B.D. Vujanovic
T.M. Atanackovic

An Introduction to Modern Variational Techniques in Mechanics and Engineering

Springer Science+Business Media, LLC

B.D. Vujanovic
University of Novi Sad
Faculty of Technical Sciences
21121 Novi Sad
Serbia and Montenegro

T.M. Atanackovic
University of Novi Sad
Faculty of Technical Sciences
21121 Novi Sad
Serbia and Montenegro

Library of Congress Cataloging-in-Publication Data

Vujanovic, B. D. (Bozidar D.). 1930-
 An introduction to modern variational techniques in mechanics and engineering / Bozidar
D. Vujanovic, Teodor M. Atanackovic.
  p. cm.
 Includes bibliographical references and index.
 ISBN 978-1-4612-6467-5   ISBN 978-0-8176-8162-3 (eBook)
 DOI 10.1007/978-0-8176-8162-3
 1. Variational principles. 2. Mechanics, Analytic. I. Atanackovic, Teodor M., 1945- II.
Title.

QA808.V85 2003
531'.01'51–dc22   2003062781

AMS Subject Classifications: 70H25, 70H30, 49-99, 74G65

ISBN 978-1-4612-6467-5   Printed on acid-free paper.

©2004 Springer Science+Business Media New York
Originally published by Birkhäuser Boston in 2004
Softcover reprint of the hardcover 1st edition 2004

9 8 7 6 5 4 3 2 1   SPIN 10951811

www.birkhauser-science.com

Contents

II   The Hamiltonian Integral Variational Principle

Preface

This book is devoted to the basic variational principles of mechanics: the Lagrange–D'Alembert differential variational principle and the Hamilton integral variational principle. These two variational principles form the main subject of contemporary analytical mechanics, and from them the whole colossal corpus of classical dynamics can be deductively derived as a part of physical theory.

In recent years students and researchers of engineering and physics have begun to realize the utility of variational principles and the vast possibilities that they offer, and have applied them as a powerful tool for the study of linear and nonlinear problems in conservative and nonconservative dynamical systems.

The present book has evolved from a series of lectures to graduate students and researchers in engineering given by the authors at the Department of Mechanics at the University of Novi Sad Serbia, and numerous foreign universities.

The objective of the authors has been to acquaint the reader with the wide possibilities to apply variational principles in numerous problems of contemporary analytical mechanics, for example, the Noether theory for finding conservation laws of conservative and nonconservative dynamical systems, application of the Hamilton–Jacobi method and the field method suitable for nonconservative dynamical systems, the variational approach to the modern optimal control theory, the application of variational methods to stability and determining the optimal shape in the elastic rod theory, among others.

In order to reach a level of practical effectiveness, numerous concrete examples are solved in order to clarify the vitality of the theory. It is hoped that this book will be useful as a text in graduate and senior undergraduate courses with an emphasis on mechanics and/or applied mathematics and in graduate engineering courses. The exposition is intended to be suggestive rather than (mathematically) rigorous. Thus, the mathematical level has been kept as elementary as possible. Each chapter starts from widely understood principles and brings the reader to the forefront of the topic in a logical way. An important part of the material presented was already published by the authors of this book in the numerous papers printed in the current literature for the last 10 or so years, and the reader is directed to these sources at the proper places in the text.

The book is divided into two parts. The first part contains four chapters. In the first chapter we consider the basic forms of the Lagrange–D'Alembert principle in the form of the central Lagrangian equation, Euler–Lagrangian differential equations of motion for holonomic and nonholonomic dynamical systems, the Hamilton canonical equations, canonical transformations, and Poisson's brackets.

The second chapter is devoted to the Hamilton–Jacobi method of integration of canonical equations. Special attention is paid to the analysis of rheolinear oscillations and quadratic conservation laws of rheolinear systems with two degrees of freedom.

In the third chapter we study methods of obtaining the conservation laws of conservative and nonconservative dynamical systems by means of Noether theory. The necessary conditions for the existence of conservation laws are obtained by studying the invariant properties of the central Lagrangian equation with respect to the infinitesimal transformations of generalized coordinates and time, in the presence of the gauge function. Generally, the generators of space and time transformations are supposed to depend upon time, generalized coordinates, and generalized velocities.

In the fourth chapter, we consider a field method suitable for applications in conservative and nonconservative dynamics. The essence of the method is the supposition that one component of the momentum vector can be represented as a field function depending on time, generalized coordinates, and the rest of the components of generalized momenta.

The second part of the book is devoted to the Hamilton integral variational principle, and its various applications. It contains four chapters.

The fifth chapter is the introductory character for this part.

The sixth chapter contains the variational problems subject to natural boundary conditions, variable end points, the Bolza problem, and the Jacobi form of the variational principle describing the trajectories of conservative dynamical systems.

Chapter 7 discusses constrained problems and the variational approach to optimal control theory. The various specified and natural boundary conditions are discussed in detail.

Chapter 8 contains applications of variational methods to the problems of elastic rod theory. The variational methods are used to estimate the critical load of elastic columns, to determine postcritical shape (the shape after buckling), and to determine the optimal shape of elastic rods and columns. By *optimal* we mean rods of minimal mass (volume) for specified buckling load.

We are grateful to Prof. Dragan Spasic and assistant Mrs. Branislava Novakovic for helping us in preparing the manuscript.

The book is gratefully dedicated to our children and grandchildren: Milica, Dragutin, Milena, Djordje Mihajlo and Bozidar (B.D.V.) and Jelena and Milica (T.M.A.).

                                                    B. D. Vujanovic
Novi Sad, June 2003   T. M. Atanackovic

An Introduction to
Modern Variational Techniques
in Mechanics and Engineering

Part I

Differential Variational Principles of Mechanics

Chapter 1

The Elements of Analytical Mechanics Expressed Using the Lagrange–D'Alembert Differential Variational Principle

1.1 Introduction

The text material of the present chapter is designed to be a more or less self-contained introduction to analytical mechanics expressed in an invariant form that is not connected to any privileged coordinate system. To accomplish this goal we turn first to the Lagrange–D'Alembert differential variational principle, whose applications are very wide and encompass holonomic and nonholonomic dynamical systems and also conservative and purely nonconservative systems as well. The elements of this part of contemporary analytical mechanics in fact, constitute the content of this chapter.

1.2 Differential Equations of Motion in Cartesian Coordinates

1.2.1 Free Dynamical Systems

We commence our considerations by regarding the simplest dynamical system consisting of N material particles that are completely free to move in a Cartesian inertial coordinate system $Oxyz$. Let us denote the position vector of the ith particle by $\mathbf{r}_i = x_i\mathbf{e}_1 + y_i\mathbf{e}_2 + z_i\mathbf{e}_3$, where $\mathbf{e}_1, \mathbf{e}_2$, and $\mathbf{e}_3$ are the unit vectors of the axes Ox, Oy, and Oz, respectively. In every problem of particle dynamics the active (applied), impressed forces $\mathbf{F}_i = F_{xi}\mathbf{e}_1 + F_{yi}\mathbf{e}_2 + F_{zi}\mathbf{e}_3$, acting on the ith particle should be given in advance. Generally, these forces are the functions of time t, position vectors $\mathbf{r}_i$, and velocity vectors $\mathbf{v}_i = d\mathbf{r}_i/dt = \dot{\mathbf{r}}_i = \dot{x}_i\mathbf{e}_1 + \dot{y}_i\mathbf{e}_2 + \dot{z}_i\mathbf{e}_3$, namely

$$\mathbf{F}_i = \mathbf{F}_i\left(t, \mathbf{r}_1, ..., \mathbf{r}_N, \mathbf{v}_1, ..., \mathbf{v}_N\right), \quad i = 1, ..., N. \tag{1.2.1}$$

Denoting by $\mathbf{a}_i = d^2\mathbf{r}_i/dt^2 = \ddot{\mathbf{r}}_i = \ddot{x}_i\mathbf{e}_1 + \ddot{y}_i\mathbf{e}_2 + \ddot{z}_i\mathbf{e}_3$ the acceleration vector of the ith particle, and applying the second Newton's law of motion, we arrive at the following simultaneous system of differential equations of motion of a free dynamical system:

$$m_i\ddot{\mathbf{r}}_i = \mathbf{F}_i\left(t, \mathbf{r}_1, ..., \mathbf{r}_N, \mathbf{v}_1, ..., \mathbf{v}_N\right), \quad i = 1, ..., N, \tag{1.2.2}$$

where m_i denotes the mass of the ith particle.

For the motion of a free dynamical system, Newton's law supplies all the dynamical information that we need. Namely, the problem of finding the motion of every particle of the dynamical system $\mathbf{r}_i = \mathbf{r}_i\left(t\right), i = 1, ..., N$, is reduced to that of integration of a set of N, vectorial (or $3N$ scalar) differential equations of the second order (1.2.2). If we are able to integrate the system (1.2.2), we find the positions $\mathbf{r}_i$ of each particle at time t if the values $\mathbf{r}_i\left(t_0\right)$ and $\mathbf{v}_i\left(t_0\right)$ are prescribed in advance at the initial moment $t = t_0$.

1.2.2 Constrained Motion. Lagrangian Equations with Multipliers

Frequently, the particles of a dynamical system are not completely free to move in the physical space, but are rather forced to be in permanent contact with some material objects that can be described in a mathematical form (for example, fixed or moving surfaces, curves, etc.). Such limitations to the freedom of motion are known as *constraints*, and they are specified by certain geometrical or kinematical relations.

Constraints may be classified in various ways,[1] and we shall use here the simplest, but very important type of constraints named *holonomic constraints*, which are of purely geometrical character and can be expressed as

$$f_s\left(t, x_1, y_1, z_1, ..., x_N, y_N, z_N\right) = 0, \quad s = 1, ..., k, \quad \text{where } k < 3N. \tag{1.2.3}$$

The explicit dependence on time in these relations means that physically the constraints are in motion. Such constraints are usually referred to as *rheonomic* or *nonstationary*, in contrast to the cases when they are fixed in space or *scleromic* or *stationary* constraints, that is, they do not depend on time t explicitly, namely $\partial f_s/\partial t = 0$. It is to be noted that the case $k = 3N$ is not of any interest

---

[1]The reader can find a rather exhaustive classification of constraints as, for example, *nonholonomic constraints*, bilateral, unilateral, etc., in the monographs of Pars [84], Santilli [95], and Papastavridis [82]. Papastavridis has also considered servo constraints.

since we could solve the complete system (1.2.3) and find all $3N$ coordinates $x_i, y_i, z_i, (i = 1, ..., N)$ as functions of time t, which means that the motion of the dynamical system is given in advance.

If the dynamical system is completely mobile without restraints, all $3N$ coordinates $x_1, y_1, z_1, ..., x_N, y_N, z_N$ can vary separately, and such a dynamical system is said to have $3N$ *degrees of freedom*. Naturally, the existence of constraints reduces the number of independent coordinates. In fact, we can use the equations of constraints to eliminate as many coordinates as there are constraints. This would bring the number of coordinates down to the number of degrees of freedom. Namely, we can eliminate k of $3N$ coordinates from (1.2.3) and express them as functions of $3N - k$ independent coordinates. Then, it is said that the dynamical system has $3N - k$ degrees of freedom. Consequently, the minimum number of the geometrical parameters that uniquely determine the position of the dynamical system at each moment of time is known as the *number of degrees of freedom*. Also, we tacitly assume that the constraints (1.2.3) are independent, that is, that they have been reduced to the least possible number, which implies that the functions $f_1, ..., f_k$ are not connected by a relation $\theta(f_1, ..., f_k) = 0$. In many practical situations the elimination of the k redundant coordinates can be tedious or difficult, and there are advantages in retaining more coordinates than the number of degrees of freedom. We will pursue this possibility in the next paragraph.

Since the particles of the dynamical system are compelled to be in permanent contact with the given constraints, we have to suppose that, as the result of interaction between the particles and constraints, there are forces of constraints $\mathbf{R}_i = R_{xi}\mathbf{e}_1 + R_{yi}\mathbf{e}_2 + R_{zi}\mathbf{e}_3, i = 1, ..., N$, acting on the particles. The differential equations of motion in the presence of holonomic constraints are

$$m_i\ddot{\mathbf{r}}_i = \mathbf{F}_i(t, \mathbf{r}_1, ..., \mathbf{r}_N, \mathbf{v}_1, ..., \mathbf{v}_N) + \mathbf{R}_i. \qquad (1.2.4)$$

In contrast to the active forces $\mathbf{F}_i$, which are fully specified, the forces of constraint are not furnished a priori. They are among the unknowns of the problem and must be obtained from the solution we seek. On the other hand, it is easy to see that the problem posed by $3N$ differential equations (1.2.4) does not constitute a sufficient set of equations for finding $3N$ unknown coordinates $x_1, y_1, z_1, ..., x_N, y_N, z_N$ as functions of time and $3N$ unknown orthogonal projections of the reaction forces $R_{xi}, R_{yi}, R_{zi}, i = 1, ..., N$.

In order to establish a consistent problem we have to introduce some additional assumptions about the character of forces of constraints $\mathbf{R}_i$. It is sufficient to require that the constraints are smooth, that is, that the reaction forces $\mathbf{R}_i$ are directed toward the normal of the hypersurfaces (1.2.3) and the magnitudes of reaction forces are not limited. In addition, it is of vital importance to underline some kinematical properties of the constraints.

Let us differentiate the expression (1.2.3) totally with respect to time

$$\sum_{i=1}^{N}\left(\frac{\partial f_s}{\partial x_i}\dot{x}_i + \frac{\partial f_s}{\partial y_i}\dot{y}_i + \frac{\partial f_s}{\partial z_i}\dot{z}_i\right) + \frac{\partial f_s}{\partial t} = 0, \quad s = 1, ..., k. \qquad (1.2.5)$$

The velocity vector of the ith particle $\mathbf{v}_i = \dot{x}_i\mathbf{e}_1 + \dot{y}_i\mathbf{e}_2 + \dot{z}_i\mathbf{e}_3$ satisfies this relation by all *possible velocities* that the dynamical system might have. However, the velocity vector that is compatible simultaneously with (1.2.4) and (1.2.5) will be referred to as the *actual velocity vector*. Equivalently, the *actual displacement vector* of the ith particle $d\mathbf{r}_i = \mathbf{v}_i dt = \dot{\mathbf{r}}_i dt = dx_i\mathbf{e}_1 + dy_i\mathbf{e}_2 + dz_i\mathbf{e}_3$, which satisfies at the same time the differential expression

$$\sum_{i=1}^{N} \left(\frac{\partial f_s}{\partial x_i}dx_i + \frac{\partial f_s}{\partial y_i}dy_i + \frac{\partial f_s}{\partial z_i}dz_i \right) + \frac{\partial f_s}{\partial t}dt = 0, \quad s = 1, ..., k, \qquad (1.2.6)$$

and the differential equations of motion (1.2.4) is said to be the *actual displacement vector* of the ith particle.

Together with the actual velocity and actual displacement vectors, we shall also introduce a new kind of infinitesimal displacement, usually referred as *virtual displacement* or simply *variation*, which we denote by

$$\delta\mathbf{r}_i = \delta x_i\mathbf{e}_1 + \delta y_i\mathbf{e}_2 + \delta z_i\mathbf{e}_3. \qquad (1.2.7)$$

This kind of displacement is introduced in such a way that of smooth constraints and notwithstanding of scleronomic or rheonomic systems, the relations

$$\sum_{i=1}^{N} \left(\frac{\partial f_s}{\partial x_i}\delta x_i + \frac{\partial f_s}{\partial y_i}\delta y_i + \frac{\partial f_s}{\partial z_i}\delta z_i \right) = 0, \quad s = 1, ..., k, \qquad (1.2.8)$$

are satisfied for the arbitrary values of the vector (1.2.7) at the *given instant of time* t. These displacements are called *virtual* to distinguish them from the actual displacements $d\mathbf{r}_i$ occurring in the time interval dt. They are the displacements that would be possible at the constraints (1.2.3) if they were petrified in the form that they have at the instant t. Note also that the virtual displacements do not satisfy the differential equations of motion, and they have purely geometrical significance since they are not influenced by the forces acting on the particles. Comparing (1.2.8) and (1.2.6) it is evident that for the case of rheonomic systems ($\partial f_s/\partial t \neq 0$) the actual displacement vector of the ith particle $d\mathbf{r}_i$ and the corresponding virtual displacement vector $\delta\mathbf{r}_i$ do not coincide. Moreover, if the constraints are scleronomic ($\partial f_s/\partial t = 0$), it follows that both vectors are belonging to the same class of displacements. Nevertheless, even in the case of scleronomic systems we will make distinctions between these two classes of displacements.

We now restrict ourselves to the dynamical systems for which the total virtual work of the forces of constraints defined as $\delta A = \sum_{i=1}^{N} \mathbf{R}_i \cdot \delta\mathbf{r}_i$ is zero:

$$\delta A = \sum_{i=1}^{N} \mathbf{R}_i \cdot \delta\mathbf{r}_i = \sum_{i=1}^{N} (R_{xi}\delta x_i + R_{yi}\delta y_i + R_{zi}\delta z_i) = 0, \qquad (1.2.9)$$

where we used $\cdot$ to denote a scalar product of vectors. The condition (1.2.9) is one of the most important properties that is fulfilled for the case in which the constraints are smooth and the system is holonomic. Namely, the particles are compelled to move on the constraints (surfaces, curves, etc.) and the reaction forces are perpendicular to those surfaces, while the virtual displacement must be tangent to them, hence the total virtual work vanishes. It should be stressed, that for the rheonomic systems, the total work done by the forces of constraints on the actual displacements $d\mathbf{r}_i$ is not zero:

$$\sum_{i=1}^{N} \mathbf{R}_i \cdot d\mathbf{r}_i \neq 0. \tag{1.2.10}$$

Comparing conditions (1.2.8) and (1.2.9), we conclude that the reaction forces can be expressed in terms of k multipliers $\lambda_s(t)$ in the following way:

$$R_{xi} = \sum_{s=1}^{k} \lambda_s \frac{\partial f_s}{\partial x_i}, \quad R_{yi} = \sum_{s=1}^{k} \lambda_s \frac{\partial f_s}{\partial y_i}, \quad R_{zi} = \sum_{s=1}^{k} \lambda_s \frac{\partial f_s}{\partial z_i}, \quad i = 1, ..., N,$$

$$\tag{1.2.11}$$

where the multipliers $\lambda_s(t)$ are related to the magnitude of the forces of constraints. Therefore, the $3N$ differential equations (1.2.4),

$$m_i \ddot{x}_i = F_{xi} + R_{xi}, \quad m_i \ddot{y}_i = F_{yi} + R_{yi}, \quad m_i \ddot{z}_i = F_{zi} + R_{zi}, \tag{1.2.12}$$

become

$$m_i \ddot{x}_i = F_{xi} + \sum_{s=1}^{k} \lambda_s \frac{\partial f_s}{\partial x_i},$$

$$m_i \ddot{y}_i = F_{yi} + \sum_{s=1}^{k} \lambda_s \frac{\partial f_s}{\partial y_i},$$

$$m_i \ddot{z}_i = F_{zi} + \sum_{s=1}^{k} \lambda_s \frac{\partial f_s}{\partial z_i}. \tag{1.2.13}$$

These differential equations should be considered together with k equations of constraints (1.2.3):

$$f_s(t, x_1, y_1, z_1, ..., x_N, y_N, z_N) = 0, \quad s = 1, ..., k, \quad k < 3N. \tag{1.2.14}$$

Note that $3N$ differential equations (1.2.13) and k equations of constraints (1.2.14) form $3N+k$ equations with $3N+k$ unknowns: $3N$ unknowns $x_1, y_1, z_1, ...,$ x_N, y_N, z_N and k unknown multipliers $\lambda_1, ..., \lambda_k$. They express the equations of motion for the general, holonomic dynamical system in a simple and conceivable form. They are known as the *Lagrangian equation of the first kind*. Note that equations (1.2.13), (1.2.14) are not quite useful in practical applications, since the process of finding the solution requires simultaneous treatment of the Cartesian coordinates x_i, y_i, z_i, and also the Lagrangian multipliers λ_s. In the text that follows, we will transform these differential equations into a different form in which the λ_i do not appear.

As a simple illustration of the preceding theory, consider the triangular prism with inclination α that moves with constant speed V on a horizontal plane Ox. A particle of mass m slides down the smooth inclined face AB under the force of gravity. The particle in motion is in contact with the moving constraint, which represents the straight line AB moving parallel to the right, as shown in Figure 1.2.1, with constant velocity V.

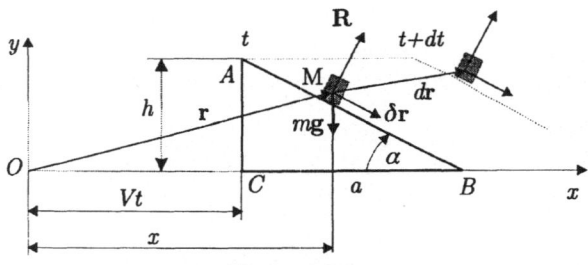

Figure 1.2.1

The equation of constraint is therefore

$$kx + y - k\left(Vt + a\right) = 0, \qquad (1.2.15)$$

where $BC = a$ and $\tan\alpha = k$. This equation represents a particular case of a holonomic, rheonomic constraint of the type (1.2.14), for $s = 1$. The particle has one degree of freedom since the projections of the virtual displacement vector are connected by the relation $\delta y + k\delta x = 0$.

From Figure 1.2.1 it is seen that the force of constraint $\mathbf{R}$ is permanently normal to the moving constraint AB and the virtual displacement vector $\delta\mathbf{r}$ is orthogonal to $\mathbf{R}$ at each moment of time. It is also seen that the actual displacement $d\mathbf{r}$ is not orthogonal to $\mathbf{R}$, and the work done by this force on the actual displacement $d\mathbf{r}$ is not zero.

The Lagrangian equations with multipliers (1.2.13) are of the form

$$m\ddot{x} = \lambda\frac{\partial f}{\partial x} = k\lambda, \quad m\ddot{y} = -mg + \lambda\frac{\partial f}{\partial y} = \lambda - mg, \qquad (1.2.16)$$

where g is gravitational acceleration.

From (1.2.15) it follows that, $k\ddot{x} = -\ddot{y}$, and combining this with (1.2.16) we easily find the components of the acceleration vector and multiplier λ:

$$\ddot{x} = \frac{kg}{1 + k^2} = M, \quad \ddot{y} = -\frac{k^2 g}{1 + k^2} = N, \quad \lambda = \frac{mg}{1 + k^2}. \qquad (1.2.17)$$

If the initial conditions are given in the form

$$x\left(0\right) = 0, \quad y\left(0\right) = h, \quad \dot{x}\left(0\right) = V, \quad \dot{y}\left(0\right) = 0, \qquad (1.2.18)$$

we find the motion of the particle

$$x = \frac{Mt^2}{2} + Vt, \quad y = h - \frac{Nt^2}{2}. \tag{1.2.19}$$

With (1.2.19) we can verify that the constraint (1.2.15) is identically satisfied. According to (1.2.11), the projections of the reaction force $\mathbf{R}$ are

$$R_x = k\lambda = mg\frac{k}{1+k^2}, \quad R_y = mg\frac{1}{1+k^2}, \tag{1.2.20}$$

and the total force of constraint is

$$R = \left(R_x^2 + R_y^2\right)^{1/2} = mg\frac{1}{\left(1+k^2\right)^{1/2}}. \tag{1.2.21}$$

1.3 An Invariant Form of Dynamics, the Lagrange–D'Alembert Differential Variational Principle for Holonomic Dynamical Systems

The differential equations of motion of holonomic dynamical systems can be described in a variety of mutually different forms, depending upon the coordinate system we employ. In the previous section we have derived the differential equations of motion in Cartesian coordinates. As mentioned already, the equations on the form (1.2.13) are not generally feasible when working with the system with many degrees of freedom. Namely, in order to find $3N - k$ independent coordinates as functions of time, we must solve $3N + k$ equations consisting of $3N$ Cartesian coordinates x_i, y_i, z_i and as many Lagrangian multipliers $\lambda_s, (s = 1, ..., k)$ as the number of holonomic constraints figuring in the system.

1.3.1 The Principle

In this section we introduce a single invariant expression known as the *Lagrange–D'Alembert differential variational principle* from which, due to its generality, we can derive analytical mechanics as a part of physical theory independently of any coordinate system we use, which is free of the unknown Lagrangian multipliers λ_s and reaction forces R_i.

Let us consider N differential equations of motion (1.2.4) together with the equations of constraints (1.2.3):

$$m_i\ddot{\mathbf{r}}_i = \mathbf{F}_i\left(t, \mathbf{r}_1, ..., \mathbf{r}_N, \mathbf{v}_1, ..., \mathbf{v}_N\right) + \mathbf{R}_i, \quad i = 1, ..., N, \tag{1.3.1}$$

and

$$f_s\left(t, x_1, y_1, z_1, ..., x_N, y_N, z_N\right) = 0, \quad s = 1, ..., k, \quad k < 3N, \tag{1.3.2}$$

where the forces of constraints $\mathbf{R}_i$ satisfy the condition of orthogonality (1.2.9):

$$\sum_{i=1}^{N} \mathbf{R}_i \cdot \delta \mathbf{r}_i = 0. \qquad (1.3.3)$$

Multiplying (1.3.1) by the virtual displacement vector (i.e., by forming a scalar product) and summing over i, we immediately arrive at the following equation which, together with the constraints equations, expresses the Lagrange–D'Alembert variational principle:

$$\sum_{i=1}^{N} (m_i \ddot{\mathbf{r}}_i - \mathbf{F}_i) \cdot \delta \mathbf{r}_i = 0, \qquad (1.3.4)$$

$$f_s (t, x_1, y_1, z_1, ..., x_N, y_N, z_N) = 0, \quad s = 1, ..., k. \qquad (1.3.5)$$

Or, written in coordinate form, the equation (1.3.4) reads

$$\sum_{i=1}^{N} [(m_i \ddot{x}_i - F_{xi}) \delta x_i + (m_i \ddot{y}_i - F_{yi}) \delta y_i + (m_i \ddot{z}_i - F_{zi}) \delta z_i] = 0. \qquad (1.3.6)$$

The significance of the Lagrange–D'Alembert principle can be summarized in the following few remarks:

(i) This principle is formulated as a scalar product, which is one of the most fundamental invariants used in physics and geometry, since the scalar product does not depend on the coordinate system used but exclusively on the vectors themselves.

(ii) By postulating scalar invariant (1.3.4) we actually replaced N vectorial differential equations of motion (1.3.1) by a *single* scalar equation.

(iii) The differential expression (1.3.4) contains the total work of active forces $\sum_{i=1}^{N} \mathbf{F}_i \cdot \delta \mathbf{r}_i$ and the unknown forces of constraints $\mathbf{R}_i$ do not figure into it. One of the important advantages of the Lagrange–D'Alembert principle is the fact that the active forces entering into equation (1.3.4) are not limited in structure. Namely, they can be potential or purely nonconservative forces.

Besides classical mechanics, the Lagrange–D'Alembert variational principle can be employed as a starting point in different branches of physics that are not intimately connected with classical mechanics. Namely, in some sense, it plays a unifying concept in physics due to its invariance and also due to the structural similarity of many physical manifestations with the models of classical mechanics. At this point it is of interest to invoke the remark of W. Heisenberg [53, p. 49], that "the concept of classical physics will always remain the basis for any exact and objective science."

For example M. A. Biot [24] extended the applications of the Lagrange–D'Alembert principle to nonlinear nonstationary heat conduction processes, and V. V. Dobronravov [39] applied this principle to the electric machine theory, to mention just two examples.

1.3.2 Generalized Coordinates and Their Variations

Since the Lagrange–D'Alembert principle is invariant with respect to the arbitrary coordinate system we can, instead of the Cartesian coordinates used so far, introduce coordinates of more general type. Namely, we shall introduce new geometrical parameters $q_1(t), ..., q_n(t)$, whose number is equal to the number of degrees of freedom, that is, $n = 3N - k$ (where N denotes the number of particles of the system and k is the number of holonomic constraints). These parameters are known as the *generalized independent coordinates* and they uniquely determine the configuration of the dynamical system at the given moment of time. By the term *independent* we understand that the set $q_1, ..., q_n$ is the minimal number of coordinates that are potentially able to specify the position of the dynamical system.

From the definition of the generalized coordinates, it follows that they must satisfy the following two requirements. First, the position vectors of each particle must be uniquely expressed in terms of generalized coordinates $q_s(t), s = 1, ..., n$, and time t,

$$\mathbf{r}_i = \mathbf{r}_i(t, q_1, ..., q_n), \quad i = 1, ..., N, \quad n = 3N - k, \tag{1.3.7}$$

or

$$x_i = x_i(t, q_1, ..., q_n), \quad y_i = y_i(t, q_1, ..., q_n), \quad z_i = z_i(t, q_1, ..., q_n). \tag{1.3.8}$$

Second, the equations of constraint (1.2.3) must be satisfied identically by the equations (1.3.8), which means that the sets of independent coordinates $q_s, s = 1, ..., n$, contain the constraint conditions implicitly in the transformation condition (1.3.7). Therefore

$$f_s(t, ..., x_i(t, q_1, ..., q_n), y_i(t, q_1, ..., q_n), z_i(t, q_1, ..., q_n), ...) \equiv 0.$$

$$i = 1, ..., N \quad s = 1, ..., k. \tag{1.3.9}$$

To illustrate this, we turn to the example considered in the previous section and shown in Figure 1.2.1. Let us introduce as the generalized coordinate the distance AM $= q$. The position of the particle in terms of this coordinate is

$$x = Vt + q\cos\alpha, \quad y = h - q\sin\alpha. \tag{1.3.10}$$

Entering with this into the constraint equation (1.2.15), $kx + y - k(Vt + a) = 0$, we verify that the identity of the type (1.3.9) is satisfied for $k = \tan\alpha$ and $h = a\tan\alpha$.

By using (1.3.7), we can transform the Lagrange–D'Alembert principle in terms of the generalized coordinates. Differentiating (1.3.7) totally with respect to time we find the following expression for the velocity vector of the ith particle in terms of generalized coordinates q_i and generalized velocities $\dot{q}_i$:

$$\dot{\mathbf{r}}_i = \mathbf{v}_i = \frac{d\mathbf{r}_i}{dt} = \sum_{s=1}^{n} \frac{\partial \mathbf{r}_i}{\partial q_s}\dot{q}_s + \frac{\partial \mathbf{r}_i}{\partial t}, \quad i = 1, ..., N; \tag{1.3.11}$$

that is, the velocity vector is the linear function of the *generalized velocities.*

Since the quantities $\partial \mathbf{r}_i / \partial t$ and $\partial \mathbf{r}_i / \partial q_s$ depend only on the generalized co-ordinates and time, it is easy to verify that the following two functional relations are valid:

$$\frac{\partial \mathbf{v}_i}{\partial \dot{q}_s} = \frac{\partial \mathbf{r}_i}{\partial q_s} = \frac{\partial \dot{\mathbf{r}}_i}{\partial \dot{q}_s}, \quad i = 1, ..., N, \quad s = 1, ..., n, \tag{1.3.12}$$

and

$$\frac{d}{dt} \frac{\partial \mathbf{r}_i}{\partial q_s} = \frac{\partial \mathbf{v}_i}{\partial q_s}, \quad i = 1, ..., N. \tag{1.3.13}$$

The virtual displacement vector (or variation) of the ith particle, according to (1.3.7), is of the form

$$\delta \mathbf{r}_i = \sum_{s=1}^{n} \frac{\partial \mathbf{r}_i}{\partial q_s} \delta q_s, \quad i = 1, ..., N, \tag{1.3.14}$$

where the variations of the generalized coordinates are denoted by the symbol δq_s.

As mentioned previously, the variation as a differential operator does not produce any infinitesimal change upon time, that is, $\delta t = 0$. The variation of velocity (1.3.11) is found to be

$$\delta \dot{\mathbf{r}}_i = \sum_{s=1}^{n} \sum_{p=1}^{n} \frac{\partial^2 \mathbf{r}_i}{\partial q_s \partial q_p} \dot{q}_s \delta q_p + \sum_{i=1}^{n} \frac{\partial \mathbf{r}_i}{\partial q_s} \delta \dot{q}_s + \sum_{i=1}^{n} \frac{\partial^2 \mathbf{r}_i}{\partial t \partial q_s} \delta q_s. \tag{1.3.15}$$

At the same time, the total time derivative of (1.3.14) (after simple interchange of dummy indices) is of the form

$$\frac{d}{dt} \delta \mathbf{r}_i = \sum_{s=1}^{n} \sum_{p=1}^{n} \frac{\partial^2 \mathbf{r}_i}{\partial q_s \partial q_p} \dot{q}_s \delta q_p + \sum_{s=1}^{n} \frac{\partial \mathbf{r}_i}{\partial q_s} \frac{d}{dt} (\delta q_s) + \sum_{s=1}^{n} \frac{\partial^2 \mathbf{r}_i}{\partial q_s \partial t} \delta q_s. \tag{1.3.16}$$

The difference between the last two expressions gives

$$\delta \dot{\mathbf{r}}_i - \frac{d}{dt} \delta \mathbf{r}_i = \sum_{s=1}^{n} \frac{\partial \mathbf{r}_i}{\partial q_s} \left(\delta \dot{q}_s - \frac{d}{dt} (\delta q_s) \right). \tag{1.3.17}$$

It is to be noted that in the classical variational calculus and integral variational principles of Hamiltonian type, it is generally accepted that the following *commutative rules* are valid:

$$\delta \dot{q}_s - \frac{d}{dt} (\delta q_s) = 0; \tag{1.3.18}$$

that is, the variation of velocity is equal to the velocity of variation. However, in the realm of the differential variational principles and especially in the formulation of the Lagrange–D'Alembert differential principle, the commutative rules (1.3.18) are not obligatory. Readers can find a very exhaustive discussion concerning the commutative and noncommutative rules in analytical mechanics in [68], [75], and [82].

1.3.3 The Lagrange–D'Alembert Variational Principle Expressed in Terms of Generalized Coordinates, Central Lagrangian Equations

Since by introducing the independent generalized coordinates the constraint equations are eliminated (see (1.3.9)), we substitute (1.3.14) into (1.3.4) and permute the sign of summation:

$$\sum_{s=1}^{n} \left[\sum_{i=1}^{N} m_i \ddot{\mathbf{r}}_i \cdot \frac{\partial \mathbf{r}_i}{\partial q_s} - \sum_{i=1}^{N} \mathbf{F}_i \cdot \frac{\partial \mathbf{r}_i}{\partial q_s} \right] \delta q_s = 0. \tag{1.3.19}$$

Transforming the first term in the following way and using relations (1.3.12) and (1.3.13), one has

$$\begin{aligned}
\sum_{i=1}^{N} m_i \ddot{\mathbf{r}}_i \cdot \frac{\partial \mathbf{r}_i}{\partial q_s} &= \frac{d}{dt} \sum m_i \dot{\mathbf{r}}_i \cdot \frac{\partial \mathbf{r}_i}{\partial q_s} - \sum_{i=1}^{N} m_i \dot{\mathbf{r}}_i \cdot \frac{d}{dt} \frac{\partial \mathbf{r}_i}{\partial q_s} \\
&= \frac{d}{dt} \sum m_i \dot{\mathbf{r}}_i \cdot \frac{\partial \dot{\mathbf{r}}_i}{\partial \dot{q}_s} - \sum_{i=1}^{N} m_i \dot{\mathbf{r}}_i \cdot \frac{\partial \dot{\mathbf{r}}_i}{\partial q_s}.
\end{aligned} \tag{1.3.20}$$

Let us introduce the kinetic energy of the dynamical system in terms of the generalized coordinates

$$T = \sum_{i=1}^{N} \left[\frac{1}{2} m_i \dot{\mathbf{r}}_i \cdot \dot{\mathbf{r}}_i \right] \quad \begin{aligned} \dot{\mathbf{r}}_1 = \dot{\mathbf{r}}_1(t, \mathbf{q}, \dot{\mathbf{q}}) \\ \dots\dots\dots\dots\dots \\ \dots\dots\dots\dots\dots \\ \dot{\mathbf{r}}_N = \dot{\mathbf{r}}_N(t, \mathbf{q}, \dot{\mathbf{q}}) \end{aligned} \quad = T(t, q_1, ..., q_n, \dot{q}_1, ..., \dot{q}_n), \tag{1.3.21}$$

where $\mathbf{q} = \{q_1, ..., q_n\}$, $\dot{\mathbf{q}} = \{\dot{q}_1, ..., \dot{q}_n\}$. Differentiating this function partially with respect to generalized velocities $\dot{q}_s$ and generalized coordinates q_s we have

$$\frac{\partial T}{\partial \dot{q}_s} = \sum_{i=1}^{N} m_i \dot{\mathbf{r}}_i \frac{\partial \dot{\mathbf{r}}_i}{\partial \dot{q}_s}, \quad \frac{\partial T}{\partial q_s} = \sum_{i=1}^{N} m_i \dot{\mathbf{r}}_i \frac{\partial \dot{\mathbf{r}}_i}{\partial q_s}. \tag{1.3.22}$$

Thus, the equation (1.3.20) reads

$$\sum_{i=1}^{N} m_i \ddot{\mathbf{r}}_i \cdot \frac{\partial \mathbf{r}_i}{\partial q_s} = \frac{d}{dt} \frac{\partial T}{\partial \dot{q}_s} - \frac{\partial T}{\partial q_s}, \quad s = 1, ..., n. \tag{1.3.23}$$

The second term in the brackets of equation (1.3.19) is usually termed the *generalized force* of the corresponding generalized coordinate $q_s, s = 1, ..., n$:

$$Q_s = \sum_{i=1}^{N} \mathbf{F}_i \cdot \frac{\partial \mathbf{r}_i}{\partial q_s} = \sum_{i=1}^{N} \left(F_{xi} \frac{\partial x_i}{\partial q_s} + F_{yi} \frac{\partial y_i}{\partial q_s} + F_{zi} \frac{\partial z_i}{\partial q_s} \right), \quad s = 1, ..., n. \tag{1.3.24}$$

Since, according to (1.2.1), the active forces $\mathbf{F}_i$ are functions of position $\mathbf{r}_i$, velocities $\mathbf{v}_i$, and time t, and in accordance with (1.3.7) and (1.3.11), these vectors depend upon q's and $\dot{q}$'s. Using the introduced notation, the virtual work of the active forces and its transformation to the generalized coordinates can be represented as

$$\delta A\left(\mathbf{F}\right) = \sum_{i=1}^{N} \mathbf{F}_i \cdot \delta \mathbf{r}_i = \sum_{s=1}^{n} Q_s \delta q_s. \qquad (1.3.25)$$

From this, it follows that a generalized force can be interpreted as a coefficient of the *independent variations* of the generalized coordinates $\delta q_1, ..., \delta q_n$ in the expression of the virtual work of the applied (impressed) forces.

Generally, the generalized forces are functions of the time t generalized coordinates q_s and generalized velocities $\dot{q}_s$:

$$Q_s = Q_s\left(t, q_1, ..., q_n, \dot{q}_1, ..., \dot{q}_n\right), \quad s = 1, ..., n, \qquad (1.3.26)$$

and they belong to the class of purely *nonconservative forces*.

As the case of special interest let the active forces $\mathbf{F}_i = F_{xi}\mathbf{e}_1 + F_{yi}\mathbf{e}_2 + F_{zi}\mathbf{e}_3$ $(i = 1, ..., N)$ not depend on the velocities $\dot{x}_i, \dot{y}_i, \dot{z}_i$ but are functions of x_i, y_i, z_i and the time t. It can happen at the same time that this type of forces is derivable from a single scalar function usually referred to as the *potential function*

$$\Pi = \Pi\left(t, x_1, y_1, z_1, ..., x_N, y_N, z_N\right), \qquad (1.3.27)$$

in the following way:

$$F_{xi} = -\frac{\partial \Pi}{\partial x_i}, \quad F_{yi} = -\frac{\partial \Pi}{\partial y_i}, \quad F_{zi} = -\frac{\partial \Pi}{\partial z_i}. \qquad (1.3.28)$$

In this case the generalized forces according to (1.3.7) and (1.3.24) can be written

$$Q_s = -\left(\frac{\partial \Pi}{\partial x_i}\frac{\partial x_i}{\partial q_s} + \frac{\partial \Pi}{\partial y_i}\frac{\partial y_i}{\partial q_s} + \frac{\partial \Pi}{\partial z_i}\frac{\partial z_i}{\partial q_s}\right), \quad i = 1, ..., N, \quad s = 1, ..., n, \qquad (1.3.29)$$

which is actually the same expression for the partial derivative of the function $-\Pi\left(t, x_1, y_1, z_1, ..., x_N, y_N, z_N\right)$ with respect to q_s:

$$Q_s = -\frac{\partial \Pi}{\partial q_s}, \qquad (1.3.30)$$

where the potential function is expressed in terms of q_s by means of the relation (1.3.8), that is,

$$\Pi = \Pi\left(t, q_1, ..., q_n\right). \qquad (1.3.31)$$

Let us suppose that the dynamical system is subjected to n nonpotential forces of the type (1.3.26) and also to n potential forces (1.3.30). The expression (1.3.19) becomes

$$\sum_{s=1}^{n} \left(\frac{d}{dt} \frac{\partial T}{\partial \dot{q}_s} - \frac{\partial T}{\partial q_s} - Q_s + \frac{\partial \Pi}{\partial q_s} \right) \delta q_s = 0. \tag{1.3.32}$$

Since, according to (1.3.31), the potential forces do not depend upon the generalized velocities $\dot{q}_s$ the last equation can be written in the form

$$\sum_{i=1}^{n} \left(\frac{d}{dt} \frac{\partial L}{\partial \dot{q}_s} - \frac{\partial L}{\partial q_s} - Q_s \right) \delta q_s = 0, \tag{1.3.33}$$

where we introduced a new function L known as the *Lagrangian function* or *kinetic potential* or simply *Lagrangian* defined as the difference between the kinetic energy and potential function

$$L = L\left(t, q_1, ..., q_n, \dot{q}_1, ..., \dot{q}_n\right) = T\left(t, q_1, ..., q_n, \dot{q}_1, ..., \dot{q}_n\right) - \Pi\left(t, q_1, ..., q_n\right). \tag{1.3.34}$$

The scalar equation (1.3.33) plays the fundamental role in analytical mechanics and is usually termed the *central Lagrangian equation*.

Note that the central Lagrangian equation can be transformed in a way that can be of interest for additional considerations. In what follows we will abandon the commutative rules, "the variation of generalized velocity is equal to the velocity of the generalized variation" shown in equation (1.3.18).

Let us calculate the variation of the Lagrangian function (1.3.34):

$$\delta L = \sum_{s=1}^{n} \left(\frac{\partial L}{\partial q_s} \delta q_s + \frac{\partial L}{\partial \dot{q}_s} \delta \dot{q}_s + \frac{\partial L}{\partial \dot{q}_s} \left(\delta q_s \right)^{\cdot} - \frac{\partial L}{\partial \dot{q}_s} \left(\delta q_s \right)^{\cdot} \right), \tag{1.3.35}$$

where we added and subtracted the term $\left(\partial L / \partial \dot{q}_s \right) \left(\delta q_s \right)^{\cdot}$. By using the identity $\left(\partial L / \partial \dot{q}_s \right) \left(\delta q_s \right)^{\cdot} = \frac{d}{dt} \left[\left(\partial L / \partial \dot{q}_s \right) \left(\delta q_s \right) \right] - \frac{d}{dt} \left(\partial L / \partial \dot{q}_s \right) \delta q_s$ the last equation becomes

$$\delta L = \sum_{s=1}^{n} \left[\left(\frac{\partial L}{\partial q_s} - \frac{d}{dt} \frac{\partial L}{\partial \dot{q}_s} \right) \delta q_s + \frac{d}{dt} \left(\frac{\partial L}{\partial \dot{q}_s} \delta q_s \right) + \frac{\partial L}{\partial \dot{q}_s} \left(\frac{d}{dt} \delta q_s - \delta \dot{q}_s \right) \right]. \tag{1.3.36}$$

Employing the central Lagrangian equation (1.3.35) we find

$$\delta L = \sum_{s=1}^{n} \left[-Q_s \delta q_s + \frac{d}{dt} \left(\frac{\partial L}{\partial \dot{q}_s} \delta q_s \right) + \frac{\partial L}{\partial \dot{q}_s} \left(\frac{d}{dt} \delta q_s - \delta \dot{q}_s \right) \right]. \tag{1.3.37}$$

Let us introduce the *generalized momentum* vector p_s defined by the equation

$$p_s = \frac{\partial L}{\partial \dot{q}_s}, \quad s = 1, ..., n. \tag{1.3.38}$$

Note that this (covariant) vector will play a very important role in the subsequent text. Finally, with (1.3.38), the equation (1.3.37) becomes

$$\frac{d}{dt} \sum_{s=1}^{n} p_s \delta q_s = \delta L + \sum_{s=1}^{n} Q_s \delta q_s + \sum_{s=1}^{n} p_s \left[(\delta q_s)^{\cdot} - \delta \dot{q}_s \right]. \tag{1.3.39}$$

This is *the second form of the central Lagrangian equation.* According to our best knowledge the form of this equation was first published by Lurie [68, p. 257] and Neimark and Fufaev [75, p. 133].

It is of interest to note that the variational equation (1.3.39) can be successfully employed if the commutativity rule (1.3.18) is accepted. For this case we have

$$\frac{d}{dt} \sum_{s=1}^{n} p_s \delta q_s = \delta L + \sum_{s=1}^{n} Q_s \delta q_s. \tag{1.3.40}$$

It is of importance to note that the derivation of the differential equations of motion do not depend on whether we accept the commutative or noncommutative variational rules.

1.4   Euler–Lagrangian Equations

The central Lagrangian equation (1.3.33) is valid for an arbitrary set of generalized coordinates $q_1(t), ..., q_n(t)$ for which we suppose that they are mutually independent. Consequently, the corresponding virtual displacements $\delta q_1, ..., \delta q_n$ are mutually independent and (1.3.33) can only be satisfied if the following n equations hold:

$$\frac{d}{dt} \frac{\partial T}{\partial \dot{q}_s} - \frac{\partial T}{\partial q_s} + \frac{\partial \Pi}{\partial q_s} - Q_s = 0, \quad s = 1, ..., n. \tag{1.4.1}$$

These equations are called the *Euler–Lagrangian equations,* or frequently *Lagrangian equations of the second kind.* They are valid for all holonomic dynamical systems. Equations (1.4.1) are the ordinary differential equations of second order with respect to the generalized coordinates q_s. To show this, we write (1.4.1) in the explicit form, taking into account that kinetic energy, according to (1.3.21), is a function of $t, q_s,$ and $\dot{q}_s$:

$$\frac{\partial^2 T}{\partial \dot{q}_s \dot{q}_m} \ddot{q}_m + \frac{\partial^2 T}{\partial \dot{q}_s \partial q_m} \dot{q}_m + \frac{\partial^2 T}{\partial \dot{q}_s \partial t} - \frac{\partial T}{\partial q_s} + \frac{\partial \Pi}{\partial q_s} - Q_s = 0, \quad s, m = 1, ..., n, \tag{1.4.2}$$

where the potential function Π and generalized forces Q_s have the structure given by (1.3.21) and (1.3.24), respectively.

Note that we have used the generally accepted *summation convention,* which means that whenever an index occurs two times in a term, it is implied that the terms are to be summed over all possible values of the index. Thus, in (1.4.2) the summation is performed with respect to the dummy index m. For the rest of this book the summation convention will be permanently assumed (applied). From the system of equations (1.4.2) it follows that the generalized accelerations $\ddot{q}_s$ enter in these equations linearly. We will suppose that the equations (1.4.2) are solvable with respect to the generalized accelerations, and to do so, the following determinant must be different from zero:

$$\det \left| \frac{\partial^2 T}{\partial \dot{q}_s \partial \dot{q}_m} \right| \neq 0. \tag{1.4.3}$$

It is easy to see that if we integrate the Euler–Lagrangian equations of motion and find the generalized coordinates as functions of time $q_1(t), ..., q_n(t)$, then the motion of the dynamical system is completely determined. Indeed, entering with $q_s(t)$ into (1.3.5) we determine the motion of each particle $\mathbf{r}_i = \mathbf{r}_i(t), i = 1, ..., N$, and the reaction forces follow from (1.3.1) as $\mathbf{R}_i = m_i \ddot{\mathbf{r}}_i(t) - \mathbf{F}_i$.[2] From the previous analysis it follows that the Euler–Lagrangian equations play the central role in the study of motion of holonomic dynamical systems, since they do not contain redundant coordinates, they are free of equations of constraints, and they are independent of unknown Lagrangian multipliers and reaction forces. In general, the Euler–Lagrangian equations can be considered as the cornerstone of the whole of analytical dynamics, and according to Pars [84, p. 76] they are "rightly regarded as one of the outstanding intellectual achievements of mankind."

For the sake of completeness, note that the Euler–Lagrangian equations (1.4.1) can be written in the form

$$\frac{d}{dt} \frac{\partial L}{\partial \dot{q}_s} - \frac{\partial L}{\partial q_s} = Q_s, \quad s = 1, ..., n, \tag{1.4.4}$$

which follows directly from (1.3.33). Here the Lagrangian function L is defined as the difference between the kinetic energy and potential function, as indicated by (1.3.34). It is of special interest to note that in the case for which the generalized nonconservative forces are equal to zero, $Q_s = 0$, that is, the case in which all active forces acting on the dynamical system are of the potential character, the Euler–Lagrangian equations of motion depend solely on *one* function L:

$$\frac{d}{dt} \frac{\partial L}{\partial \dot{q}_s} - \frac{\partial L}{\partial q_s} = 0, \quad s = 1, ..., n. \tag{1.4.5}$$

We call this type of dynamical system the *Lagrangian dynamical system*. The fact that the Lagrangian function depends explicitly on the time as indicated by (1.3.34) signifies that the dynamical system is subjected to the rheonomic constraints given by equations (1.2.3). However, if the Lagrangian function is formed as the difference between the kinetic energy and potential function which does not depend explicitly upon time t, that is, if

$$\Pi = \Pi(q_1, ..., q_n), \tag{1.4.6}$$

---

[2] Underlined indices should *not* be summed.

which we call the *potential energy function*, and also if the kinetic energy does
not depend on time t, then we call the dynamical system characterized by the
Lagrangian

$$L = L(q_1, ..., q_n, \dot{q}_1, ..., \dot{q}_n) \tag{1.4.7}$$

the *conservative dynamical system*. Note that the Lagrangian dynamical system
whose Lagrangian function is given by $L(q_1, ..., q_n, \dot{q}_1, ..., \dot{q}_n, t)$ i.e.$\partial L/\partial t \neq 0$ is
also nonconservative since the total energy is not conserved. This fact will be
discussed in the proceeding paragraphs.

1.4.1 The Structure of the Kinetic Energy. Explicit Form of Euler–Lagrangian Equations

In this section we will briefly consider the structure of the kinetic energy of a
holonomic dynamical system in terms of the generalized coordinates and the
structure of the Euler–Lagrangian equations whose form considerably depends
upon the structure of the kinetic energy.

As mentioned previously, the total kinetic energy of a dynamical system is
defined as

$$T = \frac{1}{2} \sum_{i=1}^{N} m_i \mathbf{v}_i^2. \tag{1.4.8}$$

At the same time the velocity of the ith particle in terms of the generalized
coordinates q_s and generalized velocities $\dot{q}_s$ is, according to (1.3.11),

$$\mathbf{v}_i = \frac{\partial \mathbf{r}_i}{\partial q_s} \dot{q}_s + \frac{\partial \mathbf{r}_i}{\partial t}, \quad i = 1, ..., N, \quad s = 1, ..., n. \tag{1.4.9}$$

Entering with this into (1.4.8), the kinetic energy becomes

$$T = T_0 + T_1 + T_2, \tag{1.4.10}$$

where

$$T_0 = \frac{1}{2} m_i \left(\frac{\partial \mathbf{r}_i}{\partial t} \right)^2 = F(t, q_1, ..., q_n), \quad i = 1, ..., N, \tag{1.4.11}$$

$$T_1 = \sum_{i=1}^{N} m_i \frac{\partial \mathbf{r}_i}{\partial q_s} \cdot \frac{\partial \mathbf{r}_i}{\partial t} \dot{q}_s$$

$$= K_s(t, q_1, ..., q_n) \dot{q}_s, \quad s = 1, ..., n, \tag{1.4.12}$$

$$T_2 = \frac{1}{2} a_{ks}(t, q_1, ..., q_n) \dot{q}_k \dot{q}_s, \quad k, s = 1, ..., n, \tag{1.4.13}$$

where in the last expression the coefficients $a_{sk}(t, q_1, ..., q_n)$ are given by

$$a_{sk} = a_{ks} = \sum_{i=1}^{N} m_i \frac{\partial \mathbf{r}_i}{\partial q_s} \cdot \frac{\partial \mathbf{r}_i}{\partial q_k}. \tag{1.4.14}$$

It is evident from the structure of the equations (1.4.11) and (1.4.12) that, if the dynamical system is scleronomic, that is, $\partial \mathbf{r}_i/\partial t = 0$, and if the constraints are not moving constraints, the time does not occur explicitly in the coefficients a_{sk} and T_0 and T_1 are equal to zero. In this case the kinetic energy becomes

$$T = T_2 = \frac{1}{2} a_{ks} \left(q_1, ..., q_n \right) \dot{q}_k \dot{q}_s. \tag{1.4.15}$$

From this expression it is seen that the kinetic energy for the scleronomic dynamical systems represents a homogeneous quadratic form with respect to the generalized velocities $\dot{q}_s$.

Taking into account the form of the kinetic energy (1.4.11)–(1.4.13) let us write the Euler–Lagrangian equations. First, we calculate the following expression:

$$\left(\frac{d}{dt} \frac{\partial}{\partial \dot{q}_s} - \frac{\partial}{\partial q_s} \right) \frac{1}{2} a_{km} \dot{q}_k \dot{q}_m$$
$$= \frac{d}{dt} \left(a_{ks} \dot{q}_k \right) - \frac{1}{2} \frac{\partial a_{km}}{\partial q_s} \dot{q}_k \dot{q}_m$$
$$= a_{ks} \ddot{q}_k + \left(\frac{\partial a_{ks}}{\partial q_m} - \frac{1}{2} \frac{\partial a_{km}}{\partial q_s} \right) \dot{q}_k \dot{q}_m + \frac{\partial a_{ks}}{\partial t} \dot{q}_k. \tag{1.4.16}$$

Noting that

$$\left(\frac{\partial a_{ks}}{\partial q_m} - \frac{1}{2} \frac{\partial a_{km}}{\partial q_s} \right) \dot{q}_k \dot{q}_m$$
$$= \frac{1}{2} \left(\frac{\partial a_{ks}}{\partial q_m} + \frac{\partial a_{ms}}{\partial q_k} - \frac{\partial a_{km}}{\partial q_s} \right) \dot{q}_k \dot{q}_m = [km, s] \, \dot{q}_k \dot{q}_s, \tag{1.4.17}$$

where the symbol

$$[km, s] = \frac{1}{2} \left(\frac{\partial a_{ks}}{\partial q_m} + \frac{\partial a_{ms}}{\partial q_k} - \frac{\partial a_{km}}{\partial q_s} \right) = [mk, s], \tag{1.4.18}$$

denotes the *Kristoffel symbol of the first kind*. The differential equations of motion (1.4.1) can now be written in the from

$$a_{ks} \ddot{q}_k + [km, s] \, \dot{q}_k \dot{q}_m = Q_s - \frac{\partial \Pi}{\partial q_s} - \frac{\partial F}{\partial q_s} - \frac{\partial K_s}{\partial t} - \frac{\partial a_{ks}}{\partial t} \dot{q}_k,$$
$$k, s, m = 1, ..., n, \tag{1.4.19}$$

which are the explicit form of the Euler–Lagrangian equations we have been seeking.

As mentioned previously, the generalized accelerations enter into (1.4.19) linearly, and equations (1.4.19) can be solved with respect to $\ddot{q}_s$ since according to the equation (1.4.3) the matrix $\mathbf{A}$ with elements a_{ms} is not singular; that is,

$$\det \left| \frac{\partial^2 T}{\partial \dot{q}_m \partial \dot{q}_s} \right| = \det |a_{ms}| \neq 0. \tag{1.4.20}$$

Since the matrix $\mathbf{A} = [a_{ms}]$ is a square nonsingular matrix, let us multiply equation (1.4.19) by inverse matrix $\mathbf{A}^{-1} = [a_{sr}^{-1}]$ and, taking into account that

$$a_{ks}a_{sr}^{-1} = \delta_{kr} = \begin{cases} 0 & \text{for } k \neq r, \\ 1 & \text{for } k = r, \end{cases} \tag{1.4.21}$$

where summation with respect to repeated indices goes from 1 to n, we find

$$\ddot{q}_r + a_{sr}^{-1}\,[km,s]\,\dot{q}_k\dot{q}_m = a_{sr}^{-1}\left(Q_s - \frac{\partial \Pi}{\partial q_s} - \frac{\partial F}{\partial q_s} - \frac{\partial K_s}{\partial t} - \frac{\partial a_{ks}}{\partial t}\dot{q}_k\right). \tag{1.4.22}$$

If we introduce the *Kristoffel symbols of the second kind* by the relation

$$a_{sr}^{-1}\,[km,s] = \Gamma_{km}^r = \Gamma_{mk}^r, \tag{1.4.23}$$

the equation (1.4.22) becomes

$$\ddot{q}_r + \Gamma_{km}^r\dot{q}_k\dot{q}_m = a_{sr}^{-1}\left(Q_s - \frac{\partial\,(\Pi + F)}{\partial q_s} - \frac{\partial K_s}{\partial t} - \frac{\partial a_{ks}}{\partial t}\dot{q}_k\right). \tag{1.4.24}$$

It is to be noted that the Kristoffel symbols of the first and second kind, (1.4.18) and (1.4.23), play an important role in Riemannian geometry and geometrical interpretation of classical mechanics. For an elaborate discussion of the applications of Riemannian geometry in classical mechanics, see, for example, [106] and [71].

In the case of scleronomic dynamical systems, the equations of motion (1.4.19) and (1.4.20) are reduced to the simpler forms

$$a_{ks}\ddot{q}_k + [km,s]\,\dot{q}_k\dot{q}_m = Q_s - \frac{\partial \Pi}{\partial q_s}, \qquad s = 1, ..., n, \tag{1.4.25}$$

or

$$\ddot{q}_r + \Gamma_{km}^r\dot{q}_k\dot{q}_m = a_{sr}^{-1}\left(Q_s - \frac{\partial\,(\Pi + F)}{\partial q_s}\right), \qquad r = 1, ..., n, \tag{1.4.26}$$

which are the explicit form of the Euler–Lagrangian equations of motion.

At this point it is of interest to note that as far as the Lagrangian dynamical systems are concerned, the prescription for finding the Lagrangian function $L = T - \Pi$ is a very important and reliable way to determine the corresponding Lagrangian function L. However, it is also often possible to find some alternative Lagrangian functions besides those formed by this rule. This point can be made by noting that the given Lagrangian function $L\,(t, q_1, ..., q_n, \dot{q}_1, ..., \dot{q}_n)$ can always be replaced by a new Lagrangian L^* in the following way:

$$L^* = cL + \frac{d}{dt}f\,(t, q_1, ..., q_n), \tag{1.4.27}$$

where c is an arbitrary nonzero constant and f is a function depending on time t and the generalized coordinates $q_1, ..., q_n$. The Euler–Lagrangian differential equations formed by means of the Lagrangian function L, and L^* are the same

$$\frac{d}{dt}\frac{\partial L^*}{\partial \dot{q}_s} - \frac{\partial L^*}{\partial q_s} \equiv \frac{d}{dt}\frac{\partial L}{\partial \dot{q}_s} - \frac{\partial L}{\partial q_s} = 0. \tag{1.4.28}$$

Note that the function $f(t, q_1, ..., q_n)$ figuring in equation (1.4.27) is frequently referred to as the *gauge function*.

Moreover, in many instances it is possible to find a Lagrangian function L^* satisfying the relation (1.4.28), where L^* *is not formed by the rule* (1.4.27). As an example, consider the simple harmonic oscillator whose differential equation of motion is

$$\ddot{q} + \omega^2 q = 0 \quad (\omega = \text{given constant parameter}). \tag{1.4.29}$$

Obviously, the rule $L = T - \Pi$ is equivalent to

$$L = \frac{1}{2}\left(\dot{q}^2 - \omega^2 q^2\right). \tag{1.4.30}$$

It is easy to verify that the time-dependent Lagrangian function of the form

$$L = \frac{1}{2}\left(\dot{q} + q\omega \tan \omega t\right)^2 \tag{1.4.31}$$

will generate the correct differential equation (1.4.29). For $\omega = 1$ it is also demonstrated in [92] and [96] that the following two Lagrangian functions,

$$L_1 = \frac{\dot{q}}{q}\arctan\left(\frac{\dot{q}}{q}\right) - \frac{1}{2}\ln\left[q^2\left(1 + \frac{\dot{q}^2}{q^2}\right)\right] \tag{1.4.32}$$

and

$$L_2 = \frac{1}{q^2}\left(q\cos t - \dot{q}\sin t\right)\ln\left(\frac{q\cos t - \dot{q}\sin t}{q\sin t + \dot{q}\cos t}\right), \tag{1.4.33}$$

will also generate the same differential equation (1.4.29) and all three Lagrangians (1.4.31)–(1.4.33) are not formed by the rule $L = T - \Pi$ or the prescription (1.4.27).

The fact that the form of the Lagrangian functions in dynamics is not unique raises the question of finding functions L for a given holonomic dynamical system whose differential equations of motion are given in advance. This important problem, usually referred to as the *inverse Lagrangian problem*, was first considered by Helmholtz in 1887 and later studied by numerous authors (see, for example, [95], [122], and [27]).

Finally, it is of interest to note that the Euler–Lagrangian equations are *form invariant* with respect to any one-to-one (i.e., punctual) transformation of two systems of generalized coordinates; the "old" generalized coordinates $q_1, ..., q_n$ are transformed to the "new" generalized coordinates $Q_1, ..., Q_n$ by

$$q_i = q_i\left(t, Q_1, ..., Q_n\right), \quad i = 1, ..., n. \tag{1.4.34}$$

Let us consider the given Lagrangian function $L\left(t, q_1, ..., q_n, \dot{q}_1, ..., \dot{q}_n\right)$. By using (1.4.34) one has

$$L\left(t, q_1, ..., q_n, \dot{q}_1, ..., \dot{q}_n\right) = L^*\left(t, Q_1, ..., Q_n, \dot{Q}_1, ..., \dot{Q}_n\right). \tag{1.4.35}$$

For the analysis that follows we will need the following two identities, which are similar to (1.3.12), (1.3.13):

$$\frac{\partial \dot{q}_i}{\partial Q_j} = \frac{d}{dt}\frac{\partial q_i}{\partial Q_j}, \qquad \frac{\partial \dot{q}_i}{\partial \dot{Q}_j} = \frac{\partial q_i}{\partial Q_j}. \tag{1.4.36}$$

Now we have

$$\frac{\partial L^*}{\partial \dot{Q}_i} = \frac{\partial L}{\partial \dot{q}_j}\frac{\partial \dot{q}_j}{\partial \dot{Q}_i} = \frac{\partial L}{\partial \dot{q}_j}\frac{\partial q_j}{\partial Q_i}. \tag{1.4.37}$$

Hence,

$$\frac{d}{dt}\frac{\partial L^*}{\partial \dot{Q}_i} = \left(\frac{d}{dt}\frac{\partial L}{\partial \dot{q}_j}\right)\frac{\partial q_j}{\partial Q_i} + \frac{\partial L}{\partial \dot{q}_j}\left(\frac{d}{dt}\frac{\partial q_j}{\partial Q_i}\right).$$

We also have

$$\frac{\partial L^*}{\partial Q_i} = \frac{\partial L}{\partial q_j}\frac{\partial q_j}{\partial Q_i} + \frac{\partial L}{\partial \dot{q}_j}\frac{\partial \dot{q}_j}{\partial Q_i} = \frac{\partial L}{\partial q_j}\frac{\partial q_j}{\partial Q_i} + \frac{\partial L}{\partial \dot{q}_j}\frac{d}{dt}\frac{\partial q_j}{\partial Q_i}.$$

From the last two equations, we have

$$\frac{d}{dt}\frac{\partial L^*}{\partial \dot{Q}_i} - \frac{\partial L^*}{\partial Q_i} = \left(\frac{d}{dt}\frac{\partial L}{\partial \dot{q}_j} - \frac{\partial L}{\partial q_j}\right)\frac{\partial q_j}{\partial Q_i}. \tag{1.4.38}$$

Since the determinant det $(\partial q_j/\partial Q_i) \neq 0$, the invariance of the Euler–Lagrangian equations with respect to the point transformation (1.4.34) follows.

1.4.2   Two Important Conservation Laws of the Euler–Lagrangian Equations: Momentum and Jacobi Conservation Laws

(i) *Momentum Integral*
The Euler–Lagrangian equations

$$\frac{d}{dt}\frac{\partial L}{\partial \dot{q}_s} - \frac{\partial L}{\partial q_s} = 0, \quad s = 1, ..., n, \tag{1.4.39}$$

have a very suitable form for finding conservation laws (or first integrals) of the dynamical systems whose behavior can be completely described by the Lagrangian function $L(t, q_1, ..., q_n, \dot{q}_1, ..., \dot{q}_n)$. By the term *conservation laws* or *first integrals* we understand some specific functional relations between physical and geometrical parameters figuring in dynamical systems, which are satisfied identically due to the differential equations of motion of the dynamical system in question. The existence of conservation laws can considerably simplify the integration of the differential equations of motion. Before discussing two important conservation laws that appear frequently in analytical dynamics, we introduce the *generalized momenta* p_s by the relation

$$p_s = \frac{\partial L}{\partial q_s}, \quad s = 1, ..., n. \tag{1.4.40}$$

It can happen very often that there are generalized coordinates that do not occur in the Lagrangian function L, although their time derivatives (generalized velocities) do. Such coordinates are usually referred to as *ignorable* or *cyclic* coordinates.

Let q_j be an ignorable coordinate, where j is a fixed particular integer. In this case $\partial L/\partial q_j = 0$, and from the equation (1.4.39) for $s = j$, it follows that the *momentum* or *cyclic integral*

$$p_j = \frac{\partial L}{\partial \dot{q}_j} = const. \tag{1.4.41}$$

It is important to note that when L does not depend on q_j (j-fixed), it is invariant under the translation in the jth coordinate; that is,

$$\bar{q}_j = q_j + C, \quad (C = \text{an arbitrary constant}). \tag{1.4.42}$$

Note also that the existence of a momentum integral is a privilege of a particular coordinate system in which the motion is studied.

(ii) *The Jacobi Conservation Law. Energy Integral*

To obtain the second, very important conservation law for the Euler–Lagrangian equations (1.4.5), we form the total time derivative of the Lagrangian function L and combine the result with the differential equations (1.4.5). Thus we have

$$\frac{dL}{dt} = \frac{\partial L}{\partial t} + \frac{\partial L}{\partial q_s}\dot{q}_s + \frac{\partial L}{\partial \dot{q}_s}\ddot{q}_s = \frac{\partial L}{\partial t} + \frac{d}{dt}\frac{\partial L}{\partial \dot{q}_s}\dot{q}_s + \frac{\partial L}{\partial \dot{q}_s}\ddot{q}_s. \tag{1.4.43}$$

From this, we find

$$\frac{d}{dt}\left(\frac{\partial L}{\partial \dot{q}_s}\dot{q}_s - L\right) = -\frac{\partial L}{\partial t}. \tag{1.4.44}$$

If the Lagrangian function L does not depend explicitly upon time, that is, $\partial L/\partial t = 0$, we arrive at the Jacobi conservation law (first integral)

$$\frac{\partial L}{\partial \dot{q}_s}\dot{q}_s - L = E = const. \tag{1.4.45}$$

If the dynamical system can be represented by means of the kinetic and potential energy, then the Lagrangian function is of the form

$$L = T - \Pi. \tag{1.4.46}$$

Supposing that the kinetic energy is of the form given by equation (1.4.15), $T = (1/2) \, a_{ij} \dot{q}_i \dot{q}_j$, and the potential energy according to (1.4.6) depends only on q_s, we have

$$\frac{\partial T}{\partial \dot{q}_i} \dot{q}_i = a_{ij} \dot{q}_i \dot{q}_j = 2T = \frac{\partial L}{\partial \dot{q}_i} \dot{q}_i, \qquad (1.4.47)$$

and the Jacobi conservation law (1.4.33) becomes

$$T + \Pi = E = const., \qquad (1.4.48)$$

which is a familiar expression for the conservation of the total mechanical energy: the sum of the kinetic and potential energy is constant during the motion of the system if the Lagrangian function is given in the form (1.4.46) and if it does not depend explicitly upon time, namely $\partial L/\partial t = 0$. It can also be stated that the existence of the energy integral is a consequence of the time invariance of the Lagrangian function L with respect to the time translation

$$\bar{t} = t + B, \quad (B = \text{an arbitrary constant}). \qquad (1.4.49)$$

Example 1.4.1. *Two masses on a string.* As an illustration of the foregoing theory, consider two masses m_1 and m_2 that are connected by a weightless string of length l. The mass m_1 can move freely on a horizontal plane, while the string can move frictionless through a small hole O in the plane so that the mass m_2 moves vertically. To begin, let at $t = 0$, the distance of the mass m_1 from the hole O be $r = r_0$, and the initial velocity v_0 of this particle be normal to the string while the mass m_2 does not move. The system has two degrees of freedom and we select as the generalized coordinates the polar coordinates of the mass m_1, r, and φ as, indicated in Figure 1.4.1.

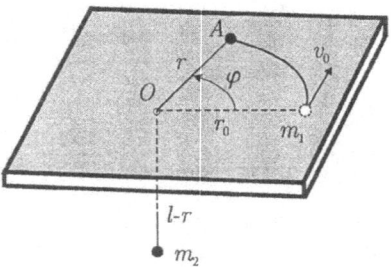

Figure 1.4.1

At the initial moment $t = 0, r = r_0$, and $\dot{r}(0) = 0$. The kinetic and potential energy of the system are

$$\begin{aligned}
T &= \frac{1}{2} \left[(m_1 + m_2) \, \dot{r}^2 + m_1 r^2 \dot{\varphi}^2 \right], \\
\Pi &= -m_2 g \, (l - r) \, m_2 g r + const.
\end{aligned} \qquad (1.4.50)$$

The Lagrangian function is given by

$$L = \frac{1}{2}\left[(m_1 + m_2)\,\dot{r}^2 + m_1 r^2 \dot{\varphi}^2\right] - m_2 g r. \tag{1.4.51}$$

It is obvious that the generalized coordinate φ is ignorable and the momentum (cyclic) integral reads

$$p_\varphi = \frac{\partial L}{\partial \dot{\varphi}} = m_1 r^2 \dot{\varphi} = C = const. \tag{1.4.52}$$

Since the Lagrangian function does not depend explicitly on time t, we have also the energy integral $T + \Pi = E$, which in our case becomes

$$\frac{1}{2}\left[(m_1 + m_2)\,\dot{r}^2 + m_1 r^2 \dot{\varphi}^2\right] + m_2 g r = E = const. \tag{1.4.53}$$

Since our dynamical system has two degrees of freedom, we can base our analysis on two conservation laws (1.4.52) and (1.4.53), ignoring the Euler–Lagrangian equations of motion.

At $t = 0$ the constant C is found to be $C = m_1 r_0 v_0$, and from (1.4.52) it follows that

$$\dot{\varphi} = \frac{r_0 v_0}{r^2}. \tag{1.4.54}$$

The total energy E is

$$E = \frac{1}{2} m_1 v_0^2 + m_2 g r_0. \tag{1.4.55}$$

Entering with (1.4.54) and (1.4.55) into (1.4.53) we find after simple calculation the following relation depending only upon variable r:

$$\dot{r}^2 = \frac{2 m_2 g}{m_1 + m_2} \frac{1}{r^2} (r_0 - r) \left[r^2 - (r + r_0)\frac{m_1 v_0^2 r_0}{2 m_2 g} \right]. \tag{1.4.56}$$

The roots of the quadratic expression with respect to r in the square brackets are found to be

$$\begin{aligned}
r_{1,2} &= \frac{m_1 v_0^2}{4 m_2 g} \pm \sqrt{\frac{m_1^2 v_0^4}{16 m_2^2 g} + \frac{m_1 v_0^2 r_0}{2 m_2 g}} \\
&= \frac{m_1 v_0^2}{4 m_2 g} \left[1 \pm \sqrt{1 + 8\frac{m_2}{m_1}\frac{g r_0}{v_0^2}} \right].
\end{aligned} \tag{1.4.57}$$

Therefore, $r_1 > 0, r_2 < 0$, and the particle m_1 will oscillate between the boundary circles r_0 and r_1. The equation (1.4.56) can be written in the form

$$\dot{r}^2 = \frac{2 m_2 g}{m_1 + m_2} \frac{1}{r^2} (r_0 - r)(r - r_1)(r - r_2) = \Phi(r). \tag{1.4.58}$$

From this equation it follows that $\Phi(r)$ must be positive during motion. Since the root r_2 is negative, the expression

$$\frac{2m_2 g}{m_1 + m_2} \frac{r - r_2}{r^2} = \Psi(r) \tag{1.4.59}$$

is always positive. The right-hand side of (1.4.58) will be positive in the following two cases:

$$\text{(a)} \quad r_0 < r < r_1; \qquad \text{(b)} \quad r_1 < r < r_0. \tag{1.4.60}$$

The trajectories of the particles m_1 for the cases (a) and (b) are depicted in Figure 1.4.2a and b.

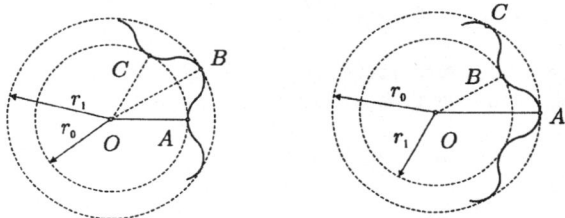

Figure 1.4.2

It is clear that the case (a) is the motion of the particle m_1 for which, at the initial moment $t = 0$, the centrifugal force of m_1 is greater than the weight of the particle m_2, that is,

$$\frac{m_1 v_0^2}{r_0} > m_2 g, \tag{1.4.61}$$

and the radius r_0 will start to increase from the initial point A if

$$v_0 > \sqrt{\frac{m_2 g r_0}{m_1}}. \tag{1.4.62}$$

Similarly, the motion of m_1 will describe the trajectory represented in Figure 1.4.2b if at $t = 0$ the weight of the particle m_2 is greater than the centrifugal force of the particle m_1, so one has the following condition for the initial velocity v_0:

$$v_0 < \sqrt{\frac{m_2 g r_0}{m_1}}, \tag{1.4.63}$$

and the distance of the particle m_1 will start to decrease from the initial distance corresponding to point A. Finally it may happen that the centrifugal force of the mass m_1 is equal to the weight $m_2 g$ of the particle m_2, and we have the motion of m_1 along the circle of the radius r_0 for which[3]

$$v_0 = \sqrt{\frac{m_2 g r_0}{m_1}}. \tag{1.4.64}$$

Let us briefly describe the motion for the case (a). During the initial increasing period from A to B the angular velocity $\dot{\varphi}$ given by (1.4.54) will decrease; the velocity and centrifugal force are going to decrease also until the point m_1 reaches the point B, where the weight $m_1 g$ is equal to the centrifugal force. In the subsequent motion from B to C the regime of motion is going to be the opposite.

From equation (1.4.64) it follows that

$$dt = \pm \frac{dr}{\sqrt{\Phi(r)}}; \tag{1.4.65}$$

since $\Phi(r) > 0$ we have to take the plus sign when r is increasing and the minus sign when r is decreasing. For the case (a) the radius is increasing from r_0 to r_1 in accordance with the equation

$$t = \int_{r_0}^{r} \frac{dr}{\sqrt{\Phi(r)}}. \tag{1.4.66}$$

The time from A to B (Figure 1.4.2a) is therefore

$$t_1 = \int_{r_0}^{r_1} \frac{dr}{\sqrt{\Phi(r)}}, \tag{1.4.67}$$

while the time needed to pass from the extreme point B to C is

$$t_2 = -\int_{r_1}^{r_0} \frac{dr}{\sqrt{\Phi(r)}} = \int_{r_0}^{r_1} \frac{dr}{\sqrt{\Phi(r)}} = t_1. \tag{1.4.68}$$

Therefore, the period of this oscillatory motion from the point A to B and back to C is finally

$$
\begin{aligned}
T &= t_1 + t_2 = 2 \int_{r_0}^{r_1} \frac{dr}{\sqrt{\Phi(r)}} \\
&= \sqrt{\frac{2(m_1 + m_2)}{m_2 g}} \int_{r_0}^{r_1} \frac{r\,dr}{\sqrt{(r_0 - r)(r - r_1)(r - r_2)}},
\end{aligned}
\tag{1.4.69}
$$

and the amplitude of the oscillation of the mass m_1 is obviously

$$a = r_1 - r_0. \tag{1.4.70}$$

---

[3]The circular motion of the particle m_1 and stability of the motion for which (1.4.64) is satisfied is considered in [68], pp. 635–638.

The integral figuring in (1.4.69) is the elliptic type, and the integration can be accomplished for given r_0, r_1, and r_2.

It is of interest to note that for this type of oscillatory motion we can use two equivalent formulas for the approximate period proposed by Pars ([84, p. 10]):

$$T_{approx} = \pi \left(r_1 - r_0\right)^{1/2} \left[\left(\frac{d\Phi}{dr}\right)_{r=r_0}^{-1/2} - \left(\frac{d\Phi}{dr}\right)_{r=r_1}^{-1/2} \right] \qquad (1.4.71)$$

or

$$T_{approx} = \pi \left[\lambda \left(r_0\right) + \lambda \left(r_1\right) \right], \qquad (1.4.72)$$

where

$$\lambda \left(r\right) = \left[\Psi \left(r\right)\right]^{-1/2} . \qquad (1.4.73)$$

For the case (a), that is, $r_0 < r < r_1$,

$$T_{approx} = \pi \left[\frac{r_0}{\sqrt{\Lambda \left(r_0 - r_2\right)}} + \frac{r_1}{\sqrt{\Lambda \left(r_1 - r_2\right)}} \right], \qquad (1.4.74)$$

with

$$\Lambda = \frac{2m_2 g}{m_1 + m_2}. \qquad (1.4.75)$$

Finally, the second generalized coordinate φ can be found from the conservation law (1.4.54)

$$\varphi = r_0 v_0 \int_{r_0}^{r} \frac{dr}{r\sqrt{\left(r_0 - r\right)\left(r - r_1\right)\left(r - r_2\right)}}. \qquad (1.4.76)$$

1.4.3   On the Disturbed Motion and Geometric Stability of the Scleronomic Potential Dynamical Systems

In this section we shall briefly discuss the problem of stability in the geometrical sense, which is based upon Synge's famous work, "On the Geometry of Dynamics" [106]. Since the exposition that follows is based strictly on the tensor calculus, we suppose that the reader is familiar with this mathematical discipline, although a broad knowledge of this subject is not essential.

As we have seen in section 1.4.1 the kinetic energy of a scleronomic dynamical system is given by equation (1.4.15): $2T = a_{ij}\dot{q}_i\dot{q}_j$. We suppose that the configuration of the dynamical system is specified by n generalized coordinates, which we now denote by $\left(q^i, ..., q^j\right)$. These coordinates define the so-called *configuration space* V_n and introduce the *metric of the configuration space* in the form

$$ds^2 = 2T dt = a_{ij}dq^i dq^j, \qquad (1.4.77)$$

which is a basic invariant of V_n. Note that the quantity $a_{ij}\left(q^1, ..., q^j\right)$ represents the *covariant metric tensor* in the configuration space V_n. Note also that the quantity ds introduced by (1.4.77) is referred to as the basic line element in V_n and configuration space defined in this way is also the Riemannian space. Together with the covariant metric tensor a_{ij} we can also introduce the *contravariant metric tensor* a^{ij} (which is denoted by a_{ij}^{-1} in equation (1.4.26)), and we have that $a_{ik}a^{kj} = \delta_i^j$, where δ_i^j denotes the Kronecker delta symbol and is equal to unity or zero when i and j are equal or different, respectively. Let us consider an arbitrary scleronomic dynamical system subject to generalized forces $Q_i\left(q^1, ..., q^n\right)$. Using the tensorial notation we can write the differential equations of motion of this system in two different forms (already indicated by (1.4.25) and (1.4.26)). In the *covariant form* the differential equations of motion are (nonconservative forces are equal to zero, $Q_s = 0$, in (1.4.25) and $Q_i = -\partial\Pi/\partial q^i$)

$$a_{ij}\ddot{q}_j + [jm, i]\,\dot{q}^j\dot{q}^m = Q_i; \quad s = 1, ..., n. \tag{1.4.78}$$

As demonstrated earlier in this section (see (1.4.19)), these equations can be derived from the corresponding Euler–Lagrangian equations.

The equations of motion can also be written in the *contravariant form*

$$\ddot{q}^i + \Gamma_{jk}^i\dot{q}^j\dot{q}^k = Q^i = a^{ij}Q_j, \quad r = 1, ..., n. \tag{1.4.79}$$

Note that the symbols $[im, j]$ and Γ_{jk}^i denote the Kristoffel symbols of the first and second kind, respectively, introduced earlier in this section.

After this introduction in the tensor notations, we now consider two motions: the *undisturbed motion* $q^i(t)$ and the infinitesimal disturbed motion $x^i(t)$. These two motions take place along the neighboring curves in the configuration space V_n. We suppose that the undisturbed motion $q^i(t)$ satisfies identically the differential equations (1.4.79). We call these trajectories, according to Synge, the *natural trajectories* C. Let $q^r, r = 1, ..., n$, be a point P of C and $q^r + x^r$ the coordinates of the corresponding (simultaneous) point P_* of disturbed natural trajectory C_*, x^r being infinitesimally small. We shall call the vector with contravariant components x^r the *disturbance vector*. If the disturbance vector between simultaneous configurations remains permanently small, we say that the undisturbed motion is *stable in the kinematical sense*.

In order to obtain the differential equations of the disturbed motion, we substitute the generalized coordinates q^i by $q^i + x^i$ in (1.4.79) and calculate the corresponding Kristoffel symbols and generalized forces in the disturbed region as

$$\ddot{q}^i + \ddot{x}^i + \Gamma_{*jk}^{\ i}\left(\dot{q}^j + \dot{x}^j\right)\left(\dot{q}^k + \dot{x}^k\right) = Q_*^{\ i}, \tag{1.4.80}$$

where the asterisk denotes the quantities that should be calculated at P_*. Therefore, expanding $\Gamma_{*jk}^{\ i}$ and $Q_*^{\ i}$ and retaining the first powers of small quantities only, we have

$$\Gamma_{*jk}^{\ i} = \Gamma_{jk}^i + x^m\frac{\partial}{\partial x^m}\Gamma_{jk}^i, \quad Q_*^{\ i} = Q^i + x^m\frac{\partial Q^i}{\partial x^m}. \tag{1.4.81}$$

Thus, the equations of the disturbances are

$$\ddot{x}^i + x^k \dot{q}^j \dot{q}^m \frac{\partial}{\partial x^k} \Gamma^i_{jm} + 2\Gamma^i_{jk} \dot{q}^j \dot{x}^k = x^j \frac{\partial Q^i}{\partial x^j}, \tag{1.4.82}$$

which are usually employed in solving concrete dynamical problems.

As demonstrated by Synge the equations of disturbed motion written in tensorial form shed light on the geometrical structure of the disturbed motion theory.[4]

Let us introduce the so called absolute derivatives of the disturbance vector. The first absolute derivative of this vector is given as

$$\frac{\overset{(a)}{\delta} x^i}{\delta t} = \dot{x}^i + \Gamma^i_{jk} \dot{q}^j x^k \tag{1.4.83}$$

where $\frac{\overset{(a)}{\delta}}{\delta t}(\cdot)$ denotes the absolute derivative with respect to time formed with respect to metric tensor a_{ij}. Differentiation with respect to time leads to

$$\begin{aligned}
\frac{d}{dt}\frac{\overset{(a)}{\delta} x^i}{\delta t} &= \ddot{x}^i + \Gamma^i_{jk}\ddot{q}^j x^k + \Gamma^i_{jk}\dot{q}^j \dot{x}^k + \frac{\partial}{\partial q^m}\Gamma^i_{jk}\dot{q}^j \dot{q}^m x^k \\
&+ \left(\dot{x}^m + \Gamma^i_{jk}\dot{q}^j x^k\right)\Gamma^i_{mr}\dot{q}^r.
\end{aligned} \tag{1.4.84}$$

Substituting $\ddot{q}^j$ from (1.4.79) in this expression we obtain the second absolute derivative of the disturbance vector in the form [68, p. 624]

$$\begin{aligned}
\frac{\overset{(a)}{\delta^2} x^i}{\delta t^2} &= \ddot{x}^i + \dot{q}^j \dot{q}^k \left(\frac{\partial}{\partial q^k}\Gamma^i_{jm} + \Gamma^i_{ks}\Gamma^s_{jm} - \Gamma^i_{mr}\Gamma^r_{jk}\right) \\
&+ 2\Gamma^i_{jk}\dot{q}^j \dot{x}^k + \Gamma^i_{jk}Q^k x^j.
\end{aligned} \tag{1.4.85}$$

Substituting $\ddot{x}^i$ from this equation into the equation of disturbed motion (1.4.82), we have after simple manipulations the following equation of disturbed motion in the contravariant form [106, p. 79], [68, p. 625]:

$$\begin{aligned}
&\frac{\overset{(a)}{\delta^2} x^i}{\delta t^2} + x^m \dot{q}^j \dot{q}^k \left[\frac{\partial}{\partial q^j}\Gamma^i_{mk} - \frac{\partial}{\partial q^m}\Gamma^i_{jk} + \Gamma^s_{mk}\Gamma^i_{js} - \Gamma^i_{ms}\Gamma^s_{jk}\right] \\
&= x^j \left(\frac{\partial Q^i}{\partial q^j} + \Gamma^i_{jk}Q^k\right).
\end{aligned} \tag{1.4.86}$$

---

[4]Concerning the geometrical interpretation of (1.4.82) which follows, Synge wrote [106, p. 78], "The use of the tensorial notation is of greatest importance. The appearance of the Riemannian curvature tensor makes it difficult to believe that similar results could be obtained without the use of this method."

Note that the expression in square brackets represents the Riemann–Kristoffel curvature tensor

$$R_{kjm}{}^j = \frac{\partial}{\partial q^j}\Gamma^i_{mk} - \frac{\partial}{\partial q^m}\Gamma^i_{jk} + \Gamma^s_{mk}\Gamma^i_{js} - \Gamma^i_{ms}\Gamma^s_{jk}. \tag{1.4.87}$$

Therefore, the equation of disturbed motion in the contravariant form becomes

$$\frac{\overset{(a)}{\delta^2 x^i}}{\delta t^2} + R_{kmj}{}^i x^m \dot{q}^j \dot{q}^k = x^j \nabla_j Q^i, \tag{1.4.88}$$

where

$$\nabla_j Q^i = \frac{\partial Q^i}{\partial q^j} + \Gamma^i_{jk} Q^k \tag{1.4.89}$$

denotes the covariant derivative of the generalized force.

It is convenient to write the equations of disturbed motion (1.4.88) in co-variant form

$$a_{ir}\frac{\overset{(a)}{\delta^2 x^i}}{\delta t^2} + R_{kmjr}\dot{q}^k \dot{q}^j x^m - x^m \nabla_m Q_r = 0, \tag{1.4.90}$$

where R_{kmjr} denotes the covariant form of the Riemann–Kristoffel tensor $R_{prmn} = a_{ps}R_{rmn}{}^s$. The explicit form of the Riemann–Kristoffel tensor is

$$R_{rsmn} = \frac{1}{2}\frac{\partial^2 a_{rn}}{\partial q^s \partial q^m} + \frac{\partial^2 a_{sm}}{\partial q^r \partial q^n} - \frac{\partial^2 a_{rm}}{\partial q^s \partial q^n} - \frac{\partial^2 a_{sn}}{\partial q^r \partial q^m}$$

$$+ a^{pq}\left([rn,p][sm,q] - [rm,p][sn,q]\right). \tag{1.4.91}$$

Note that for the case of the potential forces, that is, $Q_i = -\partial\Pi/\partial q^i$, where $\Pi = \Pi\left(q^1, ..., q^n\right)$ is the potential energy, the equations (1.4.90) become

$$a_{ir}\frac{\overset{(a)}{\delta^2 x^i}}{\delta t^2} + R_{kmjr}\dot{q}^k \dot{q}^j x^m = -x^m\left(\frac{\partial^2\Pi}{\partial q^m \partial q^r} - \Gamma^s_{rm}\frac{\partial\Pi}{\partial q^s}\right) = -x^m\nabla_m\left(\frac{\partial\Pi}{\partial q^r}\right). \tag{1.4.92}$$

We shall call the differential equations of disturbed motion (1.4.88), (1.4.90), and (1.4.92) the *Synge disturbed equations*, which are all written in tensorial (i.e., invariant) form.

The Synge disturbed equations form a basis for studying the stability of motion by geometrical means. In many practical situations the geometrical method has been used with advantage in various problems of mechanics and optimal control theory. The reason is that the Liapunov method of stability analysis, in spite of the fact that it is the strongest and most full method, is not invariant in all coordinate systems. Namely, the same motion can be stable in one coordinate system and unstable in the other. That is the reason that Pars [84, p. 174] stated that the Liapunov notion of stability "is fruitless in the classical dynamics because it demands too much."

The geometrical method for estimating the stability of motion, contrary to the Liapunov method, is direct integration of the differential equations of disturbed motion, given above. After the integration, the estimate of stability is done on the basis of the intensity of the disturbance vector, $x^2 = a_{ij}x^i x^j$.

In the text that follows, we shall show that the disturbed equations (1.4.92) together with the differential equations of motion (1.4.78) *can be derived simultaneously* from the following *modified Lagrangian function* [110]:

$$\bar{L}\left(q, \dot{q}, x, \dot{x}\right) = a_{ij}\dot{q}_i \frac{\overset{(a)}{\delta} x^i}{\delta t} - \frac{\partial \Pi}{\partial q^i}x^i.$$

(1.4.93)

Namely, it has been demonstrated that the Euler–Lagrangian equations based upon the modified Lagrangian function (1.4.93), considering the generalized coordinates q^i and corresponding disturbances x^i independently, produce the following equations:

$$\frac{d}{dt}\frac{\partial \bar{L}}{\partial \dot{q}^r} - \frac{\partial \bar{L}}{\partial q^r} = a_{ir}\frac{\overset{(a)}{\delta^2} x^i}{\delta t^2} + R_{kmjr}\dot{q}^k \dot{q}^j x^m + x^m \nabla_m \left(\frac{\partial \Pi}{\partial q^r}\right) = 0,$$

(1.4.94)

which are equal to the Synge disturbed equations (1.4.92) in covariant form. Similarly, the Euler–Lagrangian equations formed with respect to the components of the disturbance vector x^i generate the equations of motion (1.4.78), namely

$$\frac{d}{dt}\frac{\partial \bar{L}}{\partial \dot{x}^i} - \frac{\partial \bar{L}}{\partial x^i} = a_{ij}\ddot{q}^j + [jm, i]\dot{q}^j \dot{q}^m + \frac{\partial \Pi}{\partial q^i} = 0.$$

(1.4.95)

Since the modified Lagrangian function $\bar{L}$ does not depend explicitly on time t, in [110] it was shown that the energy-type first integral exists in the form $\left(\partial \bar{L}/\partial \dot{q}^i\right)\dot{q}^i + \left(\partial \bar{L}/\partial \dot{x}^i\right)\dot{x}^i - \bar{L} = const.$ This expression written explicitly leads to the following conservation law:

$$a_{ij}\frac{\overset{(a)}{\delta} x^i}{\delta t}\dot{q}^j + \frac{\partial \Pi}{\partial q^k}x^k = const.$$

(1.4.96)

Similarly, if the modified Lagrangian function (1.4.93) does not depend on a generalized coordinate q^i but does depend upon corresponding generalized velocity $\dot{q}^i$ (where i is a fixed integer), we have the cyclic conservation law

$$\frac{\partial \bar{L}}{\partial \dot{q}^i} = C_i = const.,$$

(1.4.97)

which leads to (see [110])

$$a_{ki}\frac{\overset{(a)}{\delta} x^k}{\delta t} + a_{kj}\Gamma^j_{mi}x^m\dot{q}^k = C_i = const. \tag{1.4.98}$$

This conservation law does not have the tensorial character, which one should expect, because the cyclic conservation laws are the privilege of special coordinate systems only. The reader can find more conservation laws obtained by the theory of E. Noether in [110]. Note also that applications of the Synge theory of disturbed motion can be found in [93].

Recently, in [79], almost the same modified Lagrangian function was established, the difference being that the authors of [79] allowed that the dynamical system be rheonomic, that is, the Lagrangian depend on q^i, x^i, and t.

1.5   A Brief Outline of the Nonholonomic Dynamical Systems

Thus far we have considered the dynamics of holonomic systems where the numbers of generalized coordinates $q_1, ..., q_n$ and corresponding virtual displacements $\delta q_1, ..., \delta q_n$ are precisely equal to the number of degrees of freedom of the dynamical system. Applying the Lagrange–D'Alembert principle and demanding that the virtual work of the reaction forces (1.3.3) be equal to zero, we have arrived at the central Lagrangian equation (1.3.33) in the form

$$\left(\frac{d}{dt}\frac{\partial L}{\partial \dot{q}_s} - \frac{\partial L}{\partial q_s} - Q_s\right)\delta q_s = 0, \quad s = 1, ..., n, \tag{1.5.1}$$

where the summation convention with respect to repeated indices is assumed.

In many problems, we are frequently faced with so-called *nonholonomic constraints* which are of a kinematical character. They are mostly given in the form that is linear with respect to generalized velocities:

$$A_{\alpha s}\dot{q}_s + B_\alpha = 0, \quad \alpha = 1, ..., r, \quad r < n, \tag{1.5.2}$$

where $A_{\alpha s}$ and B_α depend upon generalized coordinates q_s ($s = 1, ..., n$) and time t. It may also happen that nonholonomic constraints are the nonlinear functions of generalized velocities

$$f_\alpha(t, q_1, ..., q_n, \dot{q}_1, ..., \dot{q}_n) = 0, \quad \alpha = 1, ..., r, \quad r < n. \tag{1.5.3}$$

The term *nonholonomic* is accepted as another name for *nonintegrability* of the differential equations (1.5.2) or (1.5.3) and the impossibility of reducing them to the form

$$\theta_\alpha(t, q_1, ..., q_n) = C_\alpha = const., \tag{1.5.4}$$

since in this case we would have a holonomic constraint considered previously. Thus, the nonholonomic constraints are frequently referred to as *nonintegrable constraints*. For example, the constraint whose equation is of the form

$$q_1 \dot{q}_1 + q_2 \dot{q}_2 + q_3 \dot{q}_3 = 0 \qquad (1.5.5)$$

is holonomic, since we can integrate this relation and obtain the pure geometric holonomic constraint

$$q_1^2 + q_2^2 + q_3^2 = const., \qquad (1.5.6)$$

while the constraint of the form

$$\dot{q}_2 - q_3 \dot{q}_1 = 0 \qquad (1.5.7)$$

is nonholonomic. It is of interest to note [39, p. 9] that despite the fact that we are able to show that the particular trajectory of equation (1.5.7) satisfies the relation

$$q_1 = t^2, \quad q_2 = t^4, \quad q_3 = 2t^2, \qquad (1.5.8)$$

we are still not able to find the surface

$$f(t, q_1, q_2, q_3) = 0 \qquad (1.5.9)$$

upon which are placed *all possible trajectories* that satisfy the constraint (1.5.7).

It can be demonstrated (see, for example, [68, p. 12]) that the constraints given by equation (1.5.2) are *integrable* (*holonomic*) if the following conditions are satisfied:

$$\frac{\partial A_{\alpha s}}{\partial q_k} = \frac{\partial A_{\alpha k}}{\partial q_s}; \quad \frac{\partial A_{\alpha s}}{\partial t} = \frac{\partial B_\alpha}{\partial q_s}, \quad (k, s = 1, ..., n, \quad \alpha = 1, ..., r). \qquad (1.5.10)$$

However, if only one of the constraints (1.5.2) does not satisfy these conditions, the system must be considered nonholonomic.

It is clear that the virtual displacements δq_s $(s = 1, ..., n)$ figuring in the central Lagrangian equation (1.5.1) are not independent in the presence of nonholonomic constraints (1.5.2), (1.5.3). That means that for nonholonomic dynamical systems, it is not possible to select generalized coordinates equal in number to the degrees of freedom. Actually, in nonholonomic dynamics the number of degrees of freedom σ is always less than the number of the generalized coordinates n, namely, (number of degrees of freedom)= (number of generalized oordinates)− (number of nonholonomic constraints):

$$\sigma = n - r. \qquad (1.5.11)$$

In fact, the virtual displacements δq_s have to satisfy the following additional relations introduced by Hertz and Höllder:[5]

$$A_{\alpha s} \delta q_s = 0, \quad (\alpha = 1, ..., r, \quad s = 1, ..., n), \qquad (1.5.12)$$

---

[5]Some authors (see, for example, Mei Fengxiang [48]) call the conditions (1.5.12) Appell–Chetaev conditions.

and in the case of nonlinear nonholonomic constraints (1.5.3) these conditions read

$$\frac{\partial f_\alpha}{\partial \dot{q}_s} \delta q_s = 0 \quad (\alpha = 1, ..., r, \quad s = 1, ..., n). \tag{1.5.13}$$

We can now use (1.5.12) or (1.5.13) to reduce the number of virtual displacements to their independent number. The method for elimination of these additional displacements is the well-known procedure of Lagrange multipliers. Namely, if the equations (1.5.12) hold, then it is also true that

$$\lambda_\alpha A_{\alpha s} \delta q_s = 0, \quad (\alpha = 1, ..., r, \quad s = 1, ..., n), \tag{1.5.14}$$

where $\lambda_\alpha, \alpha = 1, ..., r$, are Lagrangian undetermined multipliers that are generally functions of generalized coordinates q_s and time t. Combining (1.5.1) and (1.5.14) we find

$$\left(\frac{d}{dt} \frac{\partial L}{\partial \dot{q}_s} - \frac{\partial L}{\partial q_s} - Q_s - \lambda_\alpha A_{\alpha s} \right) \delta q_s = 0, \quad \alpha = 1, ..., r, \quad s = 1, ..., n. \tag{1.5.15}$$

The virtual displacements δq_s are not independent as stated above, since they are also engaged in the nonholonomic constraints (1.5.12). Moreover, the values of λ_α are arbitrary. Thus, let us select the last m of the virtual displacements δq_s by accomplishing the proper choice of the λ factors, so that

$$\frac{d}{dt} \frac{\partial L}{\partial \dot{q}_s} - \frac{\partial L}{\partial q_s} - Q_s - \lambda_\alpha A_{\alpha s} = 0, \quad s = n - m + 1, ..., n. \tag{1.5.16}$$

This leaves

$$\left(\frac{d}{dt} \frac{\partial L}{\partial \dot{q}_s} - \frac{\partial L}{\partial q_s} - Q_s - \lambda_\alpha A_{\alpha s} \right) \delta q_s = 0, \quad \alpha = 1, ..., r, \quad s = 1, ..., n - m. \tag{1.5.17}$$

Now all δq_s figuring in equation (1.5.17) are independent and the coefficients of each δq_s must be equal to zero. Therefore, we finally have

$$\frac{d}{dt} \frac{\partial L}{\partial \dot{q}_s} - \frac{\partial L}{\partial q_s} - Q_s - \lambda_\alpha A_{\alpha s} = 0, \quad s = 1, ..., n, \quad \alpha = 1, ..., r. \tag{1.5.18}$$

These differential equations of motion of nonholonomic dynamical systems should be considered together with the nonholonomic constraints (1.5.2):

$$A_{\alpha s} \dot{q}_s + B_\alpha = 0, \quad \alpha = 1, ..., r, \quad r < n. \tag{1.5.19}$$

They form $n + r$ differential equations with $n + r$ unknown generalized coordinates $q_1, ..., q_n$ and $\lambda_1, ..., \lambda_r$ unknown Lagrangian multipliers.

For the case of nonlinear nonholonomic constraints (1.5.3), by repeating the similar procedure as demonstrated above, we arrive at the differential equations of motion in the form [48]

$$\frac{d}{dt} \frac{\partial L}{\partial \dot{q}_s} - \frac{\partial L}{\partial q_s} - Q_s - \lambda_\alpha \frac{\partial f_\alpha}{\partial \dot{q}_s} = 0, \quad s = 1, ..., n, \quad \alpha = 1, ..., r. \tag{1.5.20}$$

It is to be noted that the last term in (1.5.18), $Q^* = \lambda_\alpha A_{\alpha s}$, can be interpreted physically as the *generalized force* which acts on the mechanical system in order to satisfy the given nonholonomic constraints. At the same time, the nonholonomic generalized force Q^* can also be interpreted as the generalized reaction force acting on the dynamical system.

In summing up this section it is important to note that a nonholonomic dynamical system has to satisfy the following three obligatory conditions:

(a) the central Lagrangian equation (1.3.33) with specified Lagrangian function $L\,(t, \mathbf{q}, \dot{\mathbf{q}})$, which is formed in the same way as for the holonomic systems;

(b) the nonintegrable (nonholonomic) constraints in the form (1.5.2) or (1.5.3);

(c) the Hertz–Hölder conditions (1.5.12), (1.5.13).

For example, if the condition (c) is not prescribed, despite the fact that the dynamical system is subject to nonintegrable constraints (1.5.2) or (1.5.3), the system is *not nonholonomic* and must be considered as a variational problem in the presence of a given set of differential equations (1.5.2) or (1.5.3) as constraints. For equations of motion, in this case, see Chapter 4, Section 7.4 of this book.

Note that the nonholonomic dynamical systems fall into the category of purely nonconservative systems, since the noholonomic generalized force $Q^* = \lambda_\alpha A_{\alpha s}$ cannot be derived from any potential function. Therefore, there is no a single Lagrangian functions by means of which the behavior of a nonholonomic system can be completely described.

Let us underline finally that as far as the initial conditions are concerned in nonholonomic dynamics, we can select the initial position $q_1\,(0)\,, ..., q_n\,(0)$ arbitrarily, but the initial velocities $\dot{q}_1\,(0)\,, ..., \dot{q}_n\,(0)$ should be chosen in such a way that the given nonholonomic constraints are satisfied. This means that we can arbitrarily assign only $n - r$ initial generalized velocities.

The literature devoted to nonholonomic dynamical systems is very wide-reaching since there exist many nonholonomic constraints in physics and engineering. For example, an outstanding classical example whose history can be traced to the times of Hertz is the rolling solid body upon a surface without slipping. The examples of such motions are: sphere rolling on a fixed surface, and a circular homogeneous disc rolling over a plane (see example 1.5.2). The interested reader can find a variety of nonholonomic dynamical problems clearly presented in the monograph of Neimark and Fufaev [75].

Example 1.5.1. *Chaplygin sled.* A rectangular thin plate can move on the inclined plate, which makes a constant angle α with the horizon. A small knife-edge is fixed on the center of gravity C of the plate and has the direction of the line AB, as shown in Figure 1.5.1. In fact, the plate is moving in such a way that the velocity v_C of the center of gravity is always directed along the line AB. The problem is to find the motion of the plate.

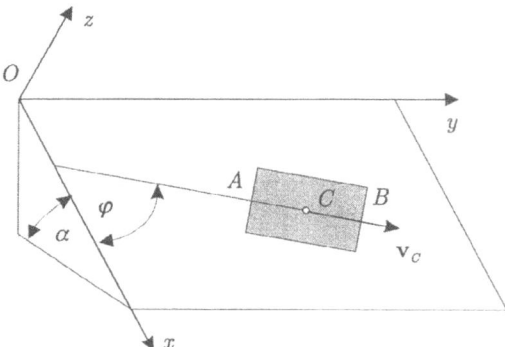

Figure 1.5.1

This example is a simplified version of the so-called Chaplygin sled. (See also [75, p. 110] and [82]). The configuration of the plate is determined by three generalized coordinates: the coordinates of the center of gravity x_c, y_c and the angle φ between the line AB and the axis x. Since the velocity $\mathbf{v}_c$ has the direction of the line AB we have

$$\dot{x}_c = v_c \cos \varphi, \quad \dot{y}_c = v_c \sin \varphi; \tag{1.5.21}$$

hence we find that the constraint

$$f = \dot{y}_c - \dot{x}_c \tan \varphi = 0 \tag{1.5.22}$$

is of the linear nonholonomic (nonintegrable) character. Thus, we see that the system has three generalized coordinates and one nonholonomic constraint, and our dynamical system has $\sigma = 3 - 1 = 2$ degrees of freedom.

Given that the mass of the plate is $m = 1$ and that the axial moment of inertia of the plate for the axis normal to the plate and passing through the mass center C is $J_C = k^2$, the Lagrangian function of the system is

$$L = \frac{1}{2}\left(\dot{x}_c^2 + \dot{y}_c^2\right) + \frac{1}{2}k^2\dot{\varphi}^2 + gx_c \sin \alpha, \tag{1.5.23}$$

where g is the acceleration of gravity. The Euler–Lagrangian equations,

$$\frac{d}{dt}\frac{\partial L}{\partial \dot{x}_c} - \frac{\partial L}{\partial x_c} = A_{11}\lambda \quad (A_{11} = -\tan \varphi),$$

$$\frac{d}{dt}\frac{\partial L}{\partial \dot{y}_c} - \frac{\partial L}{\partial y_c} = A_{12}\lambda \quad (A_{12} = 1),$$

$$\frac{d}{dt}\frac{\partial L}{\partial \dot{\varphi}} - \frac{\partial L}{\partial \varphi} = A_{13}\lambda \quad (A_{13} = 0), \tag{1.5.24}$$

read

$$\ddot{x}_c - \bar{g} = -\lambda \tan \varphi, \quad \bar{g} = g \sin \alpha = const.,$$

$$\ddot{y}_c = \lambda, \quad \ddot{\varphi} = 0. \tag{1.5.25}$$

From the third equation we find

$$\varphi = \omega t + \varphi_0, \tag{1.5.26}$$

where ω and φ_0 are arbitrary constants. From the first two equations (1.5.25) it follows that

$$\ddot{x}_c + \ddot{y}_c \tan \varphi = \bar{g}. \tag{1.5.27}$$

Differentiating the nonholonomic constraint (1.5.22) with respect to time we obtain

$$\ddot{y}_c = \ddot{x}_c \tan \varphi + \frac{\dot{x}_c \omega}{\cos^2 \varphi}; \tag{1.5.28}$$

combining (1.5.27) and (1.5.28) we arrive at the equation

$$\ddot{x}_c + \dot{x}_c \omega \tan \varphi = \bar{g} \cos^2 \varphi. \tag{1.5.29}$$

By introducing the new variable

$$\dot{x}_c = X, \tag{1.5.30}$$

the differential equation (1.5.29) is reduced to the linear differential equation of the first order, whose integration gives

$$X = \dot{x}_c = \frac{\bar{g}}{\omega} \sin \varphi \cos \varphi + D \cos \varphi, \tag{1.5.31}$$

where D is an arbitrary constant. Integrating again, we find

$$x_c = \frac{\bar{g}}{\omega^2} \frac{\sin^2 \varphi}{2} + \frac{D}{\omega} \sin \varphi + E, \tag{1.5.32}$$

where E is an arbitrary constant. Entering with (1.5.31) into nonholonomic constraint (1.5.22) and integrating, we have

$$y_c = \frac{\bar{g}}{2\omega^2} \left(\varphi - \frac{1}{2} \sin 2\varphi \right) - \frac{D}{\omega} \cos \varphi + K. \tag{1.5.33}$$

It is interesting to note that if the initial position and initial velocity are given in the form

$$x_c(0) = y_c(0) = 0, \quad \dot{x}_c(0) = \dot{y}_c(0) = 0, \quad \varphi(0) = 0, \quad \dot{\varphi}(0) = \omega, \tag{1.5.34}$$

then the motion of the plate will be given in the form

$$x_c = a^2 \sin^2 \omega t, \quad y_c = a^2 \left(\omega t - \frac{1}{2} \sin 2\omega t \right), \quad \varphi = \omega t, \tag{1.5.35}$$

where

$$a^2 = \frac{\bar{g}}{2\omega^2} \sin \alpha. \tag{1.5.36}$$

Therefore, for the given initial conditions in the form (1.5.34), the plate rotates with the constant angular velocity ω, and the center of gravity of the plate C describes the *cycloidal trajectory*.

Let us now consider the case when the plate is moving in *the horizontal plane* $x0y$ *for which* $\alpha = 0$.

For this case the Lagrangian function and the nonholonomic constraint are

$$L = \frac{1}{2} \left(\dot{x}_c^2 + \dot{y}_c^2 \right) + \frac{1}{2} k^2 \dot{\varphi}^2, \quad \dot{y}_c - \dot{x}_c \tan \varphi = 0. \tag{1.5.37}$$

The Euler–Lagangian equations are in this case

$$\ddot{x}_c = -\lambda \tan \varphi, \quad \ddot{y}_c = \lambda, \quad \ddot{\varphi} = 0. \tag{1.5.38}$$

Repeating the same procedure as in the case $\alpha \neq 0$ we find

$$\dot{x}_c = D \cos \left(\omega t + \varphi_0 \right), \quad \dot{y}_c = D \sin \left(\omega t + \varphi_0 \right), \quad \varphi = \omega t + \varphi_0, \tag{1.5.39}$$

where φ_0 and D are constants. From (1.5.39) we obtain

$$x_c - x_0 = \frac{D}{\omega} \sin \left(\omega t + \varphi_0 \right), \quad y_c - y_0 = -\frac{D}{\omega} \cos \left(\omega t + \varphi_0 \right); \tag{1.5.40}$$

namely, the plate rotates with constant angular velocity ω, while the center of gravity describes a circular orbit

$$\left(x_c - x_0 \right)^2 + \left(y_c - y_0 \right)^2 = \left(\frac{D}{\omega} \right)^2, \tag{1.5.41}$$

with the constant velocity D. Here x_0, y_0 are arbitrary constants depending upon the initial conditions.

Example 1.5.2. *The rolling disc.* Let us consider the motion of a thin, homogeneous circular disc, which rolls without slipping upon a rough horizontal plane. Let the mass of the disc be m and its radius a. Introducing a fixed coordinate system $Oxyz$, we can describe the position of the disc by the specification of five coordinates: Cartesian coordinates x, y of the point of the plane Oxy in which the disc is in contact with the plane and three Euler's angles, $\theta, \psi,$ and φ, which are depicted in Figure 1.5.2

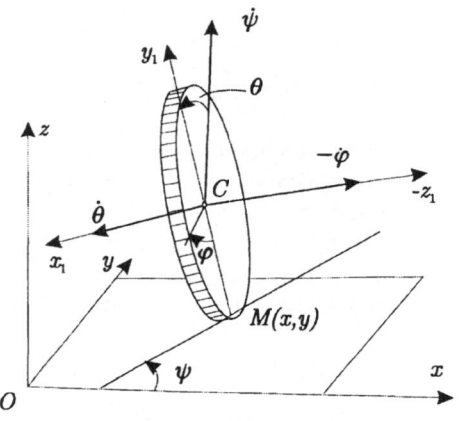

Figure 1.5.2

Ignoring for a moment the nonholonomic constraint, we first form the Lagrangian function for the disc. From Figure 1.5.2 it follows that

$$x_C = x - a \sin\theta \sin\psi, \quad y_C = y + a \sin\theta \cos\psi, \quad z_C = a \cos\theta, \qquad (1.5.42)$$

where x_C, y_C, and z_C are coordinates of the center of the disc. Therefore,

$$\begin{aligned}
\dot{x}_C &= \dot{x} - a\dot\theta \cos\theta \sin\psi - a\dot\psi \sin\theta \cos\psi, \\
\dot{y}_C &= \dot{y} + a\dot\theta \cos\theta \cos\psi - a\dot\psi \sin\theta \sin\psi, \\
\dot{z}_C &= -a\dot\theta \sin\theta.
\end{aligned} \qquad (1.5.43)$$

The kinetic energy of the disc is of the form

$$T = \frac{1}{2}m\left(\dot{x}_C^2 + \dot{y}_C^2 + \dot{z}_C^2\right) + \frac{1}{2}\left(J_{x_1}\omega_{x_1}^2 + J_{y_1}\omega_{y_1}^2 + J_{z_1}\omega_{z_1}^2\right), \qquad (1.5.44)$$

where Cx_1, Cy_1, and Cz_1 are the principal axes of the disc that are not fixed in the disc. Let $J_{x_1} = J_{y_1} = A$ and $J_{z_1} = C$ be the principal moments of inertia of the disc.

From Figure 1.5.2 we conclude that

$$\omega_{x_1} = \dot\theta, \quad \omega_{y_1} = \dot\psi \cos\theta, \quad \omega_{z_1} = \dot\varphi - \dot\psi \sin\theta, \qquad (1.5.45)$$

so that the Lagrangian function $L = T - \Pi$, with $\Pi = mga\cos\theta$, becomes (see [75, p. 305])

$$\begin{aligned}
L &= \frac{m}{2}\left(\dot{x}^2 + \dot{y}^2\right) + ma\dot{y}\left(\dot\theta \cos\theta \cos\psi - \dot\psi \sin\theta \sin\psi\right) \\
&\quad - ma\dot{x}\left(\dot\theta \cos\theta \cos\psi + \dot\psi \sin\theta \sin\psi\right) \\
&\quad + \frac{1}{2}\dot\psi^2\left(A\cos^2\theta + ma^2\sin^2\theta\right) + \frac{1}{2}\dot\theta^2\left(A + ma^2\right) \\
&\quad + \frac{1}{2}C\left(\dot\varphi - \dot\psi \sin\theta\right) - mga\cos\theta.
\end{aligned} \qquad (1.5.46)$$

The nonholonomic constraints follow from the fact that the disc is rolling without slipping upon the horizontal plane. Thus, the rolling conditions are

$$\dot{x} = a\dot{\varphi}\cos\psi, \quad \dot{y} = a\dot{\varphi}\sin\psi. \tag{1.5.47}$$

Since the problem is of the scleronomic type, the virtual displacements are the same as the possible displacements, namely

$$\delta x = a\delta\varphi\cos\psi, \quad \delta y = a\delta\varphi\sin\psi. \tag{1.5.48}$$

Noting that the generalized nonconservative forces Q's are absent and applying the equations (1.5.20) to the generalized coordinates x, y, θ, ψ, and φ, we obtain, respectively, the following differential equations of motion of the disc:

$$\frac{d}{dt}\left[m\dot{x} - ma\left(\dot{\theta}\cos\theta\sin\psi + \dot{\psi}\sin\theta\cos\psi\right)\right] = \lambda_1,$$

$$\frac{d}{dt}\left[m\dot{y} + ma\left(\dot{\theta}\cos\theta\cos\psi - \dot{\psi}\sin\theta\sin\psi\right)\right] = \lambda_2,$$

$$\frac{d}{dt}\left[ma\left(\dot{y}\cos\theta\cos\psi - \dot{x}\cos\theta\sin\psi\right) + \left(A + ma^2\right)\dot{\theta}\right]$$
$$+ma\left[\dot{y}\left(\dot{\theta}\sin\theta\cos\psi + \dot{\psi}\cos\theta\sin\psi\right)\right.$$
$$\left. - \dot{x}\left(\dot{\theta}\sin\theta\sin\psi - \dot{\psi}\cos\theta\cos\psi\right)\right]$$
$$+\dot{\psi}^2\left(A - ma^2\right)\sin\theta\cos\psi$$
$$+C\dot{\psi}\cos\theta\left(\dot{\varphi} - \dot{\psi}\sin\theta\right) - mga\sin\theta = 0,$$

$$\frac{d}{dt}\left[\left(A\cos^2\theta + ma^2\sin^2\theta\right)\dot{\psi} - ma\left(\dot{y}\sin\theta\sin\psi + \dot{x}\sin\theta\cos\psi\right)\right.$$
$$\left. -C\sin\theta\left(\dot{\varphi} - \dot{\psi}\sin\theta\right)\right] + ma\left[\dot{y}\left(\dot{\theta}\cos\theta\sin\psi + \dot{\psi}\sin\theta\cos\psi\right)\right.$$
$$\left. +\dot{x}\left(\dot{\theta}\cos\theta\cos\psi - \dot{\psi}\sin\theta\sin\psi\right)\right] = 0,$$

$$C\frac{d}{dt}\left(\dot{\varphi} - \dot{\psi}\sin\theta\right) + a\lambda_1\cos\psi + a\lambda_2\sin\psi = 0. \tag{1.5.49}$$

Entering with (1.5.47) into (1.5.49), we can easily eliminate $\lambda_1, \lambda_2, \dot{x}$, and $\dot{y}$. Thus, we arrive at the following three differential equations:

$$\left(A + ma^2\right)\ddot{\theta} + \left(C + ma^2\right)\Omega\omega\cos\theta$$
$$+A\Omega^2\sin\theta\cos\theta - mga\sin\theta = 0,$$
$$[2A\Omega\sin\theta + C\omega]\dot{\theta} - A\dot{\Omega}\cos\theta = 0,$$
$$\left(C + ma^2\right)\dot{\omega} - ma^2\Omega\dot{\theta}\cos\theta = 0, \tag{1.5.50}$$

where we introduced the notation

$$\Omega = \dot{\psi}, \quad \omega = \dot{\varphi} - \dot{\psi}\sin\theta. \tag{1.5.51}$$

Stationary States of a Rolling Disc

The differential equations of a rolling disc (1.5.50) and (1.5.51) are very convenient for the study of its stationary motions. In fact a rolling disc can perform (manifest) the following three steady state motions:

(a) *the circular motion,*

(b) *the straight line motion,*

(c) *rotation of a disc about its vertical diameter.*

The rolling disc will perform the circular motion under the following conditions:

$$\dot{\varphi} = \dot{\varphi}_0 = const., \quad \theta = \theta_0 = const., \quad \dot{\psi} = \Omega = const., \tag{1.5.52}$$

whence, taking into account (1.5.51), we conclude that $\omega = const.$ For this case, the equation $(1.5.51)_1$ becomes

$$\left(C + ma^2\right)\Omega\omega + A\Omega^2\sin\theta_0 - mga\tan\theta_0 = 0. \tag{1.5.53}$$

This equation represents the condition that has to be satisfied for the circular motion of a disc. Combining (1.5.43), (1.5.47), and (1.5.52), we have

$$\dot{x}_C = a\omega\cos\left(\Omega t + c\right), \quad \dot{y}_C = a\omega\sin\left(\Omega t + c\right), \tag{1.5.54}$$

where c is a constant of integration. Integrating these equations we find

$$x_C = \frac{a\omega}{\Omega}\sin\left(\Omega t + c\right) + x_{C_0}, \quad y_C = -\frac{a\omega}{\Omega}\cos\left(\Omega t + c\right) + y_{C_0}, \tag{1.5.55}$$

where x_{C_0} and y_{C_0} are also arbitrary constants. Therefore, the trajectory of a center of the disc is a circle:

$$\left(x_C - x_{C_0}\right)^2 + \left(y_C - y_{C_0}\right)^2 = \left(\frac{a\omega}{\Omega}\right)^2. \tag{1.5.56}$$

Readers can find the conditions of stability of circular motion of a rolling disc, in [84, p. 122] and [75, p. 305].

We determine next the first-order stability for the straight line motion (case (b)) and rotation of a disc about its vertical diameter (case (c)). Let us introduce small disturbances $\xi(t), \eta(t)$, and $\zeta(t)$ by the relations

$$\theta = \theta_0 + \xi(t), \quad \Omega = \Omega_0 + \eta(t), \quad \omega = \omega_0 + \zeta(t), \tag{1.5.57}$$

where θ_0, Ω_0, and ω_0 denote the constant stationary values (1.5.52). Entering with (1.5.57) into (1.5.50) and retaining only the first-order terms in ξ, η, and

ζ, we arrive at the following equations of the disturbed motion:

$$\left(A + ma^2\right) \ddot{\xi} - \left[\left(C + ma^2\right) \Omega_0 \omega_0 \sin\theta_0 + A\Omega_0^2 \cos 2\theta_0 + mga \cos\theta_0\right] \xi$$
$$+ \left[\left(C + ma^2\right) \omega_0 \cos\theta_0 + A\Omega_0 \sin\theta_0\right] \eta + \left(C + ma^2\right) \Omega_0 \cos\theta_0 = 0,$$
$$\left(2A\Omega_0 \sin\theta_0 + C\omega_0\right) \dot{\xi} - A\dot{\eta} \cos\theta_0 = 0,$$
$$\left(C + ma^2\right) \dot{\zeta} - ma^2 \Omega_0 \dot{\xi} \cos\theta_0 = 0. \tag{1.5.58}$$

We use now (1.5.58) for the stability analysis.

(a) *Stability of straight line motion.* The disc will perform a straight line motion in which the plane of the disc remains vertical and rotates with the constant angular velocity, that is,

$$\theta_0 = 0, \quad \Omega_0 = 0, \quad \dot{\varphi} = \omega_0 = const. \tag{1.5.59}$$

The disturbed motion satisfies equations (1.5.58) that in the present case become

$$\left(A + ma^2\right) \ddot{\xi} - mga\xi + \left(C + ma^2\right) \omega_0 \eta + \left(C + ma^2\right) = 0,$$
$$C\omega_0 \dot{\xi} - A\dot{\eta} = 0, \qquad \left(C + ma^2\right) \dot{\zeta} = 0. \tag{1.5.60}$$

From $(1.5.60)_3$ we obtain $\zeta = const.$ Integrating $(1.5.60)_2$ we find

$$\eta = \frac{C\omega_0}{A}\xi + const., \tag{1.5.61}$$

and therefore, by substituting this into $(1.5.60)_1$ we find

$$\ddot{\xi} + \left[\frac{C\left(C + ma^2\right)}{A\left(A + ma^2\right)}\omega_0^2 - \frac{mga}{A + ma^2}\right] \xi = const. \tag{1.5.62}$$

From this it follows that the disc will perform small harmonic oscillations about the vertical diameter if the expression in brackets is positive, namely, if

$$\omega_0^2 > \frac{mgaA}{C\left(C + ma^2\right)}, \tag{1.5.63}$$

which is the condition of the first-order stability of a circular motion.

(b) *Stability of the rotation about the vertical diameter.* The disc will rotate about its vertical diameter with a constant angular velocity under the conditions

$$\theta_0 = 0, \quad \omega_0 = 0, \quad \dot{\psi} = \Omega_0 = const. \tag{1.5.64}$$

The equations of distributed motions (1.5.58) become

$$\left(A + ma^2\right) \ddot{\xi} + \left(A\Omega_0^2 - mga\right) \xi + \left(C + ma^2\right) \Omega_0 = 0,$$
$$A\dot{\eta} = 0, \qquad \left(C + ma^2\right) \dot{\zeta} - ma^2\dot{\xi} = 0. \tag{1.5.65}$$

From $(1.5.65)_2$ we conclude that $\eta = const.$ Repeating the same procedure as in the previous case, we arrive at the differential equation

$$\ddot{\xi} + \frac{1}{A + ma^2} \left[A\Omega_0^2 - mga + ma^2\Omega_0^2\right] \xi = const. \tag{1.5.66}$$

Therefore, the condition of the first-order stability for the rotation about the vertical diameter is

$$\Omega_0^2 > \frac{mga}{A + ma^2}. \qquad (1.5.67)$$

1.6   Some Other Forms of the Equations of Motion

In our previous considerations we have used the Euler–Lagrangian differential equations of motion of holonomic and nonholonomic dynamical systems. In fact, these differential equations occupy the central position in all analytical mechanics due to their invariance with respect to the arbitrary selected coordinate system in which a dynamical process is taking place. However, in the evolution of the dynamics and especially in the field of nonholonomic mechanics several various forms of differential equations of motion, different from the Euler–Lagrangian equations, are discovered. Here, we briefly discuss two kinds of differential equations that are equally valid in holonomic and nonholonomic dynamics and also have invariant properties with respect to arbitrary selected coordinate systems. If considered in the realm of holonomic dynamics they are fully equivalent to the Euler–Lagrangian equations, but have quite different forms.

1.6.1   The Gibbs–Appell Equations: Holonomic Dynamical Systems

Let us consider a holonomic dynamical system consisting of N material particles, whose position in each moment of time can be specified by n generalized coordinates $q_1, ..., q_n$. The position of every particle can be expressed in terms of q_s and t:

$$\mathbf{r}_i = \mathbf{r}_i (t, q_1, ..., q_n). \qquad (1.6.1)$$

The velocity and acceleration vectors of the ith particle are

$$\mathbf{v}_i = \frac{d\mathbf{r}_i}{dt} = \frac{\partial \mathbf{r}_i}{\partial q_s} \dot{q}_s + \frac{\partial \mathbf{r}_i}{\partial t}, \quad s = 1, 2, ..., n, \qquad (1.6.2)$$

and

$$\mathbf{a}_i = \frac{d^2 \mathbf{r}_i}{dt^2} = \frac{\partial \mathbf{r}_i}{\partial q_s} \ddot{q}_s + \frac{\partial^2 \mathbf{r}_i}{\partial q_s \partial q_k} \dot{q}_s \dot{q}_k + 2 \frac{\partial^2 \mathbf{r}_i}{\partial q_s \partial t} \dot{q}_s + \frac{\partial^2 \mathbf{r}_i}{\partial t^2}. \qquad (1.6.3)$$

From these relations it is easy to verify that the following functional relations hold:

$$\frac{\partial \mathbf{r}_i}{\partial q_s} = \frac{\partial \mathbf{v}_i}{\partial \dot{q}_s} = \frac{\partial \mathbf{a}_i}{\partial \ddot{q}_s}, \quad \frac{d}{dt} \frac{\partial \mathbf{v}_i}{\partial \dot{q}_s} = \frac{\partial \mathbf{v}_i}{\partial q_s}, \quad \frac{d}{dt} \frac{\partial \mathbf{a}_i}{\partial \ddot{q}_s} = 2 \frac{\partial \mathbf{v}_i}{\partial q_s}. \qquad (1.6.4)$$

It is easy to verify that the relations $(1.6.4)_1$ and $(1.6.4)_3$ can be prolonged with respect to time derivatives

$$\frac{\partial \mathbf{r}_i}{\partial q_s} = \frac{\partial \mathbf{v}_i}{\partial \dot{q}_s} = \cdots = \frac{\partial^{(n)} \mathbf{r}_i}{\partial q_s^{(n)}} \quad \frac{d}{dt} \frac{\partial \overset{(n)}{\mathbf{r}}_i}{\partial q_s^{(n-1)}} = \frac{n}{n-1} \frac{\partial \overset{(n-1)}{\mathbf{r}}_i}{\partial q_s^{(n-2)}}. \tag{1.6.5}$$

Remembering that the kinetic energy of the dynamical system is $T = (1/2) \sum_{i=1}^{N} m_i (\mathbf{v}_i \cdot \mathbf{v}_i)$, we write the central Lagrangian equation (1.3.33) in the form

$$\left[\frac{d}{dt} \frac{\partial}{\partial \dot{q}_s} \sum_{i=1}^{N} \frac{1}{2} m_i (\mathbf{v}_i \cdot \mathbf{v}_i) - \frac{\partial}{\partial q_s} \sum_{i=1}^{N} \frac{1}{2} m_i (\mathbf{v}_i \cdot \mathbf{v}_i) - \bar{Q}_s \right] \delta q_s = 0, \tag{1.6.6}$$

where the summation with respect to the dummy index $s = 1, ..., n$ is assumed and $-\bar{Q}_s = -Q_s + \frac{\partial \Pi}{\partial q_s}$.

Performing the partial derivatives as indicated in (1.6.6) and the total differentiation with respect to time in the first term of (1.6.6), we find

$$\left[\sum_{i=1}^{N} \left(m_i \mathbf{a}_i \cdot \frac{\partial \mathbf{v}_i}{\partial \dot{q}_s} + m_i \mathbf{v}_i \cdot \frac{d}{dt} \frac{\partial \mathbf{v}_i}{\partial \dot{q}_s} - m_i \mathbf{v}_i \cdot \frac{\partial \mathbf{v}_i}{\partial q_s} \right) - \bar{Q}_s \right] \delta q_s = 0. \tag{1.6.7}$$

By using $(1.6.4)_2$ the second and third terms are equal and can be omitted. Employing $(1.6.4)_1$ we arrive at the equation

$$\sum_{i=1}^{N} \left(m_i \mathbf{a}_i \cdot \frac{\partial \mathbf{a}_i}{\partial \ddot{q}_s} - \bar{Q}_s \right) \delta q_s = 0, \tag{1.6.8}$$

whence

$$\left[\frac{\partial}{\partial \ddot{q}_s} \sum_{i=1}^{N} \frac{1}{2} m_i (\mathbf{a}_i \cdot \mathbf{a}_i) - \bar{Q}_s \right] \delta q_s = 0. \tag{1.6.9}$$

We now introduce the *Gibbs–Appell function* or *energy of acceleration*

$$S = \frac{1}{2} \sum_{i=1}^{N} m_i (\mathbf{a}_i \cdot \mathbf{a}_i), \tag{1.6.10}$$

whose structure in accordance with (1.6.1)–(1.6.3) in terms of the generalized coordinates is of the form

$$S = S(t, q_1, ..., q_n, \dot{q}_1, ..., \dot{q}_n, \ddot{q}_1, ..., \ddot{q}_n). \tag{1.6.11}$$

Thus, we arrive at the *central dynamical equation* in the Gibbs–Appell form

$$\left(\frac{\partial S}{\partial \ddot{q}_s} - \bar{Q}_s \right) \delta q_s = 0, \quad s = 1, ..., n. \tag{1.6.12}$$

Since we have supposed that the system is holonomic, all virtual displacements δq_s $(s = 1, ..., n)$ are mutually independent and arbitrary, and thus we have the following system of differential equations of motion known as the Gibbs–Appell equations:

$$\frac{\partial S}{\partial \ddot{q}_s} = \bar{Q}_s, \quad s = 1, ..., n. \tag{1.6.13}$$

From the analysis just performed it is clear that the Gibbs–Appell equations for the case of holonomic dynamical systems are fully equivalent with the Euler–Lagrangian equations (1.4.1). As we have seen the principal function in the Euler–Lagrangian equations is the kinetic energy of the dynamical system, while the principal function in the Gibbs–Appell equations is the energy of acceleration (1.6.10) or (1.6.11).

It is to be noted that in computing the energy of acceleration by means of formula (1.6.10) it is necessary to retain only those terms that contain the second derivatives of the generalized coordinates $\ddot{q}_s$.

In general, for the case of holonomic and scleronomic dynamical systems the acceleration vector given by the equation (1.6.3) is of the form

$$\mathbf{a}_i = \frac{\partial \mathbf{r}_i}{\partial q_s} \ddot{q}_s + \frac{\partial^2 \mathbf{r}_i}{\partial q_s \partial q_m} \dot{q}_s \dot{q}_m, \quad i = 1, ..., N, \quad s, k, m = 1, ..., n \tag{1.6.14}$$

since, for the scleronomic case $\partial^2 \mathbf{r}_i / \partial q_s \partial t = 0, \partial^2 \mathbf{r}_i / \partial t^2 = 0$. Entering with (1.6.14) into (1.6.10), we find after a simple but laborious calculation that the energy of acceleration can be expressed in the form (for more details see [68, p. 163])

$$S = \frac{1}{2} a_{sk} \ddot{q}_s \ddot{q}_k + [sk, m] \, \dot{q}_s \dot{q}_k \ddot{q}_m, \tag{1.6.15}$$

where

$$a_{sk} = a_{ks} = \sum_{i=1}^{N} m_i \frac{\partial \mathbf{r}_i}{\partial q_k} \cdot \frac{\partial \mathbf{r}_i}{\partial q_s} \tag{1.6.16}$$

are the coefficients already defined by equation (1.4.14) and

$$[km, s] = \frac{1}{2} \left(\frac{\partial a_{ks}}{\partial q_m} + \frac{\partial a_{ms}}{\partial q_k} - \frac{\partial a_{km}}{\partial q_s} \right) \tag{1.6.17}$$

are the Kristoffel symbols of the first kind introduced by equation (1.4.18). From the first term in equation (1.6.15) it is evident that the coefficients of the terms of the second degree in the $\ddot{q}_s$ in S are the same as the corresponding coefficients of the terms of the second degree in $\dot{q}_s$ in the kinetic energy t defined by (1.4.15). The second term in (1.6.15) which is linear in the generalized acceleration must be determined independently.

For the simple motions of a rigid body, the calculation of acceleration energy is very simple.

(a) If a rigid body is moving translatory, every particle has the same acceleration and s is found to be

$$S = \frac{1}{2}M\left[\ddot{x}_c^2 + \ddot{y}_c^2 + \ddot{z}_c^2\right], \tag{1.6.18}$$

where M is the total mass $M = \sum_{i=1}^{N} m_i$ of the body and where a_c is the acceleration of the mass center of the body C, that is, $a_c^2 = \left[\ddot{x}_c^2 + \ddot{y}_c^2 + \ddot{z}_c^2\right]$.

(b) If a rigid body rotates around a fixed axis $0z$, the energy of acceleration is

$$S = \frac{1}{2}J_z\ddot{\varphi}^2, \tag{1.6.19}$$

where J_z is the moment of inertia about the $0z$ axis and φ is the angle of rotation of the body.

(c) For a rigid lamina moving in a plane, the energy of acceleration is found to be

$$S = \frac{1}{2}M\left[\ddot{x}_c^2 + \ddot{y}_c^2\right] + \frac{1}{2}J_c\ddot{\varphi}^2, \tag{1.6.20}$$

where $a_c = \left[\ddot{x}_c^2 + \ddot{y}_c^2\right]^{1/2}$ is the acceleration of the center of gravity of the lamina and J_c is the moment of inertia of the lamina about its center of gravity C.

It is to be noted that P. Appell (see [6, p. 341]) has suggested that the differential equations (1.6.13) can be derived by introducing the function

$$R = S - (Q_1\ddot{q}_1 + Q_2\ddot{q}_2 + \cdots + Q_n\ddot{q}_n), \tag{1.6.21}$$

and that the equation of motion can be obtained from the equations

$$\frac{\partial R}{\partial \ddot{q}_1} = 0, \quad \frac{\partial R}{\partial \ddot{q}_2} = 0, ..., \frac{\partial R}{\partial \ddot{q}_n} = 0. \tag{1.6.22}$$

It was demonstrated (see also [84, p. 201]) that the expression R is minimal with respect to the generalized accelerations $\ddot{q}_s$. This fact can be connected with the so-called Gauss principle of least constraint, which we will not consider in this text. The interested reader will find more details about the Gauss principle in [84], [122], [75], and [6], to mention just a few references.

Example 1.6.1. *A particle on an elastic string.* Let us consider the motion of a material particle of the mass m that is fastened to an elastic massless spring of the spring elastic constant c whose other end is fixed as shown in the diagram. The motion of this system is situated in the vertical plane $x0y$ (see Figure 1.6.1).

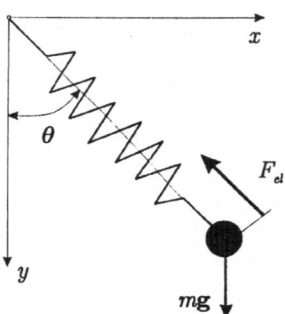

Figure 1.6.1

Let the unstretched length of the spring be l_0. In the position of (vertical) equilibrium state the weight of the particle is equal to the elastic force $mg = cf_{st}$, where f_{st} is the deformation of the spring in the equilibrium position. Hence, the length of the spring in the state of equilibrium is $l = l_0 + mg/c$. Let us denote by $z(t)$ the stretch in the spring beyond its equilibrium position. Therefore, the length of the spring at an arbitrary instant of time is

$$L(t) = l_0 + \frac{mg}{c} + z(t). \tag{1.6.23}$$

The position of the particle in the Cartesian system $x0y$ is

$$\begin{aligned} x &= \left(l_0 + \frac{mg}{c} + z\right)\sin\theta, \\ y &= \left(l_0 + \frac{mg}{c} + z\right)\cos\theta. \end{aligned} \tag{1.6.24}$$

Thus, the system has two degrees of freedom and the generalized coordinates are selected to be $q_1 = z(t)$, $q_2 = \theta(t)$.

From (1.6.24) we find

$$\begin{aligned} \ddot{x} &= \ddot{z}\sin\theta + (l+z)\ddot{\theta}\cos\theta + 2\dot{z}\dot{\theta}\cos\theta - (l+z)\dot{\theta}^2\sin\theta, \\ \ddot{y} &= \ddot{z}\cos\theta - (l+z)\ddot{\theta}\sin\theta - 2\dot{z}\dot{\theta}\sin\theta - (l+z)\dot{\theta}^2\cos\theta. \end{aligned} \tag{1.6.25}$$

The energy of acceleration is (the terms not containing second derivatives of generalized acceleration are omitted)

$$S = \frac{1}{2}m[\ddot{z}^2 + (l+z)^2\ddot{\theta}^2 - 2(l+z)\dot{\theta}^2\ddot{z} + 4(l+z)\dot{z}\dot{\theta}\ddot{\theta}]. \tag{1.6.26}$$

To find the generalized forces Q_z and Q_θ we note that the spring is linear and the elastic force is proportional to the total deformation of the spring, that is, $F_{el} = c(f_{st} + z) = c(mg/c + z)$. The direction of this force is depicted in Figure 1.6.1. The projections of elastic and gravitational forces are

$$X = -c(mg/c + z)\sin\theta, \quad Y = -c(mg/c + z)\cos\theta. \tag{1.6.27}$$

The virtual displacements δx and δy are found from (1.6.24) to be

$$\begin{aligned}
\delta x &= \sin\theta\,\delta z + (l_0 + mg/c + z)\cos\theta\,\delta\theta, \\
\delta y &= \cos\theta\,\delta z - (l_0 + mg/c + z)\sin\theta\,\delta\theta.
\end{aligned} \tag{1.6.28}$$

Therefore, the virtual work of the active forces F_{el} and mg is

$$\delta A = X\delta x + Y\delta y = [-cz - mg\,(1 - \cos\theta)]\delta z + [-mg\,(l + z)\sin\theta]\delta\theta. \tag{1.6.29}$$

The generalized forces are therefore

$$Q_z = -cz - mg\,(1 - \cos\theta)\,, \quad Q_\theta = -mg\,(l + z)\sin\theta. \tag{1.6.30}$$

The Gibbs–Appell differential equations $\partial S/\partial\ddot{z} - Q_z = 0$ and $\partial S/\partial\ddot{\theta} - Q_\theta = 0$ lead to the following system of nonlinear differential equations:

$$\begin{aligned}
m\ddot{z} - (l + z)\,\dot{\theta}^2 + cz + mg\,(1 - \cos\theta) &= 0, \\
(l + z)\,\ddot{\theta} + 2\dot{z}\dot{\theta} + g\sin\theta &= 0, \tag{1.6.31}
\end{aligned}$$

where the second equation is divided by the common factor $m\,(l + z)$. If the pendulum is performing small vibrations, the system (1.6.31) becomes

$$\ddot{z} + (c/m)\,z = 0, \quad \ddot{\theta} + (g/l)\,\theta = 0, \tag{1.6.32}$$

whose solution is easily found to be

$$z = A\sin\left(\sqrt{c/m}\,t + \alpha\right), \quad \theta = B\sin\left(\sqrt{g/l}\,t + \beta\right), \tag{1.6.33}$$

where A, α, B, and β are constants of integration.

Note that the small but still nonlinear vibrations of this problem have been studied by means of asymptotic methods in [76, pp.185–189].

1.6.2 The Gibbs–Appell Equations: Nonholonomic Dynamical Systems

The Gibbs–Appell equations play a very important role in nonholonomic dynamics since they can generate much simpler differential equations in comparison to the Euler–Lagrangian equations for nonholonomic systems (1.5.18). Namely, we shall demonstrate that the central dynamical equation (1.6.12) can be transformed into a form that leads to the differential equations which are free of the undetermined Lagrangian multipliers.

Let us consider a dynamical system whose position at an arbitrary moment of time is specified by n generalized coordinates $q_1, ..., q_n$. At the same time let us suppose that the system is subjected to r nonholonomic constraints in the form (1.5.2):

$$A_{\alpha s}\dot{q}_s + B_\alpha = 0, \quad \alpha = 1, ..., r, \quad r < n, \tag{1.6.34}$$

where the coefficients $A_{\alpha k}$ and B_α are given functions of the generalized coordinates q_s $(s = 1, ..., n)$ and time t. As demonstrated before, the virtual displacements δq_s are satisfying the Hertz–Hölder conditions (1.5.12)

$$A_{\alpha s} \delta q_s = 0. \tag{1.6.35}$$

Thus, the system has $\sigma = n - r$ degrees of freedom. Let us divide the generalized velocities and virtual displacements δq_s into two groups: dependent generalized velocities $\dot{q}_m$ and dependent virtual displacements δq_m whose number is equal to the number r of nonholonomic constraints, and independent generalized velocities $\dot{q}_p$ and independent virtual displacements δq_p whose number is equal to $\sigma = n - r$, that is, number of degrees of freedom. From (1.6.34) and (1.6.35) we can express the dependent generalized velocities and dependent virtual displacements in terms of independent ones as

$$\dot{q}_m = a_{mp}\dot{q}_p + b_m, \quad m = 1, ..., r, \quad p = r + 1, r + 2, ..., n \tag{1.6.36}$$

and

$$\delta q_m = a_{mp}\delta q_p, \quad m = 1, ..., r, \quad p = r + 1, r + 2, ..., n. \tag{1.6.37}$$

Differentiating (1.6.36) totally with respect to time, we have

$$\ddot{q}_m = a_{mp}\ddot{q}_p + \Lambda_m\left(t, q_1, ..., q_n, \dot{q}_1, ..., \dot{q}_n\right), \tag{1.6.38}$$

where Λ_m denotes the group of terms that are independent of the generalized accelerations $\ddot{q}_s$.

In analogy with the variational equation (1.6.37) we suppose that the virtual accelerations $\delta\ddot{q}_s$ satisfy the conditions

$$\delta\ddot{q}_m = a_{mp}\delta\ddot{q}_p, \quad m = 1, ..., r, \quad p = r + 1, r + 2, ..., n. \tag{1.6.39}$$

Employing (1.6.38), we can transform the Appell function S in terms of $3n - r + 1$ variables: $t, q_1, ..., q_n, \dot{q}_1, ..., \dot{q}_n, \ddot{q}_{r+1}, ..., \ddot{q}_n$, namely

$$\begin{aligned} S &= S\left(t, q_1, ..., q_n, \dot{q}_1, ..., \dot{q}_n, \ddot{q}_1, ..., \ddot{q}_n, \right) \\ &= S^*\left(t, q_1, ..., q_n, \dot{q}_1, ..., \dot{q}_n, \ddot{q}_{r+1}, ..., \ddot{q}_n\right), \end{aligned} \tag{1.6.40}$$

whence

$$\frac{\partial S}{\partial \ddot{q}_s}\delta\ddot{q}_s = \frac{\partial S^*}{\partial \ddot{q}_p}\delta\ddot{q}_p, \quad s = 1, ..., n \quad p = r + 1, ..., n. \tag{1.6.41}$$

This system of equations is equivalent to

$$\frac{\partial S}{\partial \ddot{q}_s}\delta q_s = \frac{\partial S^*}{\partial \ddot{q}_p}\delta q_p, \quad s = 1, ..., n \quad p = r + 1, ..., n \tag{1.6.42}$$

for all virtual displacements that satisfy (1.6.35).

To find transformed (independent) generalized forces Q_p^* we employ the following equation:

$$Q_s \delta q_s = Q_p^* \delta q_p, \quad s = 1, ..., n, \quad p = r + 1, ..., n, \tag{1.6.43}$$

where we employed the equation (1.6.39).

By using equations (1.6.42) and (1.6.43), the central dynamical equation (1.6.12) becomes

$$\left(\frac{\partial S^*}{\partial \ddot{q}_p} - Q_p^* \right) \delta q_p = 0, \quad p = r + 1, ..., n. \tag{1.6.44}$$

Since all δq_p are mutually independent and arbitrary, we arrive at the Gibbs–Appell equations for nonholonomic dynamical systems

$$\frac{\partial S^*}{\partial \ddot{q}_p} = Q_p^*, \quad p = r + 1, ..., n. \tag{1.6.45}$$

The number of these equations is equal to the number of degrees of freedom $\sigma = n - r$. Together with them, we must consider simultaneously the r nonholonomic constraints (1.6.34) or (1.6.36). The most important property of the Gibbs–Appell equations is the fact that they are free of the Lagrangian undetermined multipliers, and at the same time, as has been noted by L. A. Pars [84, p. 202] "The Gibbs-Appell equations provide what is probably the simplest and most comprehensive form of the equations of motion so far discovered."

As a simple illustration of the method, let us consider the example treated in section 1.5 and depicted in Figure 1.5.1, which has been solved by the Euler–Lagrangian equations with undetermined multipliers.

Since the plate moves in the plane $x0y$ the energy of accelerations, according to (1.6.20) reads

$$S = \frac{1}{2} m \left(\ddot{x}_c^2 + \ddot{y}_c^2 \right) + \frac{1}{2} k^2 \ddot{\varphi}^2. \tag{1.6.46}$$

Differentiating the nonholonomic constraint (1.5.22) with respect to time we have $\ddot{y}_c = \ddot{x}_c \tan \varphi + \dot{\varphi} \dot{x}_c / \cos^2 \varphi$. Thus, the coordinate y is the dependent coordinate and x_c and φ are independent coordinates (compare with equation (1.6.38)). Entering with this into (1.6.46) we find

$$S^* = \frac{1}{2} m \left(\frac{\ddot{x}_c^2}{\cos^2 \varphi} + 2 \dot{\varphi} \dot{x}_c \frac{\sin \varphi}{\cos^3 \varphi} \ddot{x}_s \right) + \frac{1}{2} k^2 \ddot{\varphi}^2, \tag{1.6.47}$$

where the irrelevant terms not containing the second derivatives are discarded. Since the active force mg is performing the virtual work only along the x-axis, we find that $\delta A = mg \sin \alpha \, \delta x$, and hence

$$Q_x = Q_x^* = mg \sin \alpha, \quad Q_\varphi^* = 0. \tag{1.6.48}$$

The Gibbs–Appell equations $\partial S^*/\partial \ddot{x}_c = Q_x^*, \partial S^*/\partial \varphi = Q_\varphi^*$ lead to

$$\frac{\ddot{x}_c}{\cos^2 \varphi} + \dot{\varphi}\dot{x}_c\frac{\sin \varphi}{\cos^3 \varphi} = g \sin \alpha,$$
$$\ddot{\varphi} = 0. \tag{1.6.49}$$

From the second equation we find

$$\varphi = \omega t + \varphi_0, \tag{1.6.50}$$

where ω and φ_0 are constants of integration. Then, the $(1.6.49)_1$ becomes

$$\ddot{x}_c + \omega\dot{x}_c \tan \varphi = \bar{g} \cos^2 \varphi \quad (\bar{g} = g \sin \alpha), \tag{1.6.51}$$

which is precisely the equation (1.5.29), whose solution is presented in section 1.5.

1.6.3   Kane's Equations

Euler–Lagrangian equations (1.4.1) are derived from the Lagrange–D'Alembert principle (1.3.4). Two important concepts were used in the process of derivation: the concept of *virtual displacements* introduced in section 1.2, and the *orthogonality condition* for the constraint forces (1.3.3). There is another approach to analytical mechanics, introduced by Kane in 1961 (see [57]), in which equations of motion are derived without the concept of virtual displacement. We briefly give an outline of the derivation of Kane's equations.

The starting point in Kane's approach is also D'Alembert's principle, which states that the active and inertial forces in the mechanical system are in equilibrium (see [52]). The conditions of equilibrium may be expressed in two different ways. First they could be expressed (as did Lagrange, motivated by analytical statics of Bernoulli) as a requirement that the sum of works of all forces on virtual displacements be equal to zero, or (see (1.3.1))

$$\sum_{i=1}^{N} (m_i\ddot{\mathbf{r}}_i - \mathbf{F}_i - \mathbf{R}_i) \cdot \delta\mathbf{r}_i = 0. \tag{1.6.52}$$

Equation (1.6.52) together with the definition of ideal constraint (1.3.3), that is, $\sum_{i=1}^{N} \mathbf{R}_i \cdot \delta\mathbf{r}_i = 0$, leads to Euler–Lagrangian equations (1.4.1).

The second interpretation of D'Alembert's principle is obtained if conditions of equilibrium are expressed as the requirement that the resultant force and resultant couple are equal to zero (Newton's definition of equilibrium). In this case we obtain (see [52, p. 218])

$$\sum_{i=1}^{N} (m_i\ddot{\mathbf{r}}_i - \mathbf{F}_i - \mathbf{R}_i) = 0, \quad \sum_{i=1}^{N} \mathbf{r}_i \times (m_i\ddot{\mathbf{r}}_i - \mathbf{F}_i - \mathbf{R}_i) = 0, \tag{1.6.53}$$

where $\mathbf{r}_i$ are position vectors of material points relative to an arbitrary fixed coordinate system. Kane's equations are obtained if (1.6.53) is projected on specially chosen directions (see [55, p. 416]). These directions are the *partial velocity vectors* of a point i relative to the generalized coordinate q_s and are given as (see (1.6.4))

$$\mathbf{v}_{\dot{q}_s}^{(i)} = \frac{\partial \mathbf{v}_i}{\partial \dot{q}_s}, \quad s = 1, ..., n. \tag{1.6.54}$$

Note that from (1.6.4) it follows that $\mathbf{v}_{\dot{q}_s}^{(i)} = \frac{\partial \mathbf{v}_i}{\partial \dot{q}_s} = \frac{\partial \mathbf{r}_i}{\partial q_s} = \frac{\partial \mathbf{a}_i}{\partial \ddot{q}_s}$. By taking the scalar product of $m_i \ddot{\mathbf{r}}_i$ and $\mathbf{F}_i - \mathbf{R}_i$ with $\mathbf{v}_{\dot{q}_s}^{(i)}$ for each i and summing, the following relation is obtained:

$$\sum_{i=1}^{N} m_i \ddot{\mathbf{r}}_i \cdot \mathbf{v}_{\dot{q}_s}^{(i)} = \sum_{i=1}^{N} \left(\mathbf{F}_i \cdot \frac{\partial \mathbf{v}_i}{\partial \dot{q}_s} + \mathbf{R}_i \cdot \frac{\partial \mathbf{v}_i}{\partial \dot{q}_s} \right). \tag{1.6.55}$$

The reaction forces are also assumed to satisfy $\sum_{i=1}^{N} \mathbf{R}_i \cdot \frac{\partial \mathbf{v}_i}{\partial \dot{q}_s} = \sum_{i=1}^{N} \mathbf{R}_i \cdot \frac{\partial \mathbf{r}_i}{\partial q_s} = 0$, so that from (1.6.55) we obtain

$$Q_s + Q_s^* = 0, \quad s = 1, ..., n, \tag{1.6.56}$$

where Q_s are generalized applied forces and Q_s^* are generalized inertia forces, that is

$$Q_s = \sum_{i=1}^{N} \mathbf{F}_i \cdot \frac{\partial \mathbf{v}_i}{\partial \dot{q}_s}, \quad Q_s^* = -\sum_{i=1}^{N} m_i \ddot{\mathbf{r}}_i \cdot \frac{\partial \mathbf{v}_i}{\partial \dot{q}_s} = -\frac{d}{dt}\frac{\partial T}{\partial \dot{q}_s} + \frac{\partial T}{\partial q_s}, \tag{1.6.57}$$

where $T = \frac{1}{2}\sum_{i=1}^{N} \mathbf{v}_i \cdot \mathbf{v}_i$ is the kinetic energy. In writing the last equality in (1.6.57), we used (1.3.23).

Equations (1.6.56) are *Kane's* equations. The generalized forces and generalized inertia forces in equation (1.6.56), for the case of a rigid body, could be determined either from (1.6.57) or in the following way. Suppose $\boldsymbol{\omega}$ and $\boldsymbol{\alpha}$ are angular velocity and angular acceleration of the body expressed in terms of generalized coordinates q_s and generalized velocities $\dot{q}_s$. Let $\mathbf{I}$ be the inertia tensor of the body calculated for the body's mass center and let $\boldsymbol{\omega}_{\dot{q}_s} = \partial \boldsymbol{\omega}/\partial \dot{q}_s$ be the *partial angular velocity vector*. Then the generalized forces and generalized inertia forces are given as (see [55, pp. 365–379])

$$Q_s = \mathbf{v}_{\dot{q}_s}^{C} \cdot \left(\sum_{i=1}^{N} \mathbf{F}_i \right) + \boldsymbol{\omega}_{\dot{q}_s} \cdot \sum_{i=1}^{N} \mathbf{r}_i^{C} \times \mathbf{F}_i,$$

$$Q_s^* = \mathbf{v}_{\dot{q}_s}^{C} \cdot \left(\sum_{i=1}^{N} m_i \right) \mathbf{a}^{C} + \boldsymbol{\omega}_{\dot{q}_s} \cdot \left[-\mathbf{I} \cdot \boldsymbol{\alpha} - \boldsymbol{\omega} \times (\mathbf{I} \cdot \boldsymbol{\omega}) \right], \tag{1.6.58}$$

where $\mathbf{r}_i^{C}$ denotes the position vector of the point of application of the ith force with respect to mass center, $\mathbf{a}^{C}$ is the acceleration of the mass center, and $\mathbf{v}_{\dot{q}_s}^{C}$ is the partial velocity vector of the mass center.

The procedure of forming Kane's equations will be demonstrated in the next example.

Example 1.6.2. *Double rod pendulum.* Consider a system shown in Figure 1.6.2 consisting of two equal homogeneous rods 1 and 2 having equal lengths l and mass m. The system moves in vertical plane in a constant gravity field. As generalized coordinates we take the angles between the vertical direction and the rod axes, θ_1 and θ_2.

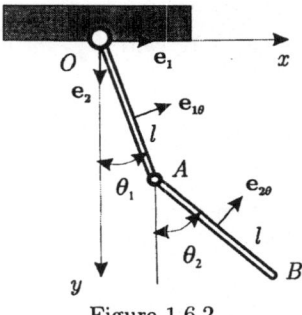

Figure 1.6.2

The position vectors of the mass centers of the rods are

$$\mathbf{r}_1 = \frac{l}{2}\sin\theta_1\mathbf{e}_1 + \frac{l}{2}\cos\theta_1\mathbf{e}_2,$$

$$\mathbf{r}_2 = \left(l\sin\theta_1 + \frac{l}{2}\sin\theta_2\right)\mathbf{e}_1 + \left(l\cos\theta_1 + \frac{l}{2}\cos\theta_2\right)\mathbf{e}_2, \quad (1.6.59)$$

where $\mathbf{e}_1$ and $\mathbf{e}_2$ are unit vectors along the x and y axes, respectively. By differentiating (1.6.59) we determine the velocities of the mass centers as

$$\mathbf{v}_1 = \frac{l}{2}\dot\theta_1\mathbf{e}_{1\theta}, \quad \mathbf{v}_2 = l\dot\theta_1\mathbf{e}_{1\theta} + \frac{l}{2}\dot\theta_2\mathbf{e}_{2\theta}, \quad (1.6.60)$$

where the vectors $\mathbf{e}_{i\theta} = -\sin\theta_i\mathbf{e}_1 + \cos\theta_i\mathbf{e}_2, i = 1, 2$, are oriented along the axis normal to the rod (see Figure 1.6.2) for the i th rod. The partial velocity vectors (1.6.54) now become

$$\mathbf{v}_{\dot\theta_1}^{(1)} = \frac{l}{2}\mathbf{e}_{1\theta} = -\frac{l}{2}\sin\theta_1\mathbf{e}_1 + \frac{l}{2}\cos\theta_1\mathbf{e}_2,$$

$$\mathbf{v}_{\dot\theta_1}^{(2)} = l\mathbf{e}_{1\theta} = -l\sin\theta_1\mathbf{e}_1 + l\cos\theta_1\mathbf{e}_2,$$

$$\mathbf{v}_{\dot\theta_2}^{(1)} = 0, \quad \mathbf{v}_{\dot\theta_2}^{(2)} = \frac{l}{2}\mathbf{e}_{2\theta} = -\frac{l}{2}\sin\theta_2\mathbf{e}_1 + \frac{l}{2}\cos\theta_2\mathbf{e}_2. \quad (1.6.61)$$

The active forces are weights of the rods so that

$$\mathbf{F}_1 = mg\mathbf{e}_1, \quad \mathbf{F}_2 = mg\mathbf{e}_1,$$

where g is acceleration of gravity. It could be shown easily that the kinetic energy of the system reads

$$T = \frac{2}{3}ml^2\dot{\theta}_1^2 + \frac{1}{2}ml^2\dot{\theta}_1\dot{\theta}_2\cos(\theta_2 - \theta_1) + \frac{1}{6}ml^2\dot{\theta}_2^2. \tag{1.6.62}$$

From (1.6.61) and by using (1.6.57), the generalized active and inertia forces are

$$\begin{aligned}
Q_{\theta_1} &= \mathbf{F}_1 \cdot \mathbf{v}_{\dot{\theta}_1}^{(1)} + \mathbf{F}_2 \cdot \mathbf{v}_{\dot{\theta}_1}^{(2)} = mg\mathbf{e}_1 \cdot \left(-\frac{l}{2}\sin\theta_1\mathbf{e}_1 + \frac{l}{2}\cos\theta_1\mathbf{e}_2\right) \\
&\quad + mg\mathbf{e}_1 \cdot (-l\sin\theta_1\mathbf{e}_1 + l\cos\theta_1\mathbf{e}_2) = -\frac{3}{2}mgl\sin\theta_1, \\
Q_{\theta_2} &= \mathbf{F}_1 \cdot \mathbf{v}_{\dot{\theta}_2}^{(1)} + \mathbf{F}_2 \cdot \mathbf{v}_{\dot{\theta}_2}^{(2)} \\
&= mg\mathbf{e}_1 \cdot \left(-\frac{l}{2}\sin\theta_2\mathbf{e}_1 + \frac{l}{2}\cos\theta_2\mathbf{e}_2\right) = -\frac{1}{2}mgl\sin\theta_2, \\
Q_{\theta_1}^* &= -\frac{d}{dt}\frac{\partial T}{\partial\dot{\theta}_1} + \frac{\partial T}{\partial\theta_1} \\
&= -\frac{4}{3}ml^2\ddot{\theta}_1 - \frac{1}{2}ml^2\ddot{\theta}_2\cos(\theta_1 - \theta_2) + \frac{1}{2}ml^2\dot{\theta}_2^2\sin(\theta_2 - \theta_1), \\
Q_{\theta_2}^* &= -\frac{d}{dt}\frac{\partial T}{\partial\dot{\theta}_2} + \frac{\partial T}{\partial\theta_2} \\
&= -\frac{1}{3}ml^2\ddot{\theta}_2 - \frac{1}{2}ml^2\ddot{\theta}_1\cos(\theta_2 - \theta_1) - \frac{1}{2}ml^2\dot{\theta}_1^2\sin(\theta_2 - \theta_1).
\end{aligned} \tag{1.6.63}$$

From (1.6.56) and (1.6.63) we obtain the Kane's form of the equation of motion as

$$\begin{aligned}
\frac{4}{3}\ddot{\theta}_1 + \frac{1}{2}\ddot{\theta}_2\cos(\theta_2 - \theta_1) - \frac{1}{2}\dot{\theta}_2^2\sin(\theta_2 - \theta_1) + \frac{3g}{2l}\sin\theta_1 &= 0, \\
\frac{1}{2}\cos(\theta_2 - \theta_1)\ddot{\theta}_1 + \frac{1}{3}\ddot{\theta}_2 - \frac{1}{2}\dot{\theta}_1^2\sin(\theta_1 - \theta_2) + \frac{g}{2l}\sin\theta_2 &= 0.
\end{aligned} \tag{1.6.64}$$

1.7   Nielsen and Mangerone–Deleanu Differential Equations

Another class of the differential equations of motion of the holonomic dynamical systems, which are also fully equivalent to the Euler–Lagrangian equations, are known as the Nielsen and Mangerone–Deleanu equations. Let us consider a dynamical system that has the Lagrangian function of the form $L = L(t, q_1, ..., q_n, \dot{q}_1, ..., \dot{q}_n)$ and at the same time is subject to the nonconservative forces (nonpotential forces) $Q_1, ..., Q_n$ depending upon time, the generalized coordinates q_s, and generalized velocities $\dot{q}_1, ..., \dot{q}_n$. The Euler–Lagrangian equations are

$$\frac{d}{dt}\frac{\partial L}{\partial \dot{q}_i} - \frac{\partial L}{\partial q_i} = Q_i, \quad i = 1, ..., n. \tag{1.7.1}$$

Let us consider the time derivative of L,

$$\dot{L} = \frac{\partial L}{\partial \dot{q}_i}\ddot{q}_i + \frac{\partial L}{\partial q_i}\dot{q}_i + \frac{\partial L}{\partial t}, \tag{1.7.2}$$

whence

$$\frac{\partial \dot{L}}{\partial \ddot{q}_i} = \frac{\partial L}{\partial \dot{q}_i}. \tag{1.7.3}$$

Finding the derivative with respect to time of this expression, we have

$$\frac{d}{dt}\frac{\partial \dot{L}}{\partial \ddot{q}_i} = \frac{d}{dt}\frac{\partial L}{\partial \dot{q}_i} = \frac{\partial L}{\partial q_i} + Q_i, \tag{1.7.4}$$

where we have used equation (1.7.1). From the facts

$$\frac{d}{dt}\frac{\partial L}{\partial \dot{q}_i} = \frac{\partial^2 L}{\partial \dot{q}_i \partial \dot{q}_j}\ddot{q}_j + \frac{\partial^2 L}{\partial \dot{q}_i \partial q_j}\dot{q}_j + \frac{\partial^2 L}{\partial \dot{q}_i \partial t} \tag{1.7.5}$$

and

$$\frac{\partial \dot{L}}{\partial \dot{q}_i} = \frac{\partial^2 L}{\partial \dot{q}_i \partial \dot{q}_j}\ddot{q}_j + \frac{\partial^2 L}{\partial q_j \partial q_i}\dot{q}_j + \frac{\partial^2 L}{\partial \dot{q}_i \partial t} + \frac{\partial L}{\partial q_i}, \tag{1.7.6}$$

we find

$$\frac{\partial \dot{L}}{\partial \dot{q}_i} = \frac{d}{dt}\frac{\partial L}{\partial \dot{q}_i} + \frac{\partial L}{\partial q_i}. \tag{1.7.7}$$

Using (1.7.3) and (1.7.4), the last equation becomes

$$\frac{\partial \dot{L}}{\partial \dot{q}_i} = 2\frac{d}{dt}\frac{\partial \dot{L}}{\partial \ddot{q}_i} - Q_i \tag{1.7.8}$$

or

$$2\frac{d}{dt}\frac{\partial \dot{L}}{\partial \ddot{q}_i} - \frac{\partial \dot{L}}{\partial \dot{q}_i} = Q_i. \tag{1.7.9}$$

These differential equations are equivalent with the classical Euler–Lagrangian equations, and we shall refer to them as the generalized Nielsen equations. They contain only the first derivative with respect to time of the Lagrangian function L.

If the dynamical system contains only the fully nonconservative forces and the Lagrangian function is just equal to the kinetic energy of the dynamical system, the equation (1.7.9) becomes

$$2\frac{d}{dt}\frac{\partial \dot{T}}{\partial \ddot{q}_i} - \frac{\partial \dot{T}}{\partial \dot{q}_i} = Q_i. \tag{1.7.10}$$

For this case, from the equation (1.7.4) we have

$$\frac{d}{dt}\frac{\partial \dot{T}}{\partial \ddot{q}_i} = \frac{\partial T}{\partial \dot{q}_i} + Q_i. \tag{1.7.11}$$

Entering with this into (1.7.10), we arrive at the differential equations known as the Nielsen differential equations (see [77])

$$\frac{\partial \dot{T}}{\partial \dot{q}_i} - 2\frac{\partial T}{\partial q_i} = Q_i, \quad i = 1, ..., n. \tag{1.7.12}$$

It is our opinion that the generalized Nielsen equations (1.7.9) derived here have some advantages in comparison to the Nielsen equations for the following two reasons: (i) The generalized Nielsen equations depend upon one function $\dot{L}$, while the Nielsen equations contain T and $\dot{T}$. (ii) For the case of Lagrangian dynamical systems for which the nonconservative forces are equal to zero, that is, $Q_i = 0$, the generalized Nielsen equations become

$$2\frac{d}{dt}\frac{\partial \dot{L}}{\partial \ddot{q}_i} - \frac{\partial \dot{L}}{\partial \dot{q}_i} = 0, \quad i = 1, ..., n, \tag{1.7.13}$$

while the Nielsen equations $\frac{\partial \dot{T}}{\partial \dot{q}_i} - 2\frac{\partial T}{\partial q_i} = 0$ describe only the inertial motion of the dynamical system, that is, the motion of a system that is not subjected to any external forces and is characterized by the kinetic energy only.

It is possible to show that by introducing consecutive time derivatives of the higher order of the Lagrangian function $\ddot{L}, \dddot{L}, ..., \overset{(p)}{L}$ and repeating a procedure similar to the one used above, we can derive differential equations of motion that depend uniformly upon the corresponding derivative of the Lagrangian function L. All those differential equations are fully equivalent to the "classical" Euler–Lagrangian equations (1.7.1).

For example, let us consider the second derivative with respect to time of the Lagrangian function $L = L(t, q, \dot{q})$[6]:

$$
\begin{aligned}
\ddot{L} =\ & \frac{\partial L}{\partial \dot{q}}\dddot{q} + \frac{\partial^2 L}{\partial \dot{q}^2}\dot{q}\ddot{q} + 2\frac{\partial^2 L}{\partial q \partial \dot{q}}\dot{q}\ddot{q} + 2\frac{\partial L}{\partial \dot{q}\partial t}\ddot{q} + 2\frac{\partial^2 L}{\partial q \partial t}\dot{q} \\
& + \frac{\partial^2 L}{\partial q^2}\dot{q}^2 + \frac{\partial^2 L}{\partial t^2} + \frac{\partial L}{\partial q}\ddot{q},
\end{aligned} \tag{1.7.14}
$$

---

[6]For the sake of simplicity, we consider here the case of the dynamical system with one degree of freedom. The transition to the case of many degrees of freedom is trivial.

whence

$$\frac{\partial \ddot{L}}{\partial \dddot{q}} = \frac{\partial L}{\partial \dot{q}}. \tag{1.7.15}$$

Differentiating this with respect to time and using the Euler–Lagrangian equations (1.7.1), one has

$$\frac{d}{dt}\frac{\partial \ddot{L}}{\partial \dddot{q}} = \frac{d}{dt}\frac{\partial L}{\partial \dot{q}} = Q + \frac{\partial L}{\partial q}, \tag{1.7.16}$$

where Q denotes the nonpotential force.

From (1.7.14) we find

$$\frac{\partial \ddot{L}}{\partial \ddot{q}} = 2\left(\frac{\partial^2 L}{\partial \dot{q}^2}\ddot{q} + \frac{\partial^2 L}{\partial \dot{q}\partial q}\dot{q} + \frac{\partial^2 L}{\partial \dot{q}\partial t}\right) + \frac{\partial L}{\partial q}. \tag{1.7.17}$$

The expression in the parentheses is equal to $\frac{d}{dt}\frac{\partial L}{\partial \dot{q}}$. Hence, employing again the Euler–Lagrangian equations, we have

$$\frac{\partial \ddot{L}}{\partial \ddot{q}} = 2\frac{d}{dt}\frac{\partial L}{\partial \dot{q}} + \frac{\partial L}{\partial q} = 2Q + 3\frac{\partial L}{\partial q} \tag{1.7.18}$$

or

$$\frac{1}{2}\left(\frac{\partial \ddot{L}}{\partial \ddot{q}} - 3\frac{\partial L}{\partial q}\right) = Q. \tag{1.7.19}$$

If we had n generalized coordinates, we would have, instead of (1.7.19), the following expression:

$$\frac{1}{2}\left(\frac{\partial \ddot{L}}{\partial \ddot{q}_i} - 3\frac{\partial L}{\partial q_i}\right) = Q_i, \quad i = 1, ..., n. \tag{1.7.20}$$

We shall refer to these differential equations as the *generalized Tzénoff equations*. In fact Tzénoff [109] derived the special form of these equations for which $L = T$, namely,

$$\frac{1}{2}\left(\frac{\partial \ddot{T}}{\partial \ddot{q}_i} - 3\frac{\partial T}{\partial q_i}\right) = Q_i, \quad i = 1, ..., n. \tag{1.7.21}$$

In order to unify the generalized Tzénoff equations with respect to the time derivatives of the Lagrangian function, we substitute $\partial L/\partial q_i$ from (1.7.16) into (1.7.20) and obtain

$$3\frac{d}{dt}\frac{\partial \ddot{L}}{\partial \dddot{q}_i} - \frac{\partial \ddot{L}}{\partial \ddot{q}_i} = Q_i, \quad i = 1, ..., n. \tag{1.7.22}$$

Finally, considering the structure of the equations (1.7.9) and (1.7.11) derived from the $\dot{L}$ and $\ddot{L}$, we can derive the general case of the differential equations of motion containing the arbitrary derivative of the Lagrangian function:

$$(p+1)\frac{d}{dt}\frac{\partial \overset{(p)}{L}}{\partial \overset{(p+1)}{q_i}} - \frac{\partial \overset{(p)}{L}}{\partial \overset{(p)}{q_i}} = Q_i, \quad i = 1,...,n, \quad p = 0,1,2,3,..., \tag{1.7.23}$$

which we shall refer to as the generalized Mangeron–Deleanu differential equations of the holonomic dynamical system possessing the given Lagrangian function L and a set of purely nonconservative generalized forces Q_i. It is obvious from (1.7.23) that for $p = 0$ we obtain the "classical" Euler–Lagrangian differential equations, for $p = 1$ the generalized Nielsen equations (1.7.9), and for $p = 2$ the generalized Tzénoff's equations (1.7.22). It is also to be noted that on the basis of the classical Nielsen differential equations (1.7.12) and (1.7.21) we can derive the case for the arbitrary derivative of the kinetic energy, namely,

$$\frac{1}{p}\left[\frac{\partial \overset{(p)}{T}}{\partial \overset{(p)}{q_i}} - (p+1)\frac{\partial T}{\partial q_i}\right] = Q_i, \quad i = 1,...,n, \quad p = 1,2,3,.... \tag{1.7.24}$$

These equations are known as the Mangeron–Deleanu differential equations [69].

As far as the holonomic dynamical systems are concerned, all differential equations derived in this paragraph are completely equivalent to the ordinary differential equations of Euler and Lagrange (1.7.1). However, some authors [39], [48], [34] have demonstrated that Nielsen, Tzénoff, and Mangeron–Deleanu equations can be advantageously used in nonholonomic mechanics, especially when we are faced with the nonlinear nonholonomic constraints and nonholonomic constraints containing the second time derivatives of the generalized coordinates.

1.8 Hamilton's Canonical Differential Equations of Motion

Thus far, we have considered the differential equations of motion of holonomic and nonholonomic dynamical systems which have been exclusively the differential equations of the second order with respect to the generalized coordinates q_i. In this section, we shall derive the famous differential equations of motion known as the Hamiltonian or canonical differential equations, which are of the first order with respect to the generalized coordinates q_i, and a new set of quantities named *generalized momenta* p_i, defined by the relations

$$p_i = \frac{\partial L}{\partial \dot{q}_i}, \quad i = 1,...,n. \tag{1.8.1}$$

If the following determinant is different from zero:

$$\det \left| \frac{\partial^2 L}{\partial \dot{q}_i \partial \dot{q}_j} \right| \neq 0, \tag{1.8.2}$$

we can solve n equations (1.8.1) with respect to generalized velocities $\dot{q}_i$ in terms of time t, generalized coordinates q_i, and generalized momenta p_i, to obtain

$$\dot{q}_i = f\left(t, q_1, ..., q_n, p_1, ..., p_n\right), \quad i = 1, ..., n. \tag{1.8.3}$$

These equations represent the first group of the canonical equations, which we will represent in a different form in the course of our analysis.

Let us now introduce a new function called the *Hamiltonian function* or simply the *Hamiltonian* defined as[7]

$$H = p_i \dot{q}_i - L\left(t, q_1, ..., q_n, \dot{q}_1, ..., \dot{q}_n\right). \tag{1.8.4}$$

Entering with (1.8.3) into (1.8.4), we can express the Hamiltonian function in terms of t, q_i, and p_i:

$$H = H\left(t, q_1, ..., q_n, p_1, ..., p_n\right). \tag{1.8.5}$$

From (1.8.4) it follows that the Lagrangian function of the transformed system can be written as

$$L = p_i \dot{q}_i - H\left(t, q_1, ..., q_n, p_1, ..., p_n\right). \tag{1.8.6}$$

Our further considerations will be based upon one of the forms of the central Lagrangian equation given by (1.3.33) or (1.3.39). For example, let us consider the central Lagrangian equation (1.3.39)

$$\frac{d}{dt}\left(p_i \delta q_i\right) = \delta L + Q_i \delta q_i + p_i \left[(\delta q_i) - \delta \dot{q}_i\right]. \tag{1.8.7}$$

Taking into account (1.8.6) this equation can be easily transformed into canonical variables

$$\frac{d}{dt}\left(p_i \delta q_i\right) = \delta \left[p_i \dot{q}_i - H\left(t, q_1, ..., q_n, p_1, ..., p_n\right)\right] + Q_i \delta q_i + p_i \left[(\delta q_i) - \delta \dot{q}_i\right]. \tag{1.8.8}$$

Note that the purely nonconservative forces $Q_i\left(t, q_1, ..., q_n, \dot{q}_1, ..., \dot{q}_n\right)$ figuring into (1.8.7) are expressed in terms of the canonical variables q_i and p_i by means of (1.8.3):

$$Q_i = Q_i\left(t, q_1, ..., q_n, p_1, ..., p_n\right), \quad i = 1, ..., n. \tag{1.8.9}$$

---

[7]The transformation from the Lagrangian position coordinates $q_1, ..., q_n$ to $2n$ canonical variables $q_1, ..., q_n, p_1, ..., p_n$ is usually referred to as the *Legendre transformation*.

From (1.8.8) we find

$$
\begin{aligned}
\dot{p}_i \delta q_i + p_i \left(\delta q_i \right)^{\cdot} &= \dot{q}_i \delta p_i + p_i \delta \dot{q}_i \\
&\quad - \frac{\partial H}{\partial q_i} \delta q_i - \frac{\partial H}{\partial p_i} \delta p_i + Q_i \delta q_i + p_i \left[(\delta q_i)^{\cdot} - \delta \dot{q}_i \right],
\end{aligned}
\tag{1.8.10}
$$

whence

$$
\left(\dot{p}_i + \frac{\partial H}{\partial q_i} - Q_i \right) \delta q_i + \left(-\dot{q}_s + \frac{\partial H}{\partial p_s} \right) \delta p_s = 0.
\tag{1.8.11}
$$

Since the generalized coordinates q_i and generalized momenta p_i should be considered mutually independent, the variations δq_i and δp_i are also mutually independent and arbitrary. Thus, from (1.8.11), it follows that

$$
\dot{q}_s = \frac{\partial H}{\partial p_s}, \quad \dot{p}_i = -\frac{\partial H}{\partial q_i} + Q_i \left(t, q_1, ..., q_n, p_1, ..., p_n \right), \quad i = 1, ..., n.
\tag{1.8.12}
$$

These are the *Hamiltonian canonical differential equations of motion*.

It is easy to demonstrate that the first group of these equations is equivalent to (1.8.3). Calculating the partial derivative of (1.8.4) with respect to p_i, we find

$$
\frac{\partial H}{\partial p_i} = f_i + p_j \frac{\partial f_j}{\partial p_i} - \frac{\partial L}{\partial \dot{q}_j} \frac{\partial f_j}{\partial p_i}.
\tag{1.8.13}
$$

Using (1.8.1) and (1.8.3), the last two terms on the right-hand side cancel, and we arrive at the first group of canonical equations (1.8.12).

If the dynamical system is not exposed to nonconservative forces, that is, if $Q_i = 0$, the most frequent form of the canonical equations is

$$
\dot{q}_s = \frac{\partial H}{\partial p_s}, \quad \dot{p}_i = -\frac{\partial H}{\partial q_i}, \quad i = 1, ..., n.
\tag{1.8.14}
$$

It is to be noted that the time derivative of the Hamiltonian function (1.8.5) can help us to understand the physical meaning of the function. From (1.8.5) we find

$$
\frac{dH}{dt} = \frac{\partial H}{\partial q_i} \dot{q}_i + \frac{\partial H}{\partial p_i} \dot{p}_i + \frac{\partial H}{\partial t}.
\tag{1.8.15}
$$

Substituting (1.8.12) into this equation one finds

$$
\frac{dH}{dt} = \frac{\partial H}{\partial p_i} Q_i + \frac{\partial H}{\partial t}.
\tag{1.8.16}
$$

Thus, if the generalized nonconservative forces are absent, $Q_i = 0$, and the dynamical system is scleronomic (i.e., $\partial L / \partial t = \partial H / \partial t = 0$), then it follows that the Hamiltonian function is a constant of motion:

$$
H = H \left(q_1, ..., q_n, p_1, ..., p_n \right) = const.
\tag{1.8.17}
$$

By using the Jacobi conservation law (1.4.45), $(\partial L/\partial \dot{q}_i)\,\dot{q}_i - L = const.$, and comparing this with (1.8.5)and (1.8.1), it is evident that the Hamiltonian function represents the Jacobi conservation law expressed in terms of $2n$ canonical variables q_i and p_i. If the dynamical system is of such a nature that $L = T - \Pi$ the Hamiltonian function (1.8.17) (under the condition $Q_i = 0$) is the constant of motion and represents the total mechanical energy $H = E = const.$

In subsection 1.4.3 we have defined a cyclic coordinate q_j as one that does not appear explicitly in the Lagrangian function. It is easy to see that if a generalized coordinate does not appear in a Lagrangian function the same coordinate will be absent from the corresponding Hamilton's function H. Therefore, for the case of the cyclic coordinate that does not occur in H, the corresponding momentum is constant: $\dot{p}_j = -\partial H/\partial q_j = 0$; that is,

$$p_j = const. \quad (j \text{ is a fixed number, and } Q_j = 0). \qquad (1.8.18)$$

It is clear that the general solution of the canonical system (1.8.12) is of the form

$$\begin{aligned} q_i &= q_i\,(t, C_1, ..., C_{2n})\,, \\ p_i &= p_i\,(t, C_1, ..., C_{2n})\,, \end{aligned} \qquad (1.8.19)$$

where $C_1, ..., C_{2n}$ are constants of integration that can be determined from the given initial conditions $q_i\,(0)$ and $p_i\,(0)\,.$

Note that from (1.8.6) the following relations follow immediately:

$$\frac{\partial L}{\partial q_i} = -\frac{\partial H}{\partial q_i}, \quad \frac{\partial L}{\partial t} = -\frac{\partial H}{\partial t}, \quad i = 1, ..., n. \qquad (1.8.20)$$

The space of $2n$ dimensions whose point is defined by $2n$ coordinates q_i and p_i $(i = 1, ..., n)$ is referred to as the *phase space*. The motion of the dynamical system can be interpreted as the motion of a point in the phase space. The structure of the phase space can be geometrically described as the Cartesian orthogonal space of the dimension $2n$. Naturally, the space defined by $2n + 1$ coordinates q_i, p_i, and t is called the *extended phase space*.

We note that the forming of canonical equations of motion is based strictly on the given Lagrangian function L and the given nonconservative forces Q_i of the dynamical problem in question. It is interesting that we can easily derive the canonical equations of motion from the Lagrangian function expressed in the form (1.8.6) and the given generalized nonconservative forces $Q_i = Q_i\,(t, q_i, p_i)\,.$ Considering the generalized coordinates q_i and generalized momenta p_i as mutually independent parameters, it is easy to see that the Euler–Lagrangian equations

$$\frac{d}{dt}\frac{\partial L}{\partial \dot{q}_i} - \frac{\partial L}{\partial q_i} = Q_i, \quad \frac{d}{dt}\frac{\partial L}{\partial \dot{p}_i} - \frac{\partial L}{\partial p_i} = 0, \qquad (1.8.21)$$

will generate exactly the canonical equations (1.8.12).

The formal procedure for forming canonical equations discussed above will be illustrated by the following example.

Let us consider a mass m that has one degree of freedom; that is, its location at any time is specified by one coordinate $q(t)$. We shall study the vibration problem for which the mass is subjected to the restoring force $-cq$ and the viscous damping force $-b\dot{q}$, and thus the differential equation of motion is of the form

$$m\ddot{q} = -cq - b\dot{q}. \tag{1.8.22}$$

We shall write this equation in the standard form

$$\ddot{q} + 2k\dot{q} + \omega^2 q = 0, \tag{1.8.23}$$

where

$$\frac{b}{m} = 2k, \quad \frac{c}{m} = \omega^2. \tag{1.8.24}$$

The parameter k is called the damping coefficient and ω is called the circular frequency of the oscillator.

Despite the fact that the system is nonconservative since the damping force is not of potential nature, a Lagrangian function exists and is of the form

$$L = \frac{1}{2}\left(\dot{q}^2 - \omega^2 q^2\right) e^{2kt}. \tag{1.8.25}$$

Namely, it is easy to verify that the Euler–Lagrangian equation $(\partial L/\partial \dot{q})' - \partial L/\partial q = 0$ will generate the differential equation (1.8.23). The generalized momentum according to (1.8.1) is found to be

$$p = \frac{\partial L}{\partial \dot{q}} = \dot{q}e^{2kt}, \tag{1.8.26}$$

thus

$$\dot{q} = pe^{-2kt}. \tag{1.8.27}$$

According to (1.8.4) the Hamiltonian function is

$$H = p\dot{q} - L(t, q, \dot{q}) = \frac{1}{2}p^2 e^{-2kt} + \frac{1}{2}\omega^2 q^2 e^{2kt}. \tag{1.8.28}$$

The canonical equations of motion according to (1.8.12) are of the form

$$\dot{q} = \frac{\partial H}{\partial p} = pe^{-2kt}, \quad \dot{p} = -\frac{\partial H}{\partial q} = -\omega^2 q e^{2kt}. \tag{1.8.29}$$

Since the Hamiltonian function (1.8.28) depends explicitly on time, this function is not a constant of motion. In fact, totally differentiating (1.8.28) with respect to time and entering into the resulting expression with $\dot{q}$ and $\dot{p}$ given by (1.8.29), we obtain

$$\dot{H} = -k\left(p^2 e^{-2kt} - \omega^2 q^2 e^{2kt}\right). \qquad (1.8.30)$$

Since in this problem the nonconservative force $Q = 0$, it is easy to verify that the equation (1.8.16), $\dot{H} = \partial H/\partial t$, holds. However, employing the canonical equations (1.8.29), the right-hand side of (1.8.30) can be written in the form

$$\dot{H} = -k\left(p\dot{q} + \dot{p}q\right), \qquad (1.8.31)$$

and we arrive at the conservation law of the form

$$H + kqp = const. = C, \qquad (1.8.32)$$

or writing explicitly using (1.8.28),

$$\frac{1}{2}p^2 e^{-2kt} + \frac{1}{2}\omega^2 q^2 e^{2kt} + kpq = C. \qquad (1.8.33)$$

If the damping coefficient is equal to zero, $k = 0$, we arrive at the *harmonic oscillator* whose canonical equations are

$$\dot{q} = p, \quad \dot{p} = -\omega^2 q, \qquad (1.8.34)$$

and the corresponding Hamiltonian

$$H = \frac{1}{2}p^2 + \frac{1}{2}\omega^2 q^2 = const., \qquad (1.8.35)$$

which follows directly from (1.8.33) by setting $k = 0$.

It is easy to express the conservation law (1.8.33) in terms of the generalized coordinate q. Substituting $p = \dot{q}e^{2kt}$ from (1.8.26) into (1.8.33) we find the following conservation law in Lagrangian form whose differential equation of motion is given by (1.8.23):

$$\left(\frac{\dot{q}^2}{2} + \frac{\omega^2 q^2}{2} + kq\dot{q}\right)e^{2kt} = C = const. \qquad (1.8.36)$$

This conservation law has been obtained by means of the Noether theorem in [111] (see also [122]).

1.9 Canonical Transformations

We have seen at the end of subsection 1.4.2 that the Euler–Lagrangian equations

$$\frac{d}{dt}\frac{\partial L}{\partial \dot{q}_i} - \frac{\partial L}{\partial q_i} = 0, \quad i = 1, ..., n, \qquad (1.9.1)$$

are invariant under the one-to-one point transformations of the generalized coordinates

$$q_i = q_i (t, Q_1, ..., Q_n) , \quad i = 1, ..., n. \tag{1.9.2}$$

Namely, by substituting (1.9.2) into the Lagrangian function $L(t, q_i, \dot{q}_i)$ we find the new Lagrangian function $L^*(t, Q_1, ..., Q_n, \dot{Q}_1, ..., \dot{Q}_n)$ whose Euler–Lagrangian equations

$$\frac{d}{dt} \frac{\partial L^*}{\partial \dot{Q}_i} - \frac{\partial L^*}{\partial Q_i} = 0, \quad i = 1, ..., n, \tag{1.9.3}$$

are generating the same dynamical trajectories of the dynamical system.

In this section we discuss the question of finding the class of transformations of the generalized coordinates q_i and generalized momenta p_i for which the canonical equations (1.8.14) will preserve their canonical form.[8]

Considering the generalized coordinates q_i and generalized momenta p_i as completely independent, we introduce the transformations from the "old" variables q_i, p_i to the "new" variables Q_i, P_i by the following relations:

$$\begin{aligned}
Q_i &= Q_i (t, q_1, ..., q_n, p_1, ..., p_n) , \\
P_i &= P_i (t, q_1, ..., q_n, p_1, ..., p_n) .
\end{aligned} \tag{1.9.4}$$

We will suppose that this transformation is reversible (nonsingular) and that we are able also to find from (1.9.4) the following relations:

$$\begin{aligned}
q_i &= q_i (t, Q_1, ..., Q_n, P_1, ..., P_n) , \\
p_i &= p_i (t, Q_1, ..., Q_n, P_1, ..., P_n) .
\end{aligned} \tag{1.9.5}$$

It is evident that the transformations (1.9.4) or (1.9.5) are much more general in comparison with the point transformations (1.9.2).

The direct substitution of the transformation (1.9.5) into the Hamiltonian function $H = H(t, q_i, p_i)$ will not preserve the Hamiltonian form of the resulting equations as was the case in the Euler–Lagrangian forms described above. Therefore, our aim is to find such transformations (1.9.4) or (1.9.5) for which the "old" canonical equations

$$\dot{q}_i = \frac{\partial H}{\partial p_i}, \quad \dot{p}_i = -\frac{\partial H}{\partial q_i} \tag{1.9.6}$$

will be transformed into the form-invariant canonical equations

$$\dot{Q}_i = \frac{\partial H^*}{\partial P_i}, \quad \dot{P}_i = -\frac{\partial H^*}{\partial Q_i}, \tag{1.9.7}$$

---

[8]Unfortunately the theory of canonical transformations is valid only for the dynamical systems that are not subject to generalized nonconservative forces, that is, $Q_i(t, q_1, ..., q_n, p_1, ..., p_n) = 0$.

where $H^* = H^*(t, Q_1, ..., Q_n, P_1, ..., P_n)$ is the new Hamiltonian function (obtained by substituting (1.9.5) into the Hamiltonian function (1.8.5)) whose form will be determined in the course of analysis. The transformations that meet these conditions are known as *canonical transformations*.

In order to determine canonical transformations we will use the fact demonstrated in the last section that the equations (1.9.6) can be derived as the Euler–Lagrange equations (see (1.8.21)) whose Lagrangian function is specified by (1.8.6):

$$L = p_i \dot{q}_i - H(t, q_1, ..., q_n, p_1, ..., p_n). \tag{1.9.8}$$

It is clear that the transformed canonical equations (1.9.7) can be derived from the Lagrangian function of the form

$$L^* = P_i \dot{Q}_i - H^*(t, Q_1, ..., Q_n, P_1, ..., P_n). \tag{1.9.9}$$

Keeping in mind that L^* is formed by means of the transformations (1.9.5), we must ensure that the equations (1.9.6) and (1.9.7) will generate the same dynamical trajectories. In section 1.4 (see (1.4.26) and (1.4.27)) we have demonstrated that two Lagrangian functions are going to produce the same differential equations if their difference is equal to a total time derivative of a gauge function. In our case, the gauge function must generally be a function of all old variables q_i, p_i, new variables Q_i, P_i, and time t, namely,

$$L - L^* = \frac{d}{dt} F(t, q_1, ..., q_n, p_1, ..., p_n, Q_1, ..., Q_n, P_1, ..., P_n), \tag{1.9.10}$$

or

$$P_i \dot{Q}_i - H^* = p_i \dot{q}_i - H - \frac{dF}{dt}. \tag{1.9.11}$$

Multiplying (1.9.11) by dt and calculating the total derivative of F, we find

$$\left(p_i - \frac{\partial F}{\partial q_i} \right) dq_i + \left(-P_i - \frac{\partial F}{\partial Q_i} \right) dQ_i$$
$$+ \left(H^* - H - \frac{\partial F}{\partial t} \right) dt - \frac{\partial F}{\partial p_i} dp_i - \frac{\partial F}{\partial P_i} dP_i = 0. \tag{1.9.12}$$

This equation will be identically satisfied if

$$p_i = \frac{\partial F}{\partial q_i}, \quad P_i = -\frac{\partial F}{\partial Q_i},$$
$$H^* = \left(H + \frac{\partial F}{\partial t} \right)_{(p_i, q_i) \to (P_i, Q_i)},$$
$$\frac{\partial F}{\partial p_i} = 0, \quad \frac{\partial F}{\partial P_i} = 0. \tag{1.9.13}$$

From $(1.9.13)_3$ it follows that the gauge function depends only upon old and new generalized coordinates and time, namely,

$$F = F\left(t, q_1, ..., q, Q_1, ..., Q_n\right). \qquad (1.9.14)$$

It should be noted that a fortunate selection of the gauge function F whose name in the theory of canonical transformation is traditionally denoted as the *generating function* can considerably simplify finding of the solution of a canonical system of differential equations of motion. Sometimes it is convenient to use generating functions whose structure is different than that given by (1.9.14).

For example, let us consider the case in which we wish that the generating function depends upon the old generalized coordinates q_i and new momenta P_i. Starting from (1.9.11) multiplied by dt, we have

$$P_i dQ_i - H^* dt = p_i dq_i - H dt - dF. \qquad (1.9.15)$$

Using the identity $P_i dQ_i = d\left(P_i Q_i\right) - Q_i dP_i$, we write (1.9.15) in the form

$$d\left(F + P_i Q_i\right) = p_i dq_i + Q_i dP_i + \left(H^* - H\right) dt. \qquad (1.9.16)$$

The expression $F + P_i Q_i$ represents the new generating function

$$F_2 = F_2\left(t, q_1, ..., q_n, P_1, ..., P_n\right), \qquad (1.9.17)$$

and we have

$$p_i = \frac{\partial F_2}{\partial q_i}, \quad Q_i = \frac{\partial F_2}{\partial P_i}, \quad H^*\left(t, Q_i, P_i\right) = \left(H + \frac{\partial F_2}{\partial t}\right)_{(q_i, p_i) \to (Q_i, P_i)}. \qquad (1.9.18)$$

Analogously we can find the following two generating functions depending upon Q_i, p_i and p_i, P_i, respectively.

(i) If

$$F_3 = F_3\left(t, Q_1, ..., Q_n, p_1, ..., p_n\right), \qquad (1.9.19)$$

we have

$$q_i = -\frac{\partial F_3}{\partial p_i}, \quad P_i = -\frac{\partial F_3}{\partial Q_i}, \quad H^* = H + \frac{\partial F_3}{\partial t}. \qquad (1.9.20)$$

(ii) If

$$F_4 = F_4\left(t, p_1, ..., p_n, P_1, ..., P_n\right), \qquad (1.9.21)$$

we have

$$q_i = -\frac{\partial F_4}{\partial p_i}, \quad Q_i = \frac{\partial F_4}{\partial P_i}, \quad H^* = H + \frac{\partial F_4}{\partial t}. \qquad (1.9.22)$$

Example 1.9.1. *Linearly damped oscillator.* Let use consider again the case of a linearly damped oscillator considered in the last section. The differential equations in canonical form are found to be (see (1.8.29))

$$\dot{q} = pe^{-2kt}, \quad \dot{p} = -\omega^2 q e^{2kt}, \tag{1.9.23}$$

with the corresponding Hamilton's function

$$H = \frac{1}{2}p^2 e^{-2kt} + \frac{1}{2}\omega^2 q^2 e^{2kt}. \tag{1.9.24}$$

Let us introduce a generating function of the form

$$F = F_1\left(t, q, Q\right) = -\frac{1}{2}q^2\left(k + \Omega \tan Q\right) e^{2kt}, \tag{1.9.25}$$

where

$$\Omega = \left(\omega^2 - k^2\right)^{1/2}. \tag{1.9.26}$$

From (1.9.13) we find

$$p = \frac{\partial F_1}{\partial q} = -q\left(k + \Omega \tan Q\right) e^{2kt}, \quad P = -\frac{\partial F_1}{\partial Q} = \frac{q^2}{2}\frac{\Omega}{\cos^2 Q}e^{2kt}. \tag{1.9.27}$$

Expressing old canonical variables in terms of new ones, we find

$$q = e^{-kt}\sqrt{\frac{2P}{\Omega}}\cos Q, \quad p = e^{kt}\left(-k\sqrt{\frac{2P}{\Omega}}\cos Q - \sqrt{2P\Omega}\sin Q\right). \tag{1.9.28}$$

Entering with this into the Hamiltonian function (1.9.24) we find

$$H_{(q,p)\to(Q,P)} = \frac{Pk^2}{\Omega}\cos^2 Q + 2Pk\cos Q \sin Q + P\Omega \sin^2 Q + \frac{P\omega^2}{\Omega}\cos^2 Q. \tag{1.9.29}$$

Similarly,

$$\left(\frac{\partial F_1}{\partial t}\right)_{(q,p)\to(Q,P)} = -\frac{2Pk^2}{\Omega}\cos^2 Q - 2Pk\sin Q \cos Q. \tag{1.9.30}$$

Therefore, according to (1.9.13)$_2$, the new Hamiltonian function is of the form

$$\begin{aligned} H^*\left(t, Q, P\right) &= \left(\frac{\partial F_1}{\partial t} + H\right)_{(q,p)\to(Q,P)} \\ &= P\Omega \sin^2 Q + \frac{P}{\Omega}\left(\omega^2 - k^2\right)\cos^2 Q. \end{aligned} \tag{1.9.31}$$

Using the relation (1.9.26) we finally have

$$H^* = P\Omega. \tag{1.9.32}$$

The canonical equations in the new coordinates are

$$\dot{Q} = \frac{\partial H^*}{\partial P} = \Omega, \quad \dot{P} = -\frac{\partial H^*}{\partial Q} = 0, \tag{1.9.33}$$

whence

$$Q = \Omega t + C, \quad P = B, \quad C = const., \quad B = const. \tag{1.9.34}$$

Now it is evident that the expressions for the canonical transformations (1.9.28) represent, at the same time, the general solution of the damped oscillator, whose canonical differential equations of motion are given by (1.9.23). Thus we have, by combining (1.9.28) and (1.9.34),

$$
\begin{aligned}
q &= \sqrt{\frac{2B}{\Omega}} e^{-kt} \cos\left(\Omega t + C\right), \\
p &= e^{-kt} \left(-k\sqrt{\frac{2B}{\Omega}} \cos\left(\Omega t + C\right) - \sqrt{2B\Omega} \sin\left(\Omega t + C\right) \right). \tag{1.9.35}
\end{aligned}
$$

It is easy to verify by repeating the same procedure that if we select the generating function in the form

$$F_1 = -\frac{1}{2} q^2 \left[k + \Omega \tan\left(Q + \Omega t\right) \right] e^{2kt}, \tag{1.9.36}$$

the new Hamiltonian function will have more a simple form than that given by (1.9.32), namely,

$$H^* = 0, \tag{1.9.37}$$

and the new canonical equations will be

$$Q = C, \quad P = B, \tag{1.9.38}$$

where C and B are constants of integration. The important question of how to find the generating function whose transformed Hamiltonian is equal to zero will be considered in the subsequent text.

1.10   Poisson Brackets, the Conditions of Canonicity of a Given Transformation

In the previous section we have seen how to find new canonical equations if one of the four generating functions is given in advance. Let us briefly discuss the question of how to know if a *given transformation* between the old and new canonical variables of the form (1.9.5) is canonical.

Before we give the answer to this question we shall introduce so-called *Poisson brackets,* which play the vital part in the Hamiltonian description of analytical mechanics.[9]

Let us consider two functions U and V depending upon time and canonical variables q_i and p_i:

$$\begin{aligned} U &= U\left(t, q_1, ...q_n, p_1, ..., p_n\right), \\ V &= V\left(t, q_1, ...q_n, p_1, ..., p_n\right). \end{aligned} \qquad (1.10.1)$$

Let us define the expression

$$\frac{\partial U}{\partial q_1}\frac{\partial V}{\partial p_1} + \cdots + \frac{\partial U}{\partial q_n}\frac{\partial V}{\partial p_n} - \left(\frac{\partial U}{\partial p_1}\frac{\partial V}{\partial q_1} + \cdots + \frac{\partial U}{\partial p_n}\frac{\partial V}{\partial q_n}\right), \qquad (1.10.2)$$

which is referred to as the Poisson bracket of the functions U and V and which is denoted as

$$(U, V)_{q,p} = \sum_{i=1}^{n}\left(\frac{\partial U}{\partial q_i}\frac{\partial V}{\partial p_i} - \frac{\partial U}{\partial p_i}\frac{\partial V}{\partial q_i}\right) = -(V, U)_{q,p}. \qquad (1.10.3)$$

The following identities can be easily verified from the definition

$$(U, U)_{q,p} = 0, \quad (U, V + K)_{q,p} = (U, V)_{q,p} + (U, K)_{q,p}, \qquad (1.10.4)$$

and

$$(q_i, q_j)_{q,p} = (p_i, p_j)_{q,p} = 0, \quad (q_i, p_j)_{q,p} = \delta_{ij}, \qquad (1.10.5)$$

where δ_{ij} is the Kronecker delta symbol. The identities (1.10.4) and (1.10.5) are usually named the basic Poisson brackets.

It is easy to verify that the canonical differential equations

$$\dot{q}_i = \frac{\partial H}{\partial p_i}, \quad \dot{p}_i = -\frac{\partial H}{\partial q_i}, \qquad (1.10.6)$$

can be expressed by means of Poisson brackets in the form

$$\dot{q}_i = (q_i, H)_{q,p}, \quad \dot{p}_i = (p_i, H)_{q,p}. \qquad (1.10.7)$$

One of the most important properties of the Poisson brackets is their invariance with respect to canonical transformations. Namely, it can be demonstrated (see, for example, [68, pp. 512–518]) that

$$(U, V)_{q,p} = (U, V)_{Q,P}, \qquad (1.10.8)$$

---

[9]For a more complete description of the Poisson bracket theory, see, for example, [68, pp. 512–518].

if the transformation from the old canonical variables q_i, p_i to the new canonical variables Q_i, P_i is canonical. This fact can be employed to verify if a *given transformation* is canonical or not canonical.

If the invariance property (1.10.8) holds, then it must be valid for the basic identities (1.10.4) and (1.10.5), namely,

$$
\begin{aligned}
(q_i, q_j)_{q,p} &= (q_i, q_j)_{Q,P} = 0, \\
(p_i, p_j)_{q,p} &= (p_i, p_j)_{Q,P} = 0, \\
(q_i, p_j)_{q,p} &= (q_i, p_j)_{Q,P} = \delta_{ij},
\end{aligned}
\tag{1.10.9}
$$

where we understand that the given transformation is of the form

$$
\begin{aligned}
q_i &= q_i\,(t, Q_1, ..., Q_n, P_1, ..., P_n)\,, \\
p_i &= p_i\,(t, Q_1, ..., Q_n, P_1, ..., P_n)\,.
\end{aligned}
\tag{1.10.10}
$$

Therefore, if the given transformation (1.10.10) is canonical it must satisfy the relations (1.10.9), and this is the test of canonicity of the given transformation.

As an example, let us consider the transformation (1.9.28):

$$
q = e^{-kt}\sqrt{\frac{2P}{\Omega}}\cos Q, \quad p = e^{kt}\left(-k\sqrt{\frac{2P}{\Omega}}\cos Q - \sqrt{2P\Omega}\sin Q\right), \tag{1.10.11}
$$

and confirm that it represents a canonical transformation. Since the dynamical system has only one degree of freedom (i.e., $n = 1$) there exists only one relation $(1.10.9)_3$, and it reads

$$
(q, p)_{Q,P} = \frac{\partial q}{\partial Q}\frac{\partial p}{\partial P} - \frac{\partial q}{\partial P}\frac{\partial p}{\partial Q}. \tag{1.10.12}
$$

Thus,

$$
(q, p)_{Q,P} = \frac{k}{\Omega}\sin Q\cos + \sin^2 Q - \frac{k}{\Omega}\sin Q\cos Q + \cos^2 Q = 1, \tag{1.10.13}
$$

and the transformation (1.10.11) is canonical.

Chapter 2

The Hamilton–Jacobi Method of Integration of Canonical Equations

2.1 Introduction

In this section we shall briefly discuss the famous Hamilton–Jacobi method, which represents a general and effective method of integration of the Hamilton canonical differential equations

$$\dot{q}_i = \frac{\partial H}{\partial p_i}, \quad \dot{p}_i = -\frac{\partial H}{\partial q_i}, \quad i = 1, ..., n, \tag{2.1.1}$$

where $H = H(t, q_1, ..., q_n, p_1, ..., p_n)$ is the Hamiltonian function. In writing (2.1.1) we assumed that the nonconservative (nonpotential) generalized forces are equal to zero:

$$Q_i = Q_i(t, q_1, ..., q_n, p_1, ..., p_n) = 0. \tag{2.1.2}$$

Note that for $Q_i \neq 0$, the Hamilton–Jacobi method is not applicable.

We will demonstrate that the integration of the system (2.1.1) can be replaced by an equivalent problem of finding a *complete solution* of a nonlinear partial differential equation of the first order, referred to as the *Hamilton–Jacobi partial differential equation*.

2.2 The Hamilton–Jacobi Partial Differential Equation

We introduce a scalar field function called *principal function*

$$S = S(t, q_1, ..., q_n), \tag{2.2.1}$$

which depends on time t and generalized position coordinates q_i. Let us *define* the generalized momentum vector p_i to be the gradient of this scalar function

$$p_i = \frac{\partial S}{\partial q_i}. \tag{2.2.2}$$

From the definition of the Hamiltonian function (1.8.6) we have

$$L\left(t, q_1, ..., q_n, \dot{q}_1, ..., \dot{q}_n\right) = p_i \dot{q}_i - H(t, q_1, ..., q_n, p_1, ..., p_n). \tag{2.2.3}$$

Using (2.2.2) and adding and subtracting the term $\partial S/\partial t$ on the right-hand side of (2.2.3), we find

$$L(t, q_1, ..., q_n, \dot{q}_1, ..., \dot{q}_n) = \frac{\partial S}{\partial t} + \frac{\partial S}{\partial q_i}\dot{q}_i - \frac{\partial S}{\partial t} - H\left(1, q_1, ..., q_n, \frac{\partial S}{\partial q_1}, ..., \frac{\partial S}{\partial q_n}\right). \tag{2.2.4}$$

We will now split this equation into two parts. The first part, taking into account the identity $dS/dt = \partial S/\partial t + (\partial S/\partial q_i)\,\dot{q}_i$, will be

$$\frac{dS}{dt} = L(t, q_1, ..., q_n, \dot{q}_1, ..., \dot{q}_n), \tag{2.2.5}$$

namely,

$$S = \int L\, dt. \tag{2.2.6}$$

This expression represents the so-called *Hamilton action integral* and plays the central role in the variational description of analytical mechanics, which will be discussed in the next part of this book. The remainder of (2.2.4) gives

$$\frac{\partial S}{\partial t} + H\left(t, q_1, ..., q_n, \frac{\partial S}{\partial q_1}, ..., \frac{\partial S}{\partial q_n}\right) = 0. \tag{2.2.7}$$

This is the Hamilton–Jacobi partial differential equation, which is formed by replacing the generalized momenta p_i in H by $\partial S/\partial q_i$ and adding the term $\partial S/\partial t$ so that the result of the equation is zero.

We shall now prove the *Jacobi theorem*, which states that if we know any *complete solution of equation* (2.2.7) containing n nonadditive arbitrary constants and one additive constant, namely,

$$S = S(t, q_1, ..., q_n, C_1, ..., C_n) + C_{n+1}, \tag{2.2.8}$$

then the general solution of canonical system (2.1.1) is given by the equations

$$\frac{\partial S}{\partial q_i} = p_i, \quad \frac{\partial S}{\partial C_i} = B_i = const., \quad i = 1, ..., n, \tag{2.2.9}$$

where B_i are new arbitrary constants.

Let us note that the additive constant C_{n+1} in (2.2.8) is not an essential parameter in the Hamilton–Jacobi method. Namely, since the *principal function* S does not enter directly into the Hamilton–Jacobi equation, and since only the partial derivatives of S figure to it, we can see that one constant should appear additively in the complete solution, as indicated in (2.2.8). However, since according to the Jacobi theorem (2.2.9) we need only the partial derivatives of S, the additive constant C_{n+1} does not play any role in our considerations. Therefore, the complete solution suitable for the application of the Hamilton–Jacobi theory is actually of the form

$$S = S(t, q_1, ..., q_n, C_1, ..., C_n), \qquad (2.2.10)$$

where C_i are nonadditive constants that are considered mutually independent parameters. The mutual independence of constants $C_i, i = 1, ..., n$, means that we have the condition that the Jacobian determinant is different from zero:

$$\det \left(\frac{\partial^2 S}{\partial C_i \partial q_j} \right) \neq 0. \qquad (2.2.11)$$

The proof of the Jacobi theorem consists in demonstrating that the solution (2.2.9) satisfies the canonical differential equations (2.1.1). To demonstrate this, we start with the Hamilton–Jacobi equation (2.2.7), into which we substitute a complete solution (2.2.10) to obtain an *identity* of the form

$$\frac{\partial S(t, \mathbf{q}, \mathbf{C})}{\partial t} + H \left(t, q_i, \frac{\partial S(t, \mathbf{q}, \mathbf{C})}{\partial q_j} \right) = 0. \qquad (2.2.12)$$

The differentiation of this equation with respect to various parameters will generate new identities.

Let us differentiate partially with respect to C_i the last expression

$$\frac{\partial S(t, \mathbf{q}, \mathbf{C})}{\partial t \partial C_i} + \frac{\partial H}{\partial p_j} \frac{\partial^2 S(t, \mathbf{q}, \mathbf{C})}{\partial q_j \partial C_i} = 0, \quad i, j = 1, ..., n. \qquad (2.2.13)$$

Taking the total time derivative of the second group of equations (2.1.1), we arrive at

$$\frac{d}{dt} \left[\frac{\partial S(t, \mathbf{q}, \mathbf{C})}{\partial C_i} \right] = \frac{\partial^2 S(t, \mathbf{q}, \mathbf{C})}{\partial t \partial C_i} + \frac{\partial^2 S(t, \mathbf{q}, \mathbf{C})}{\partial C_i \partial q_j} \dot{q}_j = 0,$$
$$i, j = 1, ..., n. \qquad (2.2.14)$$

The last two systems of equations represent a nonhomogeneous system of linear equations with respect to $\frac{\partial H}{\partial p_j}$ and $\dot{q}_j$, respectively. The coefficients of these two systems are equal, and according to (2.2.11) are different from zero. Therefore the roots of both systems have to be identically equal, namely,

$$\dot{q}_j = \frac{\partial H}{\partial p_j}, \quad j = 1, ..., n. \qquad (2.2.15)$$

This means that the first group of Hamilton's differential equations (2.1.1) are identically satisfied.

In order to demonstrate that the second group of canonical equations (2.1.1) are identically satisfied too, we find the partial derivatives with respect to q_i of the identity (2.2.12) and use (2.2.15) to get

$$\frac{\partial^2 S\left(t, \mathbf{q}, \mathbf{C}\right)}{\partial t \partial q_i} + \frac{\partial H}{\partial q_i} + \frac{\partial^2 S\left(t, \mathbf{q}, \mathbf{C}\right)}{\partial q_i \partial q_j} \dot{q}_j = 0. \qquad (2.2.16)$$

We compare this equation with the total time derivative of the first group of equations in (2.2.9):

$$\dot{p}_i = \frac{\partial^2 S\left(t, \mathbf{q}, \mathbf{C}\right)}{\partial q_i \partial q_j} \dot{q}_j + \frac{\partial^2 S\left(t, \mathbf{q}, \mathbf{C}\right)}{\partial q_i \partial t}. \qquad (2.2.17)$$

Entering with (2.2.17) into (2.2.16), we prove that the second group of canonical equations,

$$\dot{p}_i = -\frac{\partial H}{\partial q_i}, \qquad (2.2.18)$$

are also identically satisfied, which completes the proof of the Jacobi theorem (2.2.9).

Therefore, we can conclude that the problem of integration of the Hamilton equations of motion (2.1.1) is replaced by the problem of finding a complete solution of the Hamilton–Jacobi partial differential equation (2.2.7). Namely, if a complete solution of the Hamilton–Jacobi equation is known, we can find the motion of the dynamical system without any additional integration by using only the operations of simple partial differentiation and algebra.

Let us note at the end of this section that the Hamilton–Jacobi method of integration of the canonical differential equations of motion comprises one of the central pillars of analytical mechanics, theoretical physics, invariant embedding theory, and many modern branches of engineering, such as optimal control theory. Contrary to the widespread opinion that the problem of integrating a partial differential equation is usually more complicated than that of equations of motion (2.1.1), numerous authors have made it a point to emphasize that the canonical ordinary differential equations of motion "may be difficult to integrate by elementary methods, while the corresponding partial differential equation is manageable" [30, p. 107]. Similarly, Arnold [7, p. 261] made the explicit statement that the Hamilton–Jacobi method is "the most powerful method known for exact integration, and many problems which were solved by Jacobi cannot be solved by other methods."

2.3 Some Applications of the Hamilton–Jacobi Method

2.3.1 Linearly Damped Oscillator

In order to illustrate the Hamilton–Jacobi method, the linearly damped oscillator problem will again be considered. It was shown in section 1.8 that for this problem the Hamiltonian function is (see (1.8.28))

$$H = \frac{1}{2}p^2 e^{-2kt} + \frac{1}{2}\omega^2 q^2 e^{2kt}, \tag{2.3.1}$$

and the canonical equations of motion are

$$\dot{q} = pe^{-2kt}, \quad \dot{p} = -\omega^2 q e^{2kt}. \tag{2.3.2}$$

The corresponding Hamilton–Jacobi equation is

$$\frac{\partial S}{\partial t} + \frac{1}{2}\left(\frac{\partial S}{\partial q}\right)^2 e^{-2kt} + \frac{1}{2}q^2 e^{2kt} = 0. \tag{2.3.3}$$

To find a complete solution of this equation, let us seek a principal function in the form

$$S = -\frac{1}{2}f(t)q^2 e^{2kt}, \tag{2.3.4}$$

where $f(t)$ is to be determined. Substituting this into (2.3.3) we find

$$\dot{f} + 2kf + f^2 + \omega^2 = 0, \tag{2.3.5}$$

whence

$$f(t) = -k - \Omega\tan(\Omega t + C), \tag{2.3.6}$$

where C is a constant of integration and

$$\Omega^2 = \omega^2 - k^2, \quad \omega > k. \tag{2.3.7}$$

Therefore, the complete solution of (2.3.3) is found to be

$$S = -\frac{1}{2}q^2\left[k + \Omega\tan(\Omega t + C)\right]e^{2kt}. \tag{2.3.8}$$

Applying the Jacobi theorem (2.2.9), we have

$$p = \frac{\partial S}{\partial q} = -q\left[k + \Omega\tan(\Omega t + C)\right]e^{2kt} \tag{2.3.9}$$

and

$$\frac{\partial S}{\partial C} = -\frac{q^2}{2}\frac{\Omega e^{2kt}}{\cos^2(\Omega t + C)} = B = const., \tag{2.3.10}$$

namely,

$$q = (-2B/\Omega)^{1/2} e^{-kt} \cos(\Omega t + C). \tag{2.3.11}$$

This is a well-known solution of the linearly damped oscillator whose differential equation is of the form

$$\ddot{q} + 2k\dot{q} + \omega^2 q = 0. \tag{2.3.12}$$

Substituting (2.3.11) into (2.3.13), we find that the generalized momentum is

$$p = k(-2B/\Omega)^{1/2} e^{kt} \cos(\Omega t + C) - (-2B\Omega)^{1/2} e^{kt} \sin(\Omega t + C), \tag{2.3.13}$$

and the equations (2.3.11) and (2.3.13) comprise the general solution of the canonical system (2.3.2).

It is important to note that the *principal function S is at the same time the generator of a canonical transformation to constant coordinates and momenta.* Namely, if we take the constant C to be the new (constant) coordinate

$$C = Q \tag{2.3.14}$$

and consider the principal function (2.3.8) as the generator of the canonical transformation

$$F = -\frac{1}{2}q^2 \left[k + \Omega \tan(\Omega t + Q)\right], \tag{2.3.15}$$

then the new Hamiltonian function $H^*(t, P, Q) = \left(\frac{\partial F}{\partial t} + H\right)_{q,p \to Q,P}$ will be equal to zero and $Q = C$ and $P = B$ as demonstrated before by the equations (1.9.37), (1.9.38).

2.3.2   Simple Harmonic Oscillator

For the case of a simple harmonic oscillator when the damping coefficient is equal to zero, $k = 0$, the canonical equations

$$\dot{q} = p, \quad \dot{p} = -\omega^2 q \tag{2.3.16}$$

and the corresponding Hamiltonian function

$$H = \frac{1}{2}p^2 + \frac{1}{2}\omega^2 q^2 \tag{2.3.17}$$

induce the Hamilton–Jacobi equation of the form

$$\frac{\partial H}{\partial t} + \frac{1}{2}\left(\frac{\partial S}{\partial q}\right)^2 + \frac{1}{2}\omega^2 q^2 = 0. \tag{2.3.18}$$

Naturally, a complete integral of this equation can be obtained directly from (2.3.8) by putting $k = 0$, that is,

$$S = -\frac{1}{2}q^2\omega \tan(\omega t + C), \qquad (2.3.19)$$

and applying the Jacobi theorem we obtain the solution of the canonical system (2.3.16) in the form

$$q = (-2B/\omega)^{1/2}\cos(\omega t + C), \quad p = -(-2B\omega)^{1/2}\sin(\omega t + C), \qquad (2.3.20)$$

where B and C are arbitrary constants. However, since the simple harmonic oscillator represents a conservative dynamical system we can find another complete integral based on the fact that the motion is conservative. Namely, since from the Hamiltonian (2.3.17) it follows that $H = E = const.$, where E is the total mechanical energy, we will seek a complete solution of (2.3.18) in the form

$$S(t, q) = -Et + F(q), \qquad (2.3.21)$$

where $F(q)$ is to be determined. Inserting (2.3.21) into (2.3.18) we reduce the problem to an ordinary differential equation of the form

$$\left(\frac{dF}{dq}\right)^2 + \omega^2 q^2 = 2E = const., \qquad (2.3.22)$$

whence

$$F(q) = \int \sqrt{2E - \omega^2 q^2}\,dq + D, \quad D = const. \qquad (2.3.23)$$

Thus, a complete solution of (2.3.18) is found to be

$$S = -Et + \int \sqrt{2E - \omega^2 q^2}\,dq + D, \qquad (2.3.24)$$

where E is a nonadditive constant parameter and D is an additive constant that can be ignored without loss of generality. It is to be stressed that *whenever the Hamiltonian function does not depend upon t explicitly the complete solution of the Hamilton–Jacobi equation can be supposed in the form* (2.3.21). Naturally, in the case of the systems with many degrees of freedom we can take

$$S(t, q_1, ..., q_n) = -Et + F(q_1, ..., q_n). \qquad (2.3.25)$$

Applying the Jacobi theorem to the expression (2.3.24) we find

$$\frac{\partial S}{\partial E} = -t + \int \frac{dq}{\sqrt{2E - \omega^2 q^2}} = B = const. \qquad (2.3.26)$$

Integrating, we have

$$-t + \frac{1}{\omega}\arcsin\frac{\omega q}{\sqrt{2E}} = B \qquad (2.3.27)$$

or

$$q = \frac{\sqrt{2E}}{\omega} \sin\left(\omega t + \bar{B}\right), \quad \bar{B} = \omega B = const. \tag{2.3.28}$$

It is easy to see that the second equation $\partial S/\partial q = p$ gives

$$p = \sqrt{2E - \omega^2 q^2} = \sqrt{2E} \cos\left(\omega t + \bar{B}\right). \tag{2.3.29}$$

2.3.3   The Case When a Particular Solution of the Riccati Equation is Available

We have seen in this section that a very convenient form for finding a complete solution of the rheolinear dynamical systems with one degree of freedom is $S(t, q) = (1/2) f(t) q^2$, as suggested, for example, in equation (2.3.4). As a consequence of this supposition we have to solve a Riccati differential equation of the type (2.3.5).

However, in numerous problems a particular solution of the corresponding Riccati equation can be easily found, and this fact can be an important help in finding a complete integral of the corresponding Hamilton–Jacobi partial differential equation. In general, the method that follows is simple, and we shall demonstrate it by means of a few examples taken from [56].

Example 1 [56]. Let us find a general solution of the rheolinear differential equation

$$\ddot{q} + \frac{\dot{q}}{t} - \frac{q}{t^2} - a = 0, \tag{2.3.30}$$

where $a = const$. It is easy to verify that this equation can be derived from the Euler–Lagrangian equation $(d/dt)(\partial L/\partial \dot{q}) - \partial L/\partial q = 0$, where Lagrangian function is of the form

$$L = \frac{1}{2}\dot{q}^2 t + \frac{1}{2}\frac{q^2}{t} + atq. \tag{2.3.31}$$

The corresponding Hamilton's function is found to be

$$H = \frac{1}{2}\frac{p^2}{t} - \frac{1}{2}\frac{q^2}{t} - atq. \tag{2.3.32}$$

The Hamilton–Jacobi differential equation is therefore

$$\frac{\partial S}{\partial t} + \frac{1}{2t}\left(\frac{\partial S}{\partial q}\right)^2 - \frac{1}{2}\frac{q^2}{t} - atq = 0. \tag{2.3.33}$$

The central point of the method that follows is that we are going to seek a complete solution of this equation in the form

$$S(q, t) = \frac{1}{2}f_1(t) q^2 + f_2(t) q + f_3(t), \tag{2.3.34}$$

where $f_1(t)$, $f_2(t)$, and $f_3(t)$ are unknown functions of time t. Entering with (2.3.34) into (2.3.33) we find

$$\frac{q^2}{2}\left(\dot{f}_1 + \frac{1}{t}f_1^2 - \frac{1}{t}\right) + q\left(\dot{f}_2 + f_1 f_2 - at\right) + \dot{f}_3 + \frac{1}{2t}f_2^2 = 0. \qquad (2.3.35)$$

Since this relation must be satisfied for arbitrary q and t, we arrive at the following system of Riccati equations:

$$\begin{aligned}
\dot{f}_1 + \frac{1}{t}f_1^2 - \frac{1}{t} &= 0, \\
\dot{f}_2 + f_1 f_2 - at &= 0, \\
\dot{f}_3 + \frac{1}{2t}f_2^2 &= 0. \qquad (2.3.36)
\end{aligned}$$

Instead of trying to find a general solution of $(2.3.36)_1$, we try to find a particular solution of the form

$$f_1 = At^m, \qquad (2.3.37)$$

where A and m are unknown constants. Entering with (2.3.37) into $(2.3.36)_1$ we arrive at

$$mAt^{m-1} + A^2 t^{2m-1} - t^{-1} = 0. \qquad (2.3.38)$$

This equation will be identically satisfied for $m = 0$ and $A = \pm 1$. Taking the root $A = 1$, we find that

$$f_1(t) = 1. \qquad (2.3.39)$$

Entering with this f_1 into $(2.3.36)_{2,3}$ we easily find after integration that

$$f_2(t) = \frac{1}{3}at^2 + \frac{1}{t}C, \quad f_3 = -\frac{1}{72}a^2 t^4 - \frac{1}{3}Ct + \frac{1}{4}C^2 t^{-2}, \qquad (2.3.40)$$

where C is a constant of integration and where an additive constant in the expression for $f_3(t)$ is not written since it is irrelevant.

According to (2.3.34), (2.3.39), and $(2.3.40)_{1,2}$, a complete solution of the Hamilton–Jacobi equation (2.3.33) reads

$$S(t, q, C) = \frac{1}{2}q^2\left(\frac{1}{3}at^2 + \frac{1}{t}C\right)q - \frac{1}{72}a^2 t^4 - \frac{1}{3}Ct + \frac{1}{4}C^2 t^{-2}. \qquad (2.3.41)$$

Applying the Jacobi theorem $(2.2.9)_2$ (i.e., $\partial S/\partial C = B = const.$), we have $\frac{q}{t} - \frac{1}{3}at + \frac{1}{2}Ct^{-2} = B$, or

$$q = \frac{1}{3}at^2 + Bt - \frac{1}{2}t^{-1}, \qquad (2.3.42)$$

which is the general solution of the differential equation (2.3.30).

Example 2 [56]. As another example, consider the differential equation

$$\ddot{q} - \frac{3}{t}\dot{q} + 4\frac{q}{t^2} - 5t = 0. \tag{2.3.43}$$

The Lagrangian function corresponding to this equation reads

$$L = \frac{1}{2}\dot{q}^2 t^{-3} - 2q^2 t^{-5} + 5qt^{-2}, \tag{2.3.44}$$

and the corresponding Hamiltonian reads

$$H = \frac{1}{2}p^2 t^3 + 2q^2 t^{-5} - 5qt^{-2}. \tag{2.3.45}$$

The Hamilton–Jacobi equation is given as

$$\frac{\partial S}{\partial t} + \frac{1}{2}t^3 \left(\frac{\partial S}{\partial q}\right)^2 + 2q^2 t^{-5} - 5qt^{-2} = 0. \tag{2.3.46}$$

Let us suppose, as in the previous example, that $S(t,q)$ has the form (2.3.34). Entering with (2.3.34) into (2.3.46) and grouping terms with q^2, q, and free terms, we arrive at the following system of Riccati equations:

$$\begin{aligned} \dot{f}_1 + t^3 f_1^2 + 4t^{-5} &= 0, \\ \dot{f}_2 + f_1 f_2 t^3 - 5t^{-2} &= 0, \\ \dot{f}_3 + \frac{1}{2}f_2^2 t^3 &= 0. \end{aligned} \tag{2.3.47}$$

Assuming the form (2.3.37), it is easy to verify that equation (2.3.47)$_1$ has a particular solution in the form $f_1(t) = 2t^{-4}$. Substituting this into (2.3.47)$_{2,3}$ and integrating we find $f_2(t) = 5t^{-1} + Ct^{-2}$ and $f_3(t) = -\frac{25}{4}t^2 - 5Ct - \frac{1}{2}C^2 \ln t$, where C is a constant of integration and an additive constant in $f_2(t)$ is discarded. Therefore, a complete solution of the Hamilton–Jacobi equation (2.3.46) is

$$S(t,q,C) = t^{-4}q + q\left(5t^{-1} + Ct^{-2}\right) - \frac{25}{4}t^2 - 5Ct - \frac{1}{2}C^2 \ln t. \tag{2.3.48}$$

It is easy to verify that the equation $\partial S/\partial C = B = const.$ leads to

$$q = 5t^3 + Ct^2 \ln t + Bt^2, \tag{2.3.49}$$

which is a general solution of (2.3.43).

Example 3 [56]. Let us find, by applying the Hamilton–Jacobi method, a general solution of the differential equation

$$\ddot{q} + \frac{1}{t}\dot{q} - \frac{aq^2}{t^2} = 0, \quad a = \text{ given constant.} \tag{2.3.50}$$

The Lagrangian function corresponding to this equation is

$$L = \frac{1}{2}t\dot{q}^2 - \frac{1}{t}aq^2, \tag{2.3.51}$$

so that the Hamiltonian reads

$$H = \frac{1}{2t}p^2 + \frac{1}{t}aq^2. \tag{2.3.52}$$

Thus, the Hamilton–Jacobi equation becomes

$$\frac{\partial S}{\partial t} + \frac{1}{2t}\left(\frac{\partial S}{\partial q}\right)^2 + \frac{1}{t}aq^2 = 0. \tag{2.3.53}$$

Let us suppose, as in the previous two examples, that a complete solution of this equation can be represented in the form (2.3.34). Substituting (2.3.34) into (2.3.53) and equating terms with q^2, q, and free terms, we obtain the following system of Riccati equations:

$$\begin{aligned}
\dot{f}_1 + \frac{1}{t}f_1^2 + \frac{2a}{t} &= 0, \\
\dot{f}_2 + \frac{1}{t}f_1 f_2 &= 0, \\
\dot{f}_3 + \frac{1}{2t}f_2^2 &= 0.
\end{aligned} \tag{2.3.54}$$

In the analysis that follows we assume that $a^2 > 0$ so that

$$2a^2 = b^2, \tag{2.3.55}$$

with b real. Let us try to find a particular solution of $(2.3.54)_1$ in the form $f_1 = At^m$, with A and m constants. For this case it is easy to see that $(2.3.54)_1$ admits a particular solution

$$f_1(t) = ib, \quad i = \sqrt{-1}. \tag{2.3.56}$$

Integrating $(2.3.54)_{2,3}$ we find

$$f_2(t) = Ct^{-ib} = Ce^{-ib\ln t}, \quad f_3(t) = \frac{C^2}{4ib}t^{-2ib} = \frac{C^2}{4ib}e^{-2ib\ln t}, \tag{2.3.57}$$

where C is a constant and an additive constant in the expression for f_3 is discarded. Therefore, a complete solution of the Hamilton–Jacobi equation (2.3.53) becomes

$$S(t, q, C) = \frac{i}{2}bq^2 + qCe^{-ib\ln t} + \frac{C^2}{4ib}e^{-2ib\ln t}. \tag{2.3.58}$$

By applying the Jacobi theorem $\partial S/\partial C = B = const.$ and recalling the well-known relations

$$\begin{aligned}
e^{ib\ln t} &= \cos(b\ln t) + i\sin(b\ln t), \\
e^{-ib\ln t} &= \cos(b\ln t) - i\sin(b\ln t),
\end{aligned} \tag{2.3.59}$$

we find

$$q\left(t\right) = A\cos\left(b\ln t\right) + D\sin\left(b\ln t\right),\qquad(2.3.60)$$

where we introduced new constants

$$A = B - \left(\frac{C}{2bi}\right),\quad D = iB - \left(\frac{C}{2b}\right).\qquad(2.3.61)$$

The reader will easily show, repeating exactly the same procedure, that in the case when

$$2a = -k^2 < 0,\qquad(2.3.62)$$

the complete integral of the Hamilton–Jacobi equation is of the form

$$S\left(t,q,C\right) = \frac{k}{2}q^2 + Cqt^{-k} + \frac{C^2}{4k}t^{-2k}.\qquad(2.3.63)$$

From the Jacobi theorem it follows that the solution of the equation (2.3.50) for $2a = -k^2$ is

$$q\left(t\right) = Bt^k - \frac{C}{2k}t^{-k}.\qquad(2.3.64)$$

We note that the interested reader can find numerous solutions of the rheolinear dynamical systems obtained by means of the Hamilton–Jacobi method, in [21].

2.4   The Oscillatory Motion with Two Degrees of Freedom

In this section we consider the application of the Hamilton–Jacobi method in the study of oscillatory motion of a scleronomic dynamical system with two degrees of freedom, whose differential equations of motion are given in the form

$$m_1\ddot{x} + ax + by = 0,\quad m_2\ddot{y} + bx + cy = 0,\qquad(2.4.1)$$

where m_1 and m_2 are masses and a, b, and c are given constant coefficients. To ensure that the motion of this system is oscillatory, we suppose that $ac - b^2 > 0$, and $a > 0$ and $c > 0$. In order to reduce the number of physical parameters entering the system, we divide both sides of equation (2.4.1) with m_1 and obtain

$$\ddot{x} + \bar{a}x + \bar{b}y = 0,\quad \frac{1}{A}\ddot{y} + \bar{b}x + \bar{c}y = 0,\qquad(2.4.2)$$

where

$$\bar{a} = \frac{a}{m_1},\quad \bar{b} = \frac{b}{m_1},\quad \bar{c} = \frac{c}{m_1},\quad A = \frac{m_1}{m_2}.\qquad(2.4.3)$$

The system (2.4.2) can be derived from the Euler–Lagrangian equations (d/dt) $\partial L/\partial \dot{x} - \partial L/\partial x = 0, (d/dt) \partial L/\partial \dot{y} - \partial L/\partial y = 0$, whose Lagrangian function is

$$L = \frac{1}{2} \left(\dot{x}^2 + \frac{1}{A}\dot{y}^2 \right) - \frac{1}{2} \left(\bar{a}x^2 + 2\bar{b}xy + \bar{c}y^2 \right). \tag{2.4.4}$$

Accomplishing the Legendre transformation, the corresponding Hamiltonian becomes

$$H = \frac{1}{2} \left(p_x^2 + Ap_y^2 \right) - \frac{1}{2} \left(\bar{a}x^2 + 2\bar{b}xy + \bar{c}y^2 \right). \tag{2.4.5}$$

The canonical equations of motion are

$$\dot{x} = p_x, \quad \dot{y} = Ap_y, \quad \dot{p}_x = -\bar{a}x - \bar{b}y, \quad \dot{p}_y = -\bar{b}x - \bar{c}y. \tag{2.4.6}$$

The Hamilton–Jacobi partial differential equation becomes

$$\frac{\partial S}{\partial t} + \frac{1}{2} \left(\frac{\partial S}{\partial x} \right)^2 + \frac{1}{2}A \left(\frac{\partial S}{\partial y} \right)^2 + \frac{1}{2} \left(\bar{a}x^2 + 2\bar{b}xy + \bar{c}y^2 \right) = 0. \tag{2.4.7}$$

We suppose that a complete solution of this equation can be obtained by a trial principal function of the form

$$S(t, x, y) = \frac{1}{2}f_1(t) x^2 + \frac{1}{2}f_2(t) y^2 + f_3(t) xy, \tag{2.4.8}$$

where $f_1(t), f_2(t)$, and $f_3(t)$ are unknown functions of time t. Substituting (2.4.8) into (2.4.7) and equating to zero terms with x^2, y^2, and xy, we arrive at the following system of Riccati equations:

$$\begin{align} \dot{f}_1 + f_1^2 + Af_3^2 + \bar{a} &= 0, \\ \dot{f}_2 + Af_2^2 + f_3^2 + \bar{c} &= 0, \\ \dot{f}_3 + f_1 f_3 + Af_1 f_3 + \bar{b} &= 0. \end{align} \tag{2.4.9}$$

Multiplying (2.4.9)$_3$ by a constant parameter 2λ, (2.4.9)$_2$ by λ^2, and adding these equations with (2.4.9)$_1$, we obtain

$$(f_1 + \lambda f_3)\dot{\;} + \lambda (f_3 + \lambda f_2)\dot{\;} + A (f_3 + \lambda f_2)^2$$

$$+ (f_1 + \lambda f_3)^2 + \bar{a} + 2\lambda\bar{b} + \lambda^2\bar{c} = 0. \tag{2.4.10}$$

Repeating the same procedure with parameter $\mu = const.$, we also have

$$(f_1 + \mu f_3)\dot{\;} + \mu (f_3 + \mu f_2)\dot{\;} + A (f_3 + \mu f_2)^2$$

$$+ (f_1 + \mu f_3)^2 + \bar{a} + 2\mu\bar{b} + \mu^2\bar{c} = 0. \tag{2.4.11}$$

In order to be able to integrate these two Riccati equations formally, we will split them into two parts by introducing new constant parameters Ω^2 and ω^2.

Thus, we separate (2.4.10) into the following two equations:

$$(f_1 + \lambda f_3)^{\cdot} + (f_1 + \lambda f_3)^2 + \Omega^2 = 0,$$
$$\lambda (f_3 + \lambda f_2)^{\cdot} + A (f_3 + \lambda f_2)^2 + \bar{a} + 2\lambda \bar{b} + \lambda^2 \bar{c} - \Omega^2 = 0. \quad (2.4.12)$$

Similarly, we decompose (2.4.11) into the following system:

$$(f_1 + \mu f_3)^{\cdot} + (f_1 + \mu f_3)^2 + \Omega^2 = 0,$$
$$\mu (f_3 + \mu f_2)^{\cdot} + A (f_3 + \mu f_2)^2 + \bar{a} + 2\mu \bar{b} + \mu^2 \bar{c} - \Omega^2 = 0. \quad (2.4.13)$$

By integrating each equation in (2.4.12), (2.4.13) we obtain, respectively,

$$f_1 + \lambda f_3 = -\Omega \tan (\Omega t + C_1),$$

$$\lambda f_2 + f_3 = -\sqrt{\frac{\bar{a} + 2\lambda \bar{b} + \lambda^2 \bar{c} - \Omega^2}{A}}$$
$$\times \tan \left[\sqrt{\frac{A (\bar{a} + 2\lambda \bar{b} + \lambda^2 \bar{c} - \Omega^2)}{\lambda^2}} t + C_2 \right],$$

$$f_1 + \mu f_3 = -\omega \tan (\omega t + C_3),$$

$$\mu f_2 + f_3 = -\sqrt{\frac{\bar{a} + 2\mu \bar{b} + \mu^2 \bar{c} - \omega^2}{A}}$$
$$\times \tan \left[\sqrt{\frac{A (\bar{a} + 2\mu \bar{b} + \mu^2 \bar{c} - \omega^2)}{\mu^2}} t + C_4 \right], \quad (2.4.14)$$

where $C_1, ..., C_4$ are arbitrary constants.

From (2.4.14)$_1$ and (2.4.14)$_3$ we find

$$f_1 = \frac{\Omega \mu}{\lambda - \mu} \tan (\Omega t + C_1) - \frac{\lambda \omega}{\lambda - \mu} \tan (\omega t + C_3),$$

$$f_3 = -\frac{\Omega}{\lambda - \mu} \tan (\Omega t + C_1) + \frac{\omega}{\lambda - \mu} \tan (\omega t + C_3). \quad (2.4.15)$$

Solving (2.4.14)$_2$ and (2.4.14)$_4$ with respect to f_2 and f_3, we also have

$$f_2 = \sqrt{\frac{\bar{a} + 2\mu \bar{b} + \mu^2 \bar{c} - \omega^2}{A (\lambda - \mu)^2}} \tan \left[\sqrt{\frac{A (\bar{a} + 2\mu \bar{b} + \mu^2 \bar{c} - \omega^2)}{\mu^2}} t + C_4 \right]$$
$$- \sqrt{\frac{\bar{a} + 2\lambda \bar{b} + \lambda^2 \bar{c} - \Omega^2}{A (\lambda - \mu)^2}}$$
$$\times \tan \left[\sqrt{\frac{A (\bar{a} + 2\lambda \bar{b} + \lambda^2 \bar{c} - \Omega^2)}{\lambda^2}} t + C_2 \right] \quad (2.4.16)$$

and

$$
\begin{aligned}
f_3 &= \sqrt{\frac{\mu^2 \left(\bar{a} + 2\lambda\bar{b} + \lambda^2\bar{c} - \Omega^2\right)}{A\left(\lambda - \mu\right)^2}} \tan\left[\sqrt{\frac{A\left(\bar{a} + 2\lambda\bar{b} + \lambda^2\bar{c} - \Omega^2\right)}{\lambda^2}}\, t + C_2\right] \\
&\quad - \sqrt{\frac{\lambda^2 \left(\bar{a} + 2\mu\bar{b} + \mu^2\bar{c} - \omega^2\right)}{A\left(\lambda - \mu\right)^2}} \\
&\quad \times \tan\left[\sqrt{\frac{A\left(\bar{a} + 2\mu\bar{b} + \mu^2\bar{c} - \omega^2\right)}{\mu^2}}\, t + C_4\right].
\end{aligned}
\tag{2.4.17}
$$

Equating corresponding terms for f_3 from $(2.4.15)_2$ and $(2.4.17)$, we find

$$
\begin{aligned}
C_1 &= C_2; \quad C_3 = C_4, \\
\Omega &= -\sqrt{\frac{\mu^2 \left(\bar{a} + 2\lambda\bar{b} + \lambda^2\bar{c} - \Omega^2\right)}{A}}, \\
\Omega &= \sqrt{\frac{A\left(\bar{a} + 2\lambda\bar{b} + \lambda^2\bar{c} - \Omega^2\right)}{\lambda^2}}, \\
\omega &= -\sqrt{\frac{\lambda^2 \left(\bar{a} + 2\mu\bar{b} + \mu^2\bar{c} - \omega^2\right)}{A}}, \\
\omega &= \sqrt{\frac{A\left(\bar{a} + 2\mu\bar{b} + \mu^2\bar{c} - \omega^2\right)}{\mu^2}}.
\end{aligned}
\tag{2.4.18}
$$

Whence

$$
\begin{aligned}
\Omega^2 &= \frac{\mu^2 \left(\bar{a} + 2\lambda\bar{b} + \lambda^2\bar{c}\right)}{A + \mu^2} = \frac{A\left(\bar{a} + 2\lambda\bar{b} + \lambda^2\bar{c}\right)}{A + \lambda^2}, \\
\omega^2 &= \frac{\lambda^2 \left(\bar{a} + 2\mu\bar{b} + \mu^2\bar{c}\right)}{A + \lambda^2} = \frac{A\left(\bar{a} + 2\mu\bar{b} + \mu^2\bar{c}\right)}{A + \mu^2}.
\end{aligned}
\tag{2.4.19}
$$

From the last two expressions it also follows that

$$
\lambda\mu = \pm A,
\tag{2.4.20}
$$

and the sign will be determined later.

Using the relations given above, we can write expressions $(2.4.16)$ in a simpler form:

$$
f_2 = \frac{\Omega}{\mu\left(\lambda - \mu\right)} \tan\left(\Omega t + C_1\right) - \frac{\omega}{\lambda\left(\lambda - \mu\right)} \tan\left(\omega t + C_3\right).
\tag{2.4.21}
$$

We have now to ensure that the differential equations $(2.4.9)_{1,2,3}$ are satisfied for f_1, f_2, and f_3 given by $(2.4.15)_1, (2.4.17)$, and $(2.4.15)_2$. Thus, by substituting

$(2.4.15)_1$ and $(2.4.15)_2$ into $(2.4.9)_1$ we find

$$\left[\frac{\Omega^2 \mu}{\lambda - \mu} - \frac{\lambda \omega^2}{\lambda - \mu} + \bar{a} \right] + \tan^2 \psi_1 \left[\frac{\Omega^2 \mu}{\lambda - \mu} + \frac{\Omega^2 \mu^2}{(\lambda - \mu)^2} + \frac{A\Omega^2}{(\lambda - \mu)^2} \right]$$

$$+ \tan^2 \psi_2 \left[-\frac{\omega^2 \lambda}{\lambda - \mu} + \frac{\omega^2 \lambda^2}{(\lambda - \mu)^2} + \frac{A\omega^2}{(\lambda - \mu)^2} \right]$$

$$- \tan \psi_1 \tan \psi_2 \left[\frac{2\lambda \mu \omega \Omega}{(\lambda - \mu)^2} + \frac{2\omega \Omega A}{(\lambda - \mu)^2} \right] = 0, \qquad (2.4.22)$$

where

$$\psi_1 = \Omega t + C_1, \quad \psi_2 = \omega t + C_3. \qquad (2.4.23)$$

The expression (2.4.22) is identically satisfied for

$$\lambda \mu = -A, \quad \lambda + \mu = -\frac{\bar{a} - A\bar{c}}{\bar{b}}. \qquad (2.4.24)$$

In obtaining (2.4.24) from (2.4.22) we employed (2.4.19) in equating to zero the first group of terms in the square brackets.

Note also that exactly the same relations (2.4.24) follow from $(2.4.9)_2$ and $(2.4.9)_3$.

It is evident from (2.4.24) that the parameters λ and μ are roots of the quadratic equation

$$Z^2 - \frac{A\bar{c} - \bar{a}}{\bar{b}} Z - A = 0, \qquad (2.4.25)$$

namely,

$$Z_1 = \lambda = \frac{A\bar{c} - \bar{a}}{2\bar{b}} + \sqrt{\left(\frac{A\bar{c} - \bar{a}}{2\bar{b}} \right)^2 + A},$$

$$Z_2 = \mu = \frac{A\bar{c} - \bar{a}}{2\bar{b}} - \sqrt{\left(\frac{A\bar{c} - \bar{a}}{2\bar{b}} \right)^2 + A}. \qquad (2.4.26)$$

Thus, all parameters λ, μ, Ω, and ω entering into $f_1(t), f_2(t)$, and $f_3(t)$ are expressed in terms of the given quantities $\bar{a}, \bar{b}, \bar{c}$, and A defined by (2.4.3).

Therefore, a complete solution of the Hamilton–Jacobi equation (2.4.7) given by (2.4.8), (2.4.15), and (2.4.17) is of the form

$$S(t, x, y, C_1, C_3) = \frac{1}{2} \left[\frac{\Omega \mu}{\lambda - \mu} \tan \psi_1 - \frac{\lambda \omega}{(\lambda - \mu)} \tan \psi_2 \right] x^2$$

$$+ \frac{1}{2} \left[\frac{\Omega}{\mu (\lambda - \mu)} \tan \psi_1 - \frac{\omega}{\lambda (\lambda - \mu)} \tan \psi_2 \right] y^2$$

$$+ \left[-\frac{\mu}{\lambda - \mu} \tan \psi_1 + \frac{\omega}{(\lambda - \mu)} \tan \psi_2 \right] xy, \qquad (2.4.27)$$

or, collecting terms with $\tan\psi_1$ and $\tan\psi_2$, we write the principal function in more compact form:

$$S(t,x,y,C_1,C_3) = \frac{\Omega}{2\mu(\lambda-\mu)}(\mu x - y)^2 \tan(\Omega t + C_1)$$
$$-\frac{\omega}{2\lambda(\lambda-\mu)}(\lambda x - y)^2 \tan(\omega t + C_3). \quad (2.4.28)$$

To find a general solution of the canonical system (2.4.6), we apply the Jacobi theorem (2.2.9): $\partial S/\partial C_1 = B_1 = const., \partial S/\partial C_3 = B_2 = const.$, which leads to

$$\mu x - y = K_1 \cos(\Omega t + C_1), \quad \lambda x - y = K_2 \cos(\omega t + C_3), \quad (2.4.29)$$

where

$$\left[\frac{2B_1\mu(\lambda-\mu)}{\Omega}\right]^{1/2} = K_1 = const., \quad \left[-\frac{2B_2\lambda(\lambda-\mu)}{\omega}\right]^{1/2} = K_2 = const.$$
$$(2.4.30)$$

Solving (2.4.29) with respect to x and y, we obtain the general solution of the dynamical system (2.4.1). To find the momenta p_x and p_y we use $p_x = \partial S/\partial x$ and $p_y = \partial S/\partial y$, which completes the calculation of the motion of the system.

Note that the quantities $\mu x - y$ and $\lambda x - y$ figuring in (2.4.29) denote the *normal coordinates* of the oscillatory dynamical system. Thus, finding a complete solution of the Hamilton–Jacobi partial differential equation (2.4.7) represents a method for reducing the system to normal coordinates by means of the Hamilton–Jacobi method. In the next section we will present a much simpler way to study the oscillatory system by means of the Hamilton–Jacobi method in which we *previously* reduce the dynamical equation to the normal form. Still, the method of solution of such systems demonstrated here has much more mechanical significance.

Example 2.4.1. *A system with two degrees of freedom.* As an illustration we shall work out a particular example. Two masses $m_1 = m$ and $m_2 = 2m$ slide without friction on a horizontal plane. They are connected by three springs of spring constant $k_1 = k, k_2 = 2k$, and $k_3 = 2k$, as shown in Figure 2.4.1.

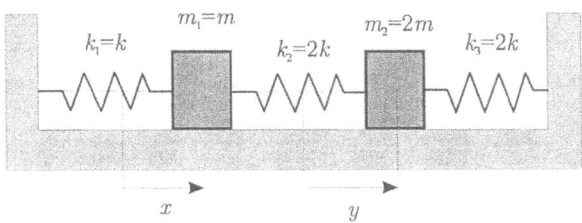

Figure 2.4.1

Let us take as coordinates two absolute displacements of the masses x and y. Then, in terms of these coordinates the kinetic and potential energies are given by

$$
\begin{aligned}
T &= \frac{1}{2}m_1\dot{x}^2 + \frac{1}{2}m_2\dot{y}^2 = \frac{1}{2}m\left(\dot{x}^2 + 2\dot{y}^2\right), \\
\Pi &= \frac{1}{2}k_1x^2 + \frac{1}{2}k_2\left(y - x\right)^2 + \frac{1}{2}k_3y^2 = \frac{1}{2}k\left(3x^2 - 4xy + 4y^2\right).
\end{aligned}
$$

$$(2.4.31)$$

Thus, in accordance with (2.4.3) we have

$$
\bar{a} = \frac{3k}{m}, \quad \bar{b} = -\frac{2k}{m}, \quad \bar{c} = \frac{4k}{m}, \quad A = \frac{1}{2}. \tag{2.4.32}
$$

The canonical differential equations of motion (2.4.6) are of the form

$$
\dot{x} = p_x, \quad \dot{y} = \frac{1}{2}p_y, \quad \dot{p}_x = -3Kx + 2Ky, \quad \dot{p}_y = -2Kx - 4Ky, \tag{2.4.33}
$$

where $K = k/m$.

The Hamilton–Jacobi partial differential equation (2.4.7) reads

$$
\frac{\partial S}{\partial t} + \frac{1}{2}\left(\frac{\partial S}{\partial x}\right)^2 + \frac{1}{4}\left(\frac{\partial S}{\partial y}\right)^2 + \frac{1}{2}\left(3Kx^2 - 4Kxy + 4Ky^2\right) = 0. \tag{2.4.34}
$$

From (2.4.26) parameters λ and μ are found to be

$$
\lambda = 1, \quad \mu = -\frac{1}{2}, \tag{2.4.35}
$$

and the frequencies Ω and ω from (2.4.19) are

$$
\Omega^2 = K, \quad \omega^2 = 4K. \tag{2.4.36}
$$

A complete solution of (2.4.34) given by (2.4.28) becomes

$$
S = -\frac{2\sqrt{K}}{3}\left[\left(\frac{x}{2} + y\right)^2\tan\left(\sqrt{K}t + C_1\right) + \left(x - y\right)^2\tan\left(2\sqrt{K}t + C_3\right)\right]. \tag{2.4.37}
$$

The equations $\partial S/\partial C_1 = B_1 = const.$ and $\partial S/\partial C_3 = B_2 = const.$ are giving

$$
\frac{x}{2} + y = \sqrt{B_1}\cos\left(\sqrt{K}t + C_1\right), \quad x - y = \sqrt{B_2}\cos\left(2\sqrt{K}t + C_3\right), \tag{2.4.38}
$$

or

$$
\begin{aligned}
x &= \frac{2}{3}\sqrt{B_1}\cos\left(\sqrt{K}t + C_1\right) + \frac{2}{3}\sqrt{B_2}\cos\left(2\sqrt{K}t + C_3\right), \\
y &= \frac{2}{3}\sqrt{B_1}\cos\left(\sqrt{K}t + C_1\right) - \frac{1}{3}\sqrt{B_2}\cos\left(2\sqrt{K}t + C_3\right). \tag{2.4.39}
\end{aligned}
$$

The second pair of equations, $p_x = \partial S/\partial x, p_y = \partial S/\partial y$, by using (2.4.38), lead to

$$
\begin{aligned}
p_x &= -\frac{2}{3}\sqrt{KB_1}\sin\left(\sqrt{K}t + C_1\right) - \frac{4}{3}\sqrt{KB_2}\cos\left(2\sqrt{K}t + C_3\right), \\
p_y &= -\frac{4}{3}\sqrt{KB_1}\sin\left(\sqrt{K}t + C_1\right) + \frac{4}{3}\sqrt{KB_2}\cos\left(2\sqrt{K}t + C_3\right).
\end{aligned}
$$
$$(2.4.40)$$

The equations (2.4.39) and (2.4.40) represent the general solution of the canonical system (2.4.33). The four arbitrary constants C_1, C_3, B_1, and B_2 should be determined from the given initial conditions $x(0), y(0), p_x(0)$, and $p_y(0)$.

Example 2.4.2. *An application of the Hamilton–Jacobi method to the vibration theory.* As the last illustrative example of application of the Hamilton–Jacobi method to the vibration theory, let us consider again the dynamical system with two degrees of freedom. Suppose that the dynamical system is subjected to the potential forces and that the Lagrangian function $L = T - \Pi$ is of the form

$$L = \frac{1}{2}\left(a_{11}\dot{q}_1^2 + 2a_{12}\dot{q}_1\dot{q}_2 + a_{22}\dot{q}_2^2\right) - \frac{1}{2}\left(c_{11}q_1^2 + 2c_{12}q_1q_2 + c_{22}q_2^2\right). \quad (2.4.41)$$

The Lagrangian equations of motion are

$$
\begin{aligned}
a_{11}\ddot{q}_1 + a_{12}\ddot{q}_2 + c_{11}q_1 + c_{12}q_2 &= 0, \\
a_{12}\ddot{q}_1 + a_{22}\ddot{q}_2 + c_{12}q_1 + c_{22}q_2 &= 0.
\end{aligned}
\quad (2.4.42)
$$

Since the kinetic and potential energies are positive definite quadratic forms, we must have

$$
\begin{aligned}
a_{11} &> 0, \quad a_{22} > 0, \quad a_{11}a_{22} - \left(a_{12}\right)^2 > 0, \\
c_{11} &> 0, \quad c_{22} > 0, \quad c_{11}c_{22} - \left(c_{12}\right)^2 > 0.
\end{aligned}
\quad (2.4.43)
$$

The direct use of the Hamilton–Jacobi method in the dynamical system given by (2.4.42) should be very complicated, and the transformation to the form convenient for the Hamilton–Jacobi analysis is necessary.

Let us multiply the second equation (2.4.42) by a constant multiplier λ and add to the first

$$\left(a_{11} + \lambda a_{12}\right)\ddot{q}_1 + \left(a_{12} + \lambda a_{22}\right)\ddot{q}_2 + \left(c_{11} + \lambda c_{12}\right)q_1 + \left(c_{12} + \lambda c_{22}\right)q_2 = 0 \quad (2.4.44)$$

or

$$\left(a_{11} + \lambda a_{12}\right)\left(\ddot{q}_1 + \frac{a_{12} + \lambda a_{22}}{a_{11} + \lambda a_{12}}\ddot{q}_2\right) + \left(c_{11} + \lambda c_{12}\right)\left(q_1 + \frac{c_{12} + \lambda c_{22}}{c_{11} + \lambda c_{12}}q_2\right) = 0. \quad (2.4.45)$$

We write (2.4.45) as

$$\left(\ddot{q}_1 + \frac{a_{12} + \lambda a_{22}}{a_{11} + \lambda a_{12}}\ddot{q}_2\right) + \frac{c_{11} + \lambda c_{12}}{a_{11} + \lambda a_{12}}\left(q_1 + \frac{c_{12} + \lambda c_{22}}{c_{11} + \lambda c_{12}}q_2\right) = 0. \qquad (2.4.46)$$

Denoting by

$$\Omega^2 = \frac{c_{11} + \lambda c_{12}}{a_{11} + \lambda a_{12}}, \qquad (2.4.47)$$

it is clear that under the condition

$$\frac{a_{12} + \lambda a_{22}}{a_{11} + \lambda a_{12}} = \frac{c_{12} + \lambda c_{22}}{c_{11} + \lambda c_{12}} = K, \qquad (2.4.48)$$

we can write the differential equation (2.4.46) in the form

$$(\ddot{q}_1 + K\ddot{q}_2) + \Omega^2(q_1 + Kq_2) = 0. \qquad (2.4.49)$$

To find the multiplier λ we have from (2.4.48) the following quadratic equation:

$$\lambda^2 + \frac{a_{22}c_{11} - a_{11}c_{22}}{a_{22}c_{12} - a_{12}c_{22}}\lambda + \frac{a_{12}c_{11} - a_{11}c_{12}}{a_{22}c_{12} - a_{12}c_{22}} = 0. \qquad (2.4.50)$$

Denoting the roots of this equation by $\lambda_{(i)}$, $i = 1, 2$, we find that the circular frequencies from (2.4.47) are

$$\Omega^2_{(i)} = \frac{c_{11} + \lambda_{(i)}c_{12}}{a_{11} + \lambda_{(i)}a_{12}}, \quad i = 1, 2, \qquad (2.4.51)$$

and the parameter K becomes

$$K_{(i)} = \frac{a_{12} + \lambda_{(i)}a_{22}}{a_{11} + \lambda_{(i)}a_{12}} = \frac{c_{12} + \lambda_{(i)}c_{22}}{c_{11} + \lambda_{(i)}c_{12}}, \quad i = 1, 2. \qquad (2.4.52)$$

Finally, if we introduce the *normal coordinates* by the relation

$$Q_{(i)} = q_1 + K_{(i)}q_2, \quad i = 1, 2, \qquad (2.4.53)$$

we have reduced the initial system of differential equations (2.4.42) to the form that follows from (2.4.49):

$$\ddot{Q}_1 + \Omega_1^2 Q_1 = 0, \quad \ddot{Q}_2 + \Omega_2^2 Q_2 = 0. \qquad (2.4.54)$$

Now it is relatively easy to analyze the uncoupled system (2.4.54) by means of the Hamilton–Jacobi method.

Since the Lagrangian function for the system (2.4.54) reads

$$L = \frac{1}{2}\left(\dot{Q}_1^2 + \dot{Q}_2^2\right) - \frac{1}{2}\left(\Omega_1^2 Q_1^2 + \Omega_2^2 Q_2^2\right), \qquad (2.4.55)$$

the corresponding Hamiltonian is

$$H = \frac{1}{2} \left(P_1^2 + P_2^2 \right) + \frac{1}{2} \left(\Omega_1^2 Q_1^2 + \Omega_2^2 Q_2^2 \right). \qquad (2.4.56)$$

The Hamilton–Jacobi partial differential equation is

$$\frac{\partial S}{\partial t} + \frac{1}{2} \left[\left(\frac{\partial S}{\partial Q_1} \right)^2 + \left(\frac{\partial S}{\partial Q_2} \right)^2 \right] + \frac{1}{2} \left(\Omega_1^2 Q_1^2 + \Omega_2^2 Q_2^2 \right) = 0. \qquad (2.4.57)$$

The Hamiltonian function (2.4.56) does not contain the time explicitly, and $H = E = const$. The Hamilton–Jacobi equation does not contain the product terms but only the squared ones, so we are faced here with a so-called *completely separable system*. Thus, we seek a complete solution of (2.4.57) in the form

$$S \left(t, Q_1, Q_2 \right) = -Et + F_1 \left(Q_1 \right) + F_2 \left(Q_2 \right), \qquad (2.4.58)$$

where E is the "total mechanical energy" of the transformed system (2.4.54). Substituting (2.4.58) into (2.4.57) one has

$$\frac{1}{2} \left[\left(\frac{dF_1}{dQ_1} \right)^2 + \Omega_1^2 Q_1^2 \right] + \frac{1}{2} \left[\left(\frac{dF_2}{dQ_2} \right)^2 + \Omega_2^2 Q_2^2 \right] = E = const. \qquad (2.4.59)$$

It is seen that the motion corresponds to two completely independent systems, each of which has only one degree of freedom. Thus

$$\frac{1}{2} \left[\left(\frac{dF_1}{dQ_1} \right)^2 + \Omega_1^2 Q_1^2 \right] = \alpha_1 = const. \qquad (2.4.60)$$

and

$$\frac{1}{2} \left[\left(\frac{dF_2}{dQ_2} \right)^2 + \Omega_2^2 Q_2^2 \right] = \alpha_2 = const., \qquad (2.4.61)$$

where the relation

$$\alpha_1 + \alpha_2 = E \qquad (2.4.62)$$

holds.

Since the complete integral of (2.4.57) contains only two nonadditive constants, we shall express one of the parameters α_1 or α_2 in terms of E and the second as $\alpha_2 = E - \alpha_1$. From (2.4.58) and (2.4.60), (2.4.61) it follows that

$$\begin{aligned} P_1 &= \frac{\partial S}{\partial Q_1} = \frac{dF_1}{dQ_1} = \sqrt{2\alpha_1 - \Omega_1^2 Q_1^2}, \\ P_2 &= \frac{\partial S}{\partial Q_2} = \frac{dF_2}{dQ_2} = \sqrt{2(E - \alpha_1) - \Omega_2^2 Q_2^2}. \end{aligned} \qquad (2.4.63)$$

Separating variables and integrating we find a complete solution of (2.4.57) as

$$
S(t, Q_1, Q_2, E, \alpha_1,) = -Et + \int \sqrt{2\alpha_1 - \Omega_1^2 Q_1^2} dQ_1
$$
$$
+ \int \sqrt{2(E - \alpha_1) - \Omega_2^2 Q_2^2} dQ_2. \qquad (2.4.64)
$$

Applying the Jacobi theorem, that is, taking the partial derivative of S with respect to α_1, we find

$$
\frac{\partial S}{\partial \alpha_1} = \int \frac{dQ_1}{(2\alpha_1 - \Omega_1^2 Q_1^2)^{1/2}} - \int \frac{dQ_2}{[2(E - \alpha_1) - \Omega_2^2 Q_2^2]^{1/2}} = B_1 = const.
$$
$$
(2.4.65)
$$

Thus, we obtain

$$
\frac{1}{\Omega_1} \arcsin \frac{\Omega_1 Q_1}{\sqrt{2\alpha_1}} - \frac{1}{\Omega_2} \arcsin \frac{\Omega_2 Q_2}{\sqrt{2(E - \alpha_1)}} = B_1. \qquad (2.4.66)
$$

This equation represents the *trajectory* in the orthogonal space of the normal coordinates (Q_1, Q_2).

The partial derivative of S with respect to E leads to

$$
\frac{\partial S}{\partial E} = -t + \int \frac{dQ_2}{[2(E - \alpha_1) - \Omega_2^2 Q_2^2]^{1/2}} = B_2 = const. \qquad (2.4.67)
$$

Integrating, we find

$$
Q_2 = \frac{\sqrt{2(E - \alpha_1)}}{\Omega_2} \sin[\Omega_2 t + B_2 \Omega_2]. \qquad (2.4.68)
$$

To find $Q_1(t)$ we enter with (2.4.68) into (2.4.66), and employing the identity $\arcsin X = X - 2n\pi, n = 0, 1, ...$, we find after simple calculation that

$$
Q_1 = \frac{\sqrt{2\alpha_1}}{\Omega_1} \sin[\Omega_1 t + (B_1 + B_2)\Omega_1]. \qquad (2.4.69)
$$

The generalized momenta (2.4.63) are of the form

$$
P_1 = \sqrt{2\alpha_1} \cos[\Omega_1 t + (B_1 + B_2)\Omega_1] = \dot{Q}_1,
$$
$$
P_2 = \sqrt{2(E - \alpha_1)} \cos[\Omega_2 t + B_2 \Omega_2] = \dot{Q}_2. \qquad (2.4.70)
$$

To find the "old" variables q_1 and q_2 from the equation (2.4.53) for $i = 1, 2$ we find

$$
q_1 = \frac{Q_1 K_2 - Q_2 K_1}{K_2 - K_1}, \qquad q_2 = \frac{Q_1 - Q_2}{K_2 - K_1}. \qquad (2.4.71)
$$

The four arbitrary constants E, α_1, B_1, and B_2 can be determined from the given initial conditions $q_1(0) = a, q_2(0) = b, \dot{q}_1(0) = c, \dot{q}_2(0) = d$.

It is interesting to note that for $\Omega_1^2 = \Omega_2^2 = \Omega^2$, that is, for the differential equations in normal coordinates

$$\ddot{Q}_1 + \Omega^2 Q_1 = 0, \quad \ddot{Q}_2 + \Omega^2 Q_2 = 0, \tag{2.4.72}$$

we can derive these equations from the Lagrangian function formed as the difference between the kinetic and potential energies, that is,

$$L = \frac{1}{2}\left(\dot{Q}_1^2 + \dot{Q}_2^2\right) - \frac{1}{2}\Omega^2\left(Q_1^2 + Q_1^2\right), \tag{2.4.73}$$

and solve the corresponding canonical equations of motion by the Hamilton–Jacobi method using as the starting point the form of the complete solution suggested by (2.4.58). However, the system of equations (2.4.72) can be equally derived from the Lagrangian function of the form

$$L = \dot{Q}_1 \dot{Q}_2 - \Omega^2 Q_1 Q_2, \tag{2.4.74}$$

whose structure has a *bilinear character*. The Hamiltonian function based upon Lagrangian (2.4.74) is

$$H_1 = P_1 P_2 + \Omega^2 Q_1 Q_2, \tag{2.4.75}$$

whose canonical equations of motion are

$$\begin{aligned} \dot{Q}_1 &= P_2, & \dot{Q}_2 &= P_1, \\ \dot{P}_1 &= -\Omega^2 Q_2, & \dot{P}_2 &= -\Omega^2 Q_1. \end{aligned} \tag{2.4.76}$$

The corresponding Hamilton–Jacobi equation reads

$$\frac{\partial S}{\partial t} + \left(\frac{\partial S}{\partial Q_1}\right)\left(\frac{\partial S}{\partial Q_2}\right) + \Omega^2 Q_1 Q_2 = 0. \tag{2.4.77}$$

To find a complete solution of this partial differential equation we select a trial function of the form

$$S = -Et + Q_2 f_1(Q_1) + f_2(Q_1), \tag{2.4.78}$$

where E is a constant. Entering with this into (2.4.77) we arrive at

$$Q_2\left[f_1\frac{df_1}{dQ_1} + \Omega^2 Q_1\right] + \frac{df_2}{dQ_1}f_1(Q_1) = E. \tag{2.4.79}$$

Equating with zero and integrating the terms in the square brackets, we find

$$f_1 = \sqrt{C - \Omega^2 Q_1^2}, \tag{2.4.80}$$

where C is an arbitrary constant. Substituting (2.4.80) into (2.4.79) and integrating, we find

$$f_2 = E\int \frac{dQ_1}{\sqrt{C - \Omega^2 Q_1^2}} = \frac{E}{\Omega}\arcsin\left(\frac{Q_1\Omega}{\sqrt{C}}\right). \tag{2.4.81}$$

Therefore, a complete solution of (2.4.77) is found to be

$$S\left(t, Q_1, Q_2, E, C\right) = Q_2\sqrt{C - \Omega^2 Q_1^2} + \frac{E}{\Omega} \arcsin\left(\frac{Q_1\Omega}{\sqrt{C}}\right) - Et. \qquad (2.4.82)$$

Applying the Jacobi theorem, we find that the equation $\partial S/\partial E = B_1 = const.$ leads to

$$Q_1 = \frac{\sqrt{C}}{\Omega} \sin\left(\Omega t + \Omega B_1\right). \qquad (2.4.83)$$

Similarly, $\partial S/\partial C = B_2 = const.$ gives the equation of the trajectory in the Q_1, Q_2 space

$$\frac{EC^{-3/2}Q_1}{2\sqrt{1 - \left(\frac{Q_1\Omega}{\sqrt{C}}\right)^2}} + \frac{Q_2}{2\sqrt{C}\sqrt{1 - \left(\frac{Q_1\Omega}{\sqrt{C}}\right)^2}} = B_2 = const. \qquad (2.4.84)$$

By simple manipulation and squaring, the form of the trajectory can be written as

$$Q_1^2\left(E^2 C^{-3} + 4B_2\Omega^2 C^{-2}\right) + 2EC^{-3/2}Q_1 Q_2 + Q_2^2 = 4B_2^2. \qquad (2.4.85)$$

Thus, the material point whose differential equations of motion are of the form (2.4.72) describe the elliptical trajectory in the plane Q_1, Q_2. Substituting (2.4.83) into (2.4.84) we find the coordinate $Q_2\left(t\right)$ in the form

$$Q_2 = \frac{E}{\Omega\sqrt{C}} \sin\left[\Omega\left(t + B_1\right)\right] + 2\sqrt{C}B_2 \cos\left[\Omega\left(t + B_1\right)\right]. \qquad (2.4.86)$$

Finally, it is easy to verify that the second pair of equations $P_1 = \partial S/\partial Q_1$ and $P_2 = \partial S/\partial Q_2$ give

$$
\begin{aligned}
P_1 &= \frac{E}{\sqrt{C}} \cos\left[\Omega\left(t + B_1\right)\right] - 2\Omega\sqrt{C}B_2 \sin\left[\Omega\left(t + B_1\right)\right], \\
P_2 &= \sqrt{C} \cos\left[\Omega\left(t + B_1\right)\right]. \qquad (2.4.87)
\end{aligned}
$$

2.5 Application of the Hamilton–Jacobi Method to the Study of Rheolinear Oscillations

In this section we study the problem of finding the quadratic conservation laws and the motion of a rheolinear (i.e., time-dependent) harmonic oscillator with a single degree of freedom whose Hamiltonian function is given in the form

$$H = \frac{1}{2}a\left(t\right)p^2 + b\left(t\right)px + \frac{1}{2}m\left(t\right)x^2, \qquad (2.5.1)$$

where $x\left(t\right)$ is a position coordinate, p is the momentum, and $a\left(t\right), b\left(t\right),$ and $m\left(t\right)$ are arbitrary sufficiently smooth functions of time t. We will suppose that these functions are of such a nature that the solution of the canonical differential equations of motion, which follow from (2.5.1) via Hamilton's equations

$$\dot{x} = \frac{\partial H}{\partial p} = a\left(t\right)p + b\left(t\right)x, \quad \dot{p} = -\frac{\partial H}{\partial x} = -b\left(t\right)p - m\left(t\right)x, \tag{2.5.2}$$

are oscillatory when a, b, and m are held constant.

Our objective is to find a complete integral of the Hamilton–Jacobi equation

$$\frac{\partial S}{\partial t} + \frac{1}{2}\left(\frac{\partial S}{\partial x}\right)^2 + b\left(t\right)\frac{\partial S}{\partial x}x + \frac{1}{2}m\left(t\right)x^2 = 0. \tag{2.5.3}$$

Since the dynamical system has one degree of freedom the principal function is of the form

$$S = S\left(t, x, I\right), \tag{2.5.4}$$

where I is a nonadditive constant. According to the Jacobi theorem

$$\frac{\partial S}{\partial x} = p, \quad \frac{\partial S}{\partial I} = K = const. \tag{2.5.5}$$

In order to reduce the problem to a simpler form we introduce a canonical transformation that transforms the "old" variables x, p into the "new" variables X, P whose generating function is

$$F_1 = \frac{1}{2}A\left(t\right)x^2 - B\left(t\right)xX, \tag{2.5.6}$$

where the unknown functions of time $A\left(t\right)$ and $B\left(t\right)$ are going to be determined in the course of analysis. According to $(1.9.13)_{1,2}$ the function F_1 implies the following transformation rules $p = \partial F_1/\partial x, P = -\partial F_1/\partial X$, whence

$$P = B\left(t\right)x, \quad X = -\frac{1}{B\left(t\right)}p + \frac{A\left(t\right)}{B\left(t\right)}x, \tag{2.5.7}$$

and the inverse transformation is of the form

$$x = \frac{1}{B\left(t\right)}P, \quad p = -B\left(t\right)X + \frac{A\left(t\right)}{B\left(t\right)}P. \tag{2.5.8}$$

By means of this canonical transformation we find the new Hamiltonian function $\bar{H} = \left(\partial F_1/\partial t + H\right)_{(x,p)\to(X,P)}$, which is of the form

$$\bar{H} = \frac{1}{2B^2}\left(\dot{A} + aA^2 + 2bA + m\right)P^2 + \left(-aA - b - \frac{\dot{B}}{B}\right)PX + \frac{1}{2}aB^2X^2. \tag{2.5.9}$$

It is easy to see that under the conditions

$$\dot{A} + aA^2 + 2bA + m - aB^4 = 0,$$
$$\dot{B} + aAB + bB = 0, \tag{2.5.10}$$

the new Hamiltonian function will be reduced to a much simpler form,

$$\bar{H} = \frac{1}{2}aB^2 \left(P^2 + X^2\right),\tag{2.5.11}$$

whose canonical equations are

$$\dot{X} = aB^2 P, \quad \dot{P} = -aB^2 X.\tag{2.5.12}$$

Note that this system admits the following quadratic conservation law:

$$X^2 + P^2 = I = const.\tag{2.5.13}$$

The new Hamilton–Jacobi equation, which corresponds to the new Hamiltonian function (2.5.11), is

$$\frac{\partial \bar{S}}{\partial t} + aB^2 \left[\frac{1}{2}\left(\frac{\partial \bar{S}}{\partial X}\right)^2 + \frac{1}{2}X^2\right] = 0,\tag{2.5.14}$$

where $\bar{S}$ is new principal function.

To find a complete integral of (2.5.14) we suppose that it is of the form

$$\bar{S} = -\frac{C}{2}\int aB^2 dt + \varphi(X),\tag{2.5.15}$$

where C is a constant. Entering this into (2.5.14) we have

$$\frac{d\varphi}{dX} = \left(C - X^2\right)^{1/2}.\tag{2.5.16}$$

Integrating, we find that a complete solution of the new Hamilton–Jacobi equation (2.5.14) is

$$\bar{S} = -\frac{C}{2}\int aB^2 dt + \frac{X}{2}\left(C - X^2\right)^{1/2} + \frac{C}{2}\arcsin\frac{X}{\sqrt{C}}.\tag{2.5.17}$$

Applying the Jacobi theorem we easily verify that the equation

$$\frac{\partial \bar{S}}{\partial C} = \frac{K}{2} = const.\tag{2.5.18}$$

gives

$$X = \sqrt{C}\sin\left(\int aB^2 dt + K\right).\tag{2.5.19}$$

The second equation arising from the Jacobi theorem, that is, $P = \partial\bar{S}/\partial X = d\varphi/dX$, gives

$$P = \sqrt{C}\cos\left(\int aB^2 dt + K\right).\tag{2.5.20}$$

To find the old coordinates x and p in terms of time t we enter with (2.5.19) and (2.5.20) into (2.5.8) to get

$$x = \frac{1}{B(t)} \sqrt{C} \cos \left(\int aB^2 dt + K \right) \tag{2.5.21}$$

and

$$p = -B\sqrt{C} \sin \left(\int aB^2 dt + K \right) + \frac{A}{B} \sqrt{C} \cos \left(\int aB^2 dt + K \right), \tag{2.5.22}$$

where $A(t)$ and $B(t)$ are *any* solution of the system of auxiliary differential equations $(2.5.10)_{1,2}$ subject to arbitrary initial conditions.

We can reduce this system of nonlinear equations to a single (nonlinear) equation of the second order by introducing a new function $w(t)$ by the relation

$$B(t) = \frac{1}{w(t) a^{1/2}}. \tag{2.5.23}$$

Thus, we obtain

$$A(t) = \frac{\dot{w}}{aw} + \frac{1}{2} \frac{\dot{a}}{a^2} - \frac{b}{a}. \tag{2.5.24}$$

Entering with (2.5.23) and (2.5.24) into (2.5.10) we get the second-order auxiliary equation

$$\ddot{w} + \left(\frac{1}{2} \frac{\ddot{a}}{a} - \frac{3}{4} \frac{\dot{a}^2}{a^2} - \dot{b} + \frac{\dot{a}b}{a} - b^2 + am \right) w - \frac{1}{w^3} = 0. \tag{2.5.25}$$

The general solution of the canonical system (2.5.2) can be expressed in the form

$$x = \sqrt{C} w a^{1/2} \cos \left(\int \frac{dt}{w^2} + K \right),$$

$$p = -\sqrt{C} \frac{1}{w a^{1/2}} \sin \left(\int \frac{dt}{w^2} + K \right)$$

$$+ \sqrt{C} \left(\frac{\dot{w}}{a^{1/2}} + \frac{1}{2} \frac{\dot{a}w}{a^{2/3}} - \frac{bw}{a^{1/2}} \right) \cos \left(\int \frac{dt}{w^2} + K \right), \tag{2.5.26}$$

where $w(t)$ is any solution of the auxiliary equation (2.5.25).

If the expressions for A and B given by (2.5.24) and (2.5.23) are substituted into (2.5.7), the conservation law (2.5.13) can be written in the form

$$I = \left[w a^{1/2} p - \left(\frac{\dot{w}}{a^{1/2}} + \frac{1}{2} \frac{\dot{a}w}{a^{3/2}} - \frac{bw}{a^{1/2}} \right) x \right]^2 + \frac{x^2}{aw^2} = const. \tag{2.5.27}$$

It is evident from (2.5.19), (2.5.20), and (2.5.13) that the constant I is equal to the constant C figuring in (2.5.17).

Remark. It is of interest to note that we can also find a complete integral of the Hamilton–Jacobi equation (2.5.3) directly, without passing to the transformed Hamilton–Jacobi equation (2.5.14) (see also [124]). In order to accomplish this, we nominate the generating function F_1 given by (2.5.6) as the principal function, that is, $F_1 = S$. Then, from (2.5.6), it follows that

$$p = \frac{\partial F_1}{\partial x} = \frac{\partial S}{\partial x} = A(t)x - B(t)X. \tag{2.5.28}$$

Combining the first equation (2.5.7) with the conservation law (2.5.13) we have

$$p = \frac{\partial S}{\partial x} = Ax - B\left(I - B^2x^2\right)^{1/2}. \tag{2.5.29}$$

Integrating this partially with respect to x one arrives at the equation

$$S(t,x,I,A,B) = \frac{1}{2}Ax^2 - \frac{1}{2}Bx\sqrt{I - (Bx)^2} - \frac{IB}{2}\arcsin\frac{x}{\sqrt{I}} + U(t), \tag{2.5.30}$$

where $U(t)$ is an arbitrary function of time. Substituting this expression into (2.5.3) we find

$$\frac{x}{2}\left(\dot{A} + aA^2 + 2bA + m - aB^4\right)$$
$$-x\sqrt{I - (Bx)^2}\left(\dot{B} + aAB - bB\right) - \left(\frac{1}{2}IaB^2 + \dot{U}\right) = 0. \tag{2.5.31}$$

This expression has to be satisfied identically for all values of x, and the terms in the parentheses must vanish. Therefore, we arrive again at the equations $(2.5.10)_{1,2}$ and

$$U(t) = -\frac{I}{2}\int aB^2 dt. \tag{2.5.32}$$

Employing the same transformations to the new function $w(t)$ as in equations (2.5.23)–(2.5.25), we finally arrive at a complete integral of the Hamilton–Jacobi equation (2.5.3) in the form

$$S(t,x,I,w) = \frac{x^2}{2}\left(\frac{\dot{w}}{aw} + \frac{\dot{a}}{2a^2} - \frac{b}{a}\right) - \frac{x}{2wa^{1/2}}\sqrt{I - \left(\frac{x}{wa^{1/2}}\right)^2}$$
$$-\frac{I}{2}\arcsin\left(\frac{x}{wa^{1/2}I^{1/2}}\right) - \frac{I}{2}\int\frac{dt}{w^2}, \tag{2.5.33}$$

where $w(t)$ is any solution of the auxiliary equation (2.5.25). It is easy to verify that the Jacobi theorem (2.5.5) generates the same solution of the canonical system (2.5.2) as those indicated by equation (2.5.26).

It is of interest to note some special cases following these general considerations.

(a) *The Lewis invariant.* As a case of special interest, let us consider the harmonic rheolinear oscillator for which $a(t) = 1, b(t) = 0, m(t) = \omega^2(t)$, where $\omega(t)$ is the time-dependent circular frequency of the oscillator.

The differential equations in the canonical form are

$$\dot{x} = p, \quad \dot{p} = -\omega^2\left(t\right)x. \qquad (2.5.34)$$

The general solution of this system follows from (2.5.26), namely,

$$x = \sqrt{I}w\cos\left(\int\frac{dt}{w^2} + K\right),$$

$$p = -\sqrt{I}\frac{1}{w}\sin\left(\int\frac{dt}{w^2} + K\right) + \sqrt{I}\dot{w}\cos\left(\int\frac{dt}{w^2} + K\right), \quad (2.5.35)$$

where $w\left(t\right)$ is any solution of the auxiliary differential equation

$$\ddot{w} + \omega^2\left(t\right)w - \frac{1}{w^3} = 0, \qquad (2.5.36)$$

which follows directly from equation (2.5.25).

The conservation law (2.5.27) now becomes

$$I = \left(wp - \dot{w}x\right)^2 + \left(\frac{x}{w}\right)^2 = const. \qquad (2.5.37)$$

This conservation law has been obtained by Lewis [66], and the general form of this conservation law was found by Symon [105] using quite a different approach, which is not connected with the Hamilton–Jacobi method.

It is important to note that, if the circular frequency $\omega\left(t\right)$ is slowly varying, we can take $\ddot{w} \approx 0$ and a solution of (2.5.36) to the zeroth order in time derivative as

$$w \approx \frac{1}{\sqrt{\omega}}, \qquad (2.5.38)$$

and the conservation law of the slowly varying harmonic oscillator takes the form

$$\frac{1}{\omega}\left(\frac{p^2}{2} + \frac{\omega^2}{2}x^2\right) = \frac{E}{\omega} = const., \qquad (2.5.39)$$

where $E = \left(p^2/2 + \omega^2 x^2/2\right) = H$ denotes the total mechanical energy of the harmonic oscillator.

The expression (2.5.39) is usually referred to as the *adiabatic invariant* of the slowly varying harmonic oscillator.

Note also that for the case when the circular frequency is constant, the auxiliary differential equation

$$\ddot{w} + \omega^2 w - \frac{1}{w^3} = 0 \qquad (2.5.40)$$

has the general solution in the form[10]

$$w^2 = \left(\frac{1}{\omega^2} + A^2 + B^2 \right)^{1/2} + A \cos\left(2\omega t\right) + B \sin\left(2\omega t\right), \qquad (2.5.41)$$

where A and B are arbitrary constants. If the constants A and B are equal to zero we arrive again to the approximation (2.5.38).

We note that the conservation law (2.5.21) and corresponding auxiliary equation (2.5.36) have been discussed also by Courant [30].

(b) *Bessel pendulum.* As an example, let us consider the so-called Bessel pendulum whose Hamiltonian function is given by

$$H = \frac{p^2}{2t} + \left(t - \frac{1}{4t} \right) \frac{x^2}{2} \qquad (2.5.42)$$

and whose canonical equations are of the form

$$\dot{x} = \frac{p}{t}, \quad \dot{p} = -\left(t - \frac{1}{4t} \right) x. \qquad (2.5.43)$$

Comparing these equations with (2.5.1) and (2.5.2), we have

$$a\left(t\right) = \frac{1}{t}, \quad b\left(t\right) = 0, \quad m\left(t\right) = t - \frac{1}{4t}. \qquad (2.5.44)$$

Thus, the auxiliary equation (2.5.25) becomes

$$\ddot{w} + w - \frac{1}{w^3} = 0. \qquad (2.5.45)$$

The general solution (2.5.41) of this equation for $A = B = 0$ is $w = 1$ and the conservation law (2.5.27) is of the form

$$I = \frac{1}{t} \left(p + \frac{x}{2} \right)^2 + t x^2 = const. \qquad (2.5.46)$$

The motion of the Bessel pendulum is, according to the first equation (2.5.26), given by

$$x = \sqrt{\frac{I}{t}} \cos\left(t + K\right), \qquad (2.5.47)$$

where I and K are arbitrary constants. This solution is a general solution of the Bessel differential equation of the second order,

$$t\ddot{x} + \dot{x} + \left(t - \frac{1}{4t} \right) x = 0, \qquad (2.5.48)$$

---

[10]The general solution of (2.5.40) in the form (2.5.41) is found in the section 3.8, pp. 148–149.

which is obtained by elimination of momenta p from the canonical system (2.5.43). The momentum $p(t)$ can be obtained from the second equation (2.5.26) or by means of the conservation law (2.5.46) by substituting (2.5.47) into it.

(c) *Elevated (inclined) pendulum.* As another example let us find the adiabatic invariant of a pendulum of a constant length l and the bob of mass m mounted upon an elevated (inclined) frictionless plane, whose angle of elevation α is a slowly variable function of time during many periods of small oscillations of the pendulum (see Figure 2.5.1). The position of the point is given by the equations

$$x = l \cos \Theta, \quad y = h - l \cos \Theta \cos \alpha, \quad z = h - l \cos \Theta \sin \alpha. \tag{2.5.49}$$

Considering h, l, and α as constants, the kinetic energy of mass m is

$$T = \frac{1}{2} m \left(\dot{x}^2 + \dot{y}^2 + \dot{z}^2 \right) = \frac{1}{2} m l^2 \dot{\Theta}^2. \tag{2.5.50}$$

Since the pendulum performs small oscillations, that is, the angle Θ remains small, the potential energy of the point is

$$\Pi = mg \left(h - l \cos \Theta \right) \sin \alpha \approx \frac{1}{2} mgl \Theta^2 \sin \alpha, \tag{2.5.51}$$

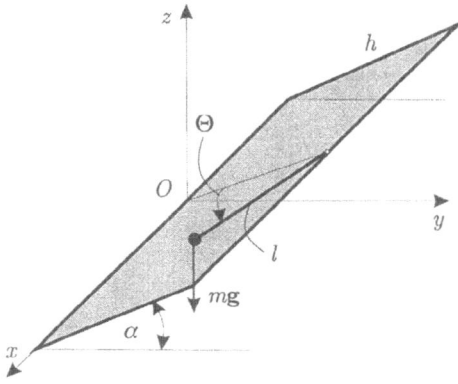

Figure 2.5.1

where g is the gravitational acceleration. The constant term $mgh \sin \alpha$ is discarded and due to the assumption of small oscillations, we used $\cos \Theta \approx 1 - \Theta^2/2$.

Forming the Lagrangian function $L = T - \Pi$, the Euler–Lagrangian equation with respect to the generalized coordinate Θ leads to

$$\frac{d}{dt} \frac{\partial L}{\partial \dot{\Theta}} - \frac{\partial L}{\partial \Theta} = \ddot{\Theta} + \frac{g}{l} \Theta \sin \alpha = 0 \tag{2.5.52}$$

or

$$\ddot{\Theta} + \omega^2 \Theta = 0, \tag{2.5.53}$$

where $\omega = \sqrt{\frac{g}{l} \sin \alpha}$. Thus, we suppose that the circular frequency is a slowly varying function of time. The solution of the equation (2.5.53) is

$$\Theta = C \cos{(\omega t + K)}, \tag{2.5.54}$$

where C is the amplitude and K is the initial phase. Entering with this solution into the expression for the total energy $E = E + \Pi$, we find

$$E = \frac{1}{2} mglC^2 \sin \alpha. \tag{2.5.55}$$

According to equation (2.5.39) the adiabatic invariant is of the form $E/\omega = const.$, namely,

$$\frac{E}{\omega} = C^2 \sqrt{\sin \alpha} = const. \tag{2.5.56}$$

From this expression we conclude that the amplitude of the oscillations of the inclined (elevated) pendulum whose angle of elevation α is a slowly varying function of time, and is changing according to

$$C \approx const. \ \sin^{-1/4} \alpha. \tag{2.5.57}$$

Finally, let us consider the *vertical pendulum whose length l is changing adiabatically* (see Figure 2.5.2) and find the corresponding adiabatic invariant.

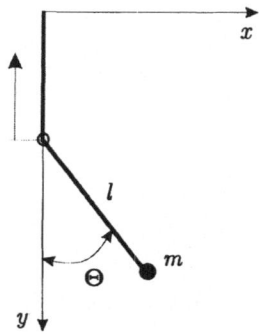

Figure 2.5.2

Repeating the same reasoning as in the previous example we see that the frequency and total energy of the small oscillations of the pendulum follows directly from $(2.5.53)_2$ and (2.5.55) for $\alpha = \pi/2$. Namely,

$$\omega = \sqrt{\frac{g}{l}}, \quad E = \frac{1}{2} mglC^2. \tag{2.5.58}$$

Therefore, the adiabatic invariant reads

$$\frac{E}{\omega} = \frac{1}{2}mg^{1/2}l^{3/2}C^2 = const.,\qquad(2.5.59)$$

whence

$$C \approx const.\ l^{-3/4},\qquad(2.5.60)$$

where C is the amplitude of the oscillations of the vertical pendulum whose length l is a slowly varying function of time.

(d) *The case of viscous dissipation.* The general time-dependent Hamiltonian (2.5.1) contains the special case of the dynamical system subject to time-dependent viscous forces that will be discussed here.

Let us write the canonical system of differential equations (2.5.2) in Lagrangian form, that is, by eliminating the momentum p we obtain

$$\ddot{x} + \dot{x}\left(-\frac{\dot{a}}{a}\right) + x\left(\frac{\dot{a}b}{a} - \dot{b} - b^2 + am\right) = 0.\qquad(2.5.61)$$

By taking

$$-\frac{\dot{a}}{a} = 2k\left(t\right),\quad b = 0,\quad am = \omega^2\left(t\right),\qquad(2.5.62)$$

we have the differential equation of a rheolinear dissipative dynamical system in the form

$$\ddot{x} + 2k\left(t\right)\dot{x} + \omega^2\left(t\right)x = 0,\qquad(2.5.63)$$

where $k\left(t\right)$ is the time-dependent damping coefficient and $\omega\left(t\right)$ is the circular frequency.

From (2.5.62) we find

$$a\left(t\right) = C\exp\left[-2\int^t k\left(u\right)du\right],\quad b\left(t\right) = 0,$$
$$m\left(t\right) = \frac{1}{C}\omega^2\left(t\right)\exp\left[2\int^t k\left(u\right)du\right].\qquad(2.5.64)$$

Therefore, for given $k\left(t\right),\omega\left(t\right)$, and $C = 1$, the Hamiltonian (2.5.1) becomes

$$H = \frac{1}{2}p^2\exp\left[-2\int^t k\left(u\right)du\right] + \frac{1}{2}\omega^2\left(t\right)x^2\exp\left[2\int^t k\left(u\right)du\right].\qquad(2.5.65)$$

The canonical equations of motion are of the form

$$\dot{x} = p\exp\left[-2\int^t k\left(u\right)du\right],\quad \dot{p} = -\omega^2\exp\left[2\int^t k\left(u\right)du\right]x.\qquad(2.5.66)$$

To find a general solution of this system by means of the Hamilton–Jacobi method, we note that the corresponding partial differential equation is

$$\frac{\partial S}{\partial t} + \frac{1}{2}\left(\frac{\partial S}{\partial x}\right)^2 \exp\left[-2\int^t k(u)\,du\right] + \frac{1}{2}\omega^2(t)\,x^2 \exp\left[2\int^t k(u)\,du\right] = 0,$$

$$(2.5.67)$$

and using the general form of the principal function given by (2.5.33) and (2.5.64), we find that the complete integral of (2.5.67) becomes

$$S(t,x,,I,w) = \frac{x^2}{2}\left(\frac{\dot{w}}{w}-k\right)\exp\left[2\int^t k(u)\,du\right]$$

$$-\frac{1}{2}\frac{x}{w}\sqrt{I-\left(\frac{x}{w}\right)^2\exp\left[2\int^t k(u)\,du\right]}$$

$$\times\sqrt{I-(x/w)^2\left[2\int^t k(u)\,du\right]}$$

$$-\frac{I}{2}\arcsin\left[\frac{x\exp\left(2\int^t k(u)\,du\right)}{wI^{1/2}}\right] - \frac{I}{2}\int^t \frac{du}{w^2},$$

$$(2.5.68)$$

where $w(t)$ is any solution of the auxiliary equation (2.5.25) which is, subject to (2.5.64), of the form

$$\ddot{w} + \left(\omega^2 - k^2 - \dot{k}\right)w - \frac{1}{w^3} = 0.\qquad(2.5.69)$$

It is easy to verify that the equation $\partial S/\partial I = K/2 = const.$ generates $x(t)$ in the form

$$x = -I^{1/2}w\exp\left[-\int^t k(u)\,du\right]\sin\left(\int^t \frac{du}{w^2(u)} + K\right),\qquad(2.5.70)$$

while $p = \partial S/\partial x$ leads to

$$p = -I^{1/2}\,(\dot{w}-kw)\exp\left[\int^t k(u)\,du\right]\sin\left(\int^t \frac{du}{w^2(u)} + K\right)$$

$$-\frac{I^{1/2}}{w}\exp\left[\int^t \frac{du}{w^2(u)}\right]\cos\left[\int^t \frac{du}{w^2(u)} + K\right].\qquad(2.5.71)$$

Finally, the conservation law (2.5.27) for the case considered here is of the form

$$I = \left\{pw\exp\left[-\int^t k(u)\,du\right]\right.$$

$$\left. -x\,(\dot{w}-kw)\exp\left[\int^t k(u)\,du\right]\right\}^2$$

$$+ \left(\frac{x}{w}\right)^2\exp\left[2\int^t k(u)\,du\right] = const.\qquad(2.5.72)$$

Note that this quadratic conservation law of the rheolinear dissipative oscillator, together with the auxiliary equation (2.5.69), was obtained in [85] and [123] by using a completely different approach.

(e) *The case of viscous dissipation and the forcing term.* Let us consider the linear time-dependent system whose Hamiltonian function is of the form

$$H(t, x, p) = \frac{1}{2}p^2 e^{-2F} + \left[\frac{1}{2}q(t)x^2 - h(t)x\right]e^{2F}, \qquad (2.5.73)$$

where $F(t) = \int^t k(u)du$ and $q(t)$ and $h(t)$ are given functions of time. The canonical system of differential equations of motion are

$$\dot{x} = pe^{-2F}, \quad \dot{p} = \left[-q(t)x + h(t)\right]e^{2F}. \qquad (2.5.74)$$

Under certain conditions, the equations (2.5.74) can represent a time-dependent oscillator whose dissipation is characterized by the damping coefficient $k(t)$, and an external force is denoted by $h(t)$.

The Hamilton–Jacobi partial differential equation is

$$\frac{\partial S}{\partial t} + \frac{1}{2}\left(\frac{\partial S}{\partial x}\right)^2 e^{-2F} + \left[\frac{1}{2}q(t)x^2 - h(t)x\right]e^{2F} = 0. \qquad (2.5.75)$$

We shall suppose that a complete solution can be sought in the form (see [127])

$$S(t, x, I) = \frac{1}{2}A(t)x^2 + E(t)x + \frac{1}{3}\left[I + K(t) + 2C(t)x\right]^{3/2} + \Lambda(t, I), \qquad (2.5.76)$$

where $A(t)$, $E(t)$, $K(t)$, and $C(t)$ are unknown functions of time, I is an arbitrary constant $I = const.$, and $\Lambda(t, I)$ is an unknown function of t and I. Substituting (2.5.76) into the Hamilton–Jacobi equation (2.5.75) we find

$$\frac{1}{2}x^2 \left(\dot{A} + A^2 e^{-2F} + qe^{2F}\right) + x\left(\dot{E} + ACe^{-2F} + C^3 e^{-2F} - he^{2F}\right)$$
$$+ \frac{1}{2}\sqrt{\theta}\left(\dot{K} + 2ECe^{-2F}\right) + x\sqrt{\theta}\left(\dot{C} + ACe^{-2F}\right)$$
$$+ \left(\dot{\Lambda} + \frac{1}{2}E^2 e^{-2F} + \frac{1}{2}IC^2 e^{-2F} + \frac{1}{2}KC^2 e^{-2F}\right) = 0, \qquad (2.5.77)$$

where $\theta = I + K + 2Cx$.

Equation (2.5.77) will be satisfied for each x and t if the terms in round brackets are equal to zero. Thus, we arrive at the following system of equations:

$$
\begin{aligned}
\dot{A} + A^2 + qe^{2F} &= 0, \\
\dot{E} + AEe^{-2F} + C^3 e^{-2F} - he^{2F} &= 0, \\
\dot{K} + 2ECe^{-2F} &= 0, \\
\dot{C} + ACe^{-2F} &= 0, \\
\dot{\Lambda} + \frac{1}{2}e^{-2F}\left(E^2 + IC^2 + KC^2\right) &= 0. \qquad (2.5.78)
\end{aligned}
$$

In order to analyze this system we first introduce two unknown functions $w\,(t)$ and $\Psi\,(t)$ and connect them with functions $A\,(t)$ by the relation

$$\int^t \left[A\,(u)\,e^{-2F} + k\,(u)\right] du = \ln\left[w\,(t)\cos\Psi\,(t)\right].$$ (2.5.79)

Differentiating (2.5.79) with respect to time we obtain

$$A(t) = \left(\frac{\dot{w}}{w} - \dot{\Psi}\tan\Psi - k\right) e^{2F}.$$ (2.5.80)

Entering with this into $(2.5.78)_1$ we have

$$\frac{\ddot{w}}{w} + q - k^2 - \dot{k} - \dot{\Psi}^2 - \left(\ddot{\Psi} + 2\frac{\dot{w}}{w}\dot{\Psi}\right)\tan\Psi = 0.$$ (2.5.81)

Let us select the function Ψ in such a way that $\ddot{\Psi} + 2\frac{\dot{w}}{w}\dot{\Psi} = 0$. Integrating this condition twice we find $\Psi = C_1 \int^t \frac{du}{w^2} + C_2$, where C_1 and C_2 are arbitrary constants. However, for the sake of simplicity, we can take $C_1 = 1$ and $C_2 = 0$. Therefore

$$\Psi = \int^t \frac{du}{w^2(u)},$$ (2.5.82)

and the equation (2.5.81) becomes

$$\ddot{w} + \left[q(t) - k^2(t) - \dot{k}(t)\right] w - \frac{1}{w^3} = 0.$$ (2.5.83)

This equation plays the basic role in our consideration and we will refer to it as the auxiliary equation. The function $A(t)$ given by (2.5.80) now becomes

$$A(t) = \left(\frac{\dot{w}}{w} - \frac{1}{w^2}\tan\Psi - k\right) e^{2F},$$ (2.5.84)

where Ψ is given by (2.5.82).

The equation $(2.5.78)_4$ leads now to

$$\frac{dC}{C} = -\frac{dw}{w} + \tan\Psi d\Psi + dF.$$ (2.5.85)

Integrating, and setting the arbitrary integration constant equal to 1, we find

$$C(t) = \frac{e^F}{w\cos\Psi},$$ (2.5.86)

where the integration constant C^* is taken to be unity.

Repeating the same procedure of integration we obtain from $(2.5.78)_2$ and $(2.5.78)_3$

$$
E(t) = \frac{e^F}{w(t)\cos\Psi(t)} \int^t \left(h(u)w(u)\cos\Psi(u)e^F - \frac{1}{w^2(u)\cos\Psi(u)} \right) du,
$$

$$
K(t) = -2\int^t \frac{1}{w^2(\xi)\cos^2\Psi(\xi)}
$$
$$
\times \left[\int^\xi \left(h(u)w(u)\cos\Psi(u)e^F - \frac{1}{w^2(u)\cos^2\Psi(u)} \right) du \right] d\xi.
$$

(2.5.87)

It is to be noted that the functions $F(t), E(t), C(t)$, and $K(t)$ are functions of time, and they do not contain the constant I, so that the integral expression

$$
\frac{1}{2}\int^t e^{-2F} \left(E^2(u) + C^2(u)K(u) \right) du = \text{function of time} \tag{2.5.88}
$$

will be called *irrelevant term* $= (i.t.)$, since in the further analysis the Jacobi theorem is not influenced by this term. Finally, integrating $(2.5.78)_5$ we find

$$
\Lambda(t, I) = -\frac{I}{2}\int^t e^{-2F} C^2(u) du + i.t. \tag{2.5.89}
$$

Therefore, the complete solution of the Hamilton–Jacobi partial differential equation (2.5.75) in the form (2.5.76) is

$$
S(t, x, I) = \frac{1}{2}A(t)x^2 + E(t)x + \frac{1}{3}\left[I + K(t) + 2C(t)x \right]^{3/2}
$$
$$
-\frac{I}{2}\int^t e^{-2F} C^2(u) du + i.t., \tag{2.5.90}
$$

where $A(t), E(t), K(t)$, and $C(t)$ are given by (2.5.84), $(2.5.87)_{1,2}$, and (2.5.86), respectively.

Applying the Jacobi theorem we find

$$
\frac{\partial S}{\partial I} = \frac{1}{2}\left(I + K + 2Cx \right)^{1/2} - \frac{1}{2}\int^t C^2(u)e^{-2F} du = \frac{M}{2} = \text{const.}, \tag{2.5.91}
$$

where $M/2$ is a new arbitrary constant. Therefore

$$
x(t) = -\frac{I + K(t)}{2C(t)} + \frac{1}{2C(t)}\left[\int^t C^2(u)e^{-2F} du + M \right]^2. \tag{2.5.92}
$$

The momentum is fully determined from the second equation of the Jacobi theorem:

$$
p = \frac{\partial S}{\partial x} = A(t)x + E(t) + C(t)\left[I + K(t) + 2C(t)x \right]^{1/2}. \tag{2.5.93}
$$

Example 2.5.1. Let us consider the Hamiltonian function [127]

$$H = \frac{p^2}{2f^2(t)} + \frac{1}{2}f(t)\ddot{f}(t)x^2 - Q(t)f(t)x, \tag{2.5.94}$$

where $f(t)$ and $Q(t)$ are arbitrary functions of time t. The canonical equations are

$$\dot{x} = \frac{p}{f^2}, \quad \dot{p} = -f(t)\ddot{f}(t)x + Q(t)f(t). \tag{2.5.95}$$

Comparing this system with the general canonical form (2.5.74) we find that the dynamical parameters are

$$q(t) = \frac{\ddot{f}(t)}{f(t)}, \quad k(t) = \frac{\dot{f}(t)}{f(t)}, \quad h(t) = \frac{Q(t)}{f(t)}. \tag{2.5.96}$$

The auxiliary equation (2.5.83) becomes $\ddot{w} - 1/w^3 = 0$, whose solution is found to be $w(t) = (1 + t^2)^{1/2}$. Therefore, $\Psi = \int^t \frac{du}{w^2(u)} = \arctan t, \sin \Psi = t/(1 + t^2)^{1/2}, \cos \Psi = 1/(1 + t^2)^{1/2}$. We also find $e^F = f(t)$. Note that

$$\int^t \left([h(u)w(u)\cos \Psi(u)] e^F - \frac{1}{w^2(u)\cos^2 \Psi(u)} \right) du = \int^t Q(u)du - t, \tag{2.5.97}$$

and

$$\int^t \frac{1}{w^2(\xi)\cos^2 \Psi(\xi)}$$
$$\times \left[\int^\xi \left(h(u)w(u)\cos \Psi(u)e^F - \frac{1}{w^2(u)\cos^2 \Psi(u)} \right) du \right] d\xi$$
$$= \int^t \left(\int^\xi Q(u)du \right) d\xi - \frac{t^2}{2}. \tag{2.5.98}$$

It is easy to verify that

$$C(t) = \frac{e^F}{\cos \Psi} = f(t); \quad K(t) = -2 \left[\int^t \left(\int^\xi Q(u)du \right) d\xi - \frac{t^2}{2} \right],$$
$$t = \int^t C^2(u)e^{-2F} du. \tag{2.5.99}$$

Thus, from (2.5.92) we have

$$x(t) = \frac{R + Mt}{f(t)} + \frac{1}{f(t)} \int^t \left(\int^\xi Q(u)du \right) d\xi, \tag{2.5.100}$$

where $R = M^2 - I = const.$ The linear momentum $p(t)$ we can calculate from (2.5.93).

2.6 A Conjugate Approach to Hamilton–Jacobi Theory. The Case of Rheolinear Systems

In the study of rheolinear dynamical systems it can be of interest to apply the so-called Legendre's dual transformation to the principal function $S(t, q_1, ..., q_n)$ and form a new principal function $V(t, p_1, ..., p_n)$ and corresponding conjugated Hamilton–Jacobi partial differential equation. The Legendre transformation will be accomplished from the generalized coordinates q_i to conjugated variables p_i considering time t as a passive parameter that does not participate in the following transformation.

Let us suppose that the principal function S is given and let us introduce new coordinates p_i by the relations

$$p_i = \frac{\partial S}{\partial q_i}. \tag{2.6.1}$$

We suppose that we can solve these equations with respect to q_i:

$$q_i = q_i(t, p_1, ..., p_n), \tag{2.6.2}$$

and we introduce the new the function $V(t, p_1, ..., p_n)$ by the relation

$$V = q_i p_i - S. \tag{2.6.3}$$

Substituting (2.6.2) into this relation we express the new function V in terms of t and p_i alone as

$$V = V(t, p_1, ..., p_n). \tag{2.6.4}$$

Differentiating (2.6.3) and taking into account the last equation we have

$$\frac{\partial V}{\partial t} + \frac{\partial V}{\partial p_i} dp_i = q_i dp_i + dq_i \left(p_i - \frac{\partial S}{\partial q_i} \right) - \frac{\partial S}{\partial t} dt, \tag{2.6.5}$$

whence

$$\frac{\partial S}{\partial t} = -\frac{\partial V}{\partial t}, \quad q_i = \frac{\partial V}{\partial p_i}. \tag{2.6.6}$$

Therefore, the Hamilton–Jacobi partial differential equation (2.2.9),

$$\frac{\partial S}{\partial t} + H \left(t, q_1, ..., q_n, \frac{\partial S}{\partial q_1}, ..., \frac{\partial S}{\partial q_n} \right) = 0, \tag{2.6.7}$$

can be transformed to its conjugated form:

$$\frac{\partial V}{\partial t} - H \left(t, p_1, ..., p_n, \frac{\partial V}{\partial p_1}, ..., \frac{\partial V}{\partial p_n} \right) = 0. \tag{2.6.8}$$

If we are able to find a complete solution to (2.6.8) in terms of t, p_i, and n nonadditive constants C_i,

$$V = V\left(, p_1, ..., p_n, C_1, ..., C_n\right),\qquad(2.6.9)$$

then the general solution of the dynamical system

$$\dot{q}_i = \frac{\partial H}{\partial p_i}, \quad \dot{p}_i = -\frac{\partial H}{\partial q_i},\qquad(2.6.10)$$

can be obtained by means of the Jacobi theorem whose form is given by the relations

$$q_i = \frac{\partial V}{\partial p_i}, \quad \frac{\partial V}{\partial C_1} = K_i = const.\qquad(2.6.11)$$

Let us apply our considerations to the rheolinear dynamical system whose Hamiltonian is of the form

$$H = \frac{1}{2}p^2 + \frac{1}{2}\omega^2\left(t\right) x^2,\qquad(2.6.12)$$

and corresponding differential equations in canonical form are given by (1.8.14).

The Hamilton–Jacobi equation in conjugated form is

$$\frac{\partial V}{\partial t} - \frac{1}{2}\omega^2\left(t\right)\left(\frac{\partial V}{\partial p}\right)^2 - \frac{1}{2}p^2 = 0.\qquad(2.6.13)$$

To find a general solution we introduce a canonical transformation F_2 in the form of an incomplete quadratic function which is conjugate to the generator F_1 introduced in (1.9.36):

$$F_2\left(t, x, X\right) = -B\left(t\right) xX + \frac{1}{2}C\left(t\right) X^2,\qquad(2.6.14)$$

where $B\left(t\right)$ and $C\left(t\right)$ are unknown function of time. By applying $p = \partial F_2/\partial x$, $P = -\partial F_2/\partial X$, we find

$$X = -\frac{p}{B}; \quad P = Bx + \frac{C}{B}p.\qquad(2.6.15)$$

As previously, we can show that the new Hamiltonian $H\left(t, X, P\right)$ admits a conservation law of the form (2.5.37), namely,

$$I = X^2 + P^2, \quad P = \left(I - X^2\right)^{1/2}.\qquad(2.6.16)$$

Expressing (2.6.16) in terms of the old coordinates by means of (2.6.15), we find

$$x = \frac{\partial V}{\partial p} = -\frac{C}{B^2}p + \frac{1}{B}\sqrt{I - \left(\frac{p}{B}\right)^2}.\qquad(2.6.17)$$

Integrating this expression partially with respect to p, one has

$$V = -\frac{C}{2B^2}p^2 + \frac{p}{2B^2}\sqrt{IB^2 - p^2} + \frac{I}{2}\arcsin\left(\frac{p}{I^{1/2}B}\right) + f(t), \qquad (2.6.18)$$

where $f(t)$ is an arbitrary function of time. Substituting (2.6.18) into (2.6.13) and equating to zero the like powers of p, we obtain the following system of ordinary differential equations:

$$\frac{\dot{C}}{B^2} - \frac{2C\dot{B}}{B^3} + \frac{C^2 - 1}{B^4}\omega^2 + 1 = 0,$$

$$\frac{B\dot{B}}{\omega^2} - C = 0,$$

$$\dot{f} + \frac{I\omega^2}{2B^2} = 0. \qquad (2.6.19)$$

The structure of these equations indicates that the function $B(t)$ can be taken as a basic auxiliary variable. Putting C, given by $(2.6.19)_2$, into $(2.6.19)_1$, we obtain the auxiliary equation

$$\frac{d}{dt}\left(\frac{\dot{B}}{\omega^2}\right) + B - \frac{\omega^2}{B^3} = 0. \qquad (2.6.20)$$

Therefore, the complete solution of the Hamilton–Jacobi equation (2.6.13) is found to be

$$\begin{aligned}
V(t, p, B, I) &= -\frac{\dot{B}}{2B\omega^2} + \frac{p}{2B}\sqrt{IB^2 - p^2} \\
&\quad + \frac{I}{2}\arcsin\left(\frac{p}{I^{1/2}B}\right) - \frac{I}{2}\int^t \frac{\omega^2(u)}{B^2(u)}du. \qquad (2.6.21)
\end{aligned}$$

Applying the Jacobi theorem, we find from equation $x = \partial V/\partial p$ after squaring the following quadratic conservation law:

$$I = I_2 = \left(\frac{p}{B}\right)^2 + \left(Bx + \frac{\dot{B}}{\omega^2}p\right)^2 = const., \qquad (2.6.22)$$

where $B(t)$ is any solution of the auxiliary equation (2.6.20).

Equation $\partial V/\partial I = K$ yields

$$p = I^{1/2}B\sin\left(\int^t \frac{\omega^2(u)}{B^2(u)}du + K\right). \qquad (2.6.23)$$

Combining this with (2.6.22), we find that the position coordinate is given by

$$x = -\frac{\dot{B}I^{1/2}}{\omega^2}\sin\left(\int^t \frac{\omega^2(u)}{B^2(u)}du + K\right) + \frac{I^{1/2}}{B}\cos\left(\int^t \frac{\omega^2(u)}{B^2(u)}du + K\right), \qquad (2.6.24)$$

which completes the general solution of the canonical system (2.5.34).

Since the Lewis invariant (2.5.37)

$$I_1 = (wp - \dot{w}x)^2 + \left(\frac{x}{w}\right)^2 = const., \qquad (2.6.25)$$

and the conservation law (2.6.22) of the time-dependent oscillator

$$\dot{x} = p, \quad \dot{p} = -\omega^2(t)x, \qquad (2.6.26)$$

are of different structures, it is of interest to examine under what conditions these two quadratic first integrals will be equivalent.

Forming the difference $I_1 - I_2$ we have

$$
I_1 - I_2 = \left(\frac{1}{B^2} + \frac{\dot{B}}{\omega^4} - w^2\right)p^2 + 2\left(w\dot{w} + \frac{B\dot{B}}{\omega^2}\right)px
$$
$$
+ \left(B^2 - \dot{w}^2 - \frac{1}{w^2}\right)x^2. \qquad (2.6.27)
$$

Therefore, the conservation laws I_1 and I_2 will be equivalent if the following system of ordinary differential equations is satisfied:

$$\dot{B} = \omega^2\sqrt{w^2 - \frac{1}{B^2}}, \quad \dot{w} = -\sqrt{B^2 - \frac{1}{w^2}}, \qquad (2.6.28)$$

and

$$w\dot{w} = -\frac{B\dot{B}}{\omega^2}. \qquad (2.6.29)$$

It is easy to verify that the last equation can be considered as a direct consequence of the system (2.6.28). It is also easy to show that by eliminating B from (2.6.28) we arrive at the Lewis auxiliary equation (2.5.36)

$$\ddot{w} + \omega^2 w - \frac{1}{w^3} = 0. \qquad (2.6.30)$$

Similarly, by eliminating w from (2.6.28), the resulting differential equation is identical with (2.6.20).

Substituting $\dot{w}$ and $\dot{B}$ given by (2.6.28) we arrive at a new "hybrid" quadratic conservation law of the dynamical system (2.6.26) in the form (see [85])

$$I_3 = w^2p^2 + 2\left(w^2B^2 - 1\right)^{1/2}px + B^2x^2, \qquad (2.6.31)$$

where w and B are any solution of the auxiliary system (2.6.28).

Therefore, the conservation laws (2.6.22) and (2.6.25) are fully equivalent if the auxiliary functions $w(t)$ and $B(t)$ satisfy the system of ordinary differential equations (2.6.28). Otherwise, they can be considered as mutually independent. Thus, if $w(t)$ and $B(t)$ are not limited by the differential equations (2.6.28) then from two conservation laws I_1 and I_2 we are able to find the solution of the dynamical system (2.6.26) by purely algebraic means, which will now be demonstrated. Namely, if I_1 and I_2 are independent, these conservation laws can be employed for finding the motion of the dynamical system without integrating the canonical differential equations (2.6.26) by calculating $x(t)$ and $p(t)$ from independent conservation laws I_1 and I_2. However, to do this, we have to know some solutions of auxiliary equations (2.6.20) and (2.6.30). Since both auxiliary equations are nonlinear, it is difficult to find solutions of this system in a closed form for arbitrary initial conditions. Thus, we will employ numerical procedures by means of a computer.

Let us consider the following time-dependent oscillator formulated as an initial-value problem

$$\dot{x} = p, \quad \dot{p} = -\omega(t)^2\, x, \quad x(0) = 1, \quad p(0) = 2, \quad 0 \le t \le 5, \quad \omega(t) = e^{t/2}. \tag{2.6.32}$$

The problem is to find $x(t)$ and $p(t)$ from the expressions I_1 and I_2 given by (2.6.22) and (2.6.25), namely,

$$I_1 = (wp - \dot{w}x)^2 + \left(\frac{x}{w}\right)^2, \quad I_2 = \left(\frac{p}{B}\right)^2 + \left(Bx + \frac{\dot{B}}{\omega^2}p\right)^2. \tag{2.6.33}$$

Here, w and Y are solutions of the auxiliary system, given as an initial value problem,

$$\dot{w} = Y, \quad w(0) = 1, \quad \dot{Y} = \frac{1}{w} - \omega^2(t)\,w, \quad Y(0) = \dot{w}(0) = 0. \tag{2.6.34}$$

Furthermore,

$$I_2 = \frac{p^2}{B^2} + (Zp + wx)^2, \tag{2.6.35}$$

where B and Z are solutions of the auxiliary system, given also as an initial-value problem

$$\dot{B} = Z\omega^2(t), \quad B(0) = 2, \quad \dot{Z} = -B + \frac{\omega^2}{B^3}, \quad Z(0) = 1, \quad \text{(i.e., } \dot{B}(0) = 1\text{)}. \tag{2.6.36}$$

Note that the initial conditions $w(0), \dot{w}(0), B(0)$, and $\dot{B}(0)$ listed in (2.6.34) and (2.6.36) are selected arbitrarily. The only restriction is that they do *not* satisfy the differential equations (2.6.28) and (2.6.29) at $t = 0$.

The system of differential equations (2.6.34) and (2.6.36) has been integrated numerically using the Runge–Kutta method. The time step is taken to be $\Delta t = 10^{-6}$. In each step of the calculation the values of x and p have been found from (2.6.33) and (2.6.35).

The results of calculations, that is, x and p versus time are depicted in Figure 2.6.1 in the time interval $0 \le t \le 5$. In order to check the accuracy of the calculations the dynamical system (2.6.32) has been directly integrated numerically and L_2 norms have been calculated for the difference of the so obtained solutions, that is,

$$\|x - x_1\|_{L_2} = \left[\int_0^5 \left(x\left(t\right) - x_1\left(t\right)\right)^2 dt \right]^{1/2} \qquad (2.6.37)$$

and

$$\|x - p_1\|_{L_2} = \left[\int_0^5 \left(x\left(t\right) - p_1\left(t\right)\right)^2 dt \right]^{1/2}, \qquad (2.6.38)$$

where x_1 and p_1 are the values of x and p obtained from the direct integration of (2.6.34) and (2.6.36) and x_2 and p_2 are obtained as the algebraic solution of the conservation laws (2.6.33) and (2.6.35). Both norms (2.6.37) and (2.6.38) are practically equal to zero (the largest values of both norms are found to be at $t = 5$ and are of the order 4×10^{-7}).

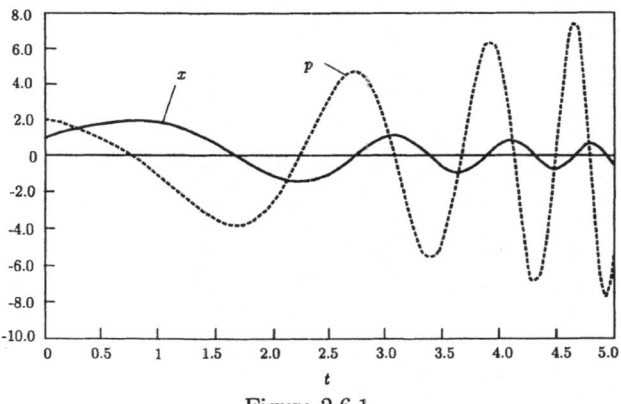

Figure 2.6.1

2.7 Quadratic Conservation Laws of Rheolinear Dynamical Systems with Two Degrees of Freedom

In this section we outline a method for finding the quadratic conservation laws of the dynamical system described by the Hamiltonian of the form

$$H = \frac{1}{2}\Lambda(t)p_x^2 + \frac{1}{2}\Theta(t)p_y^2 + \frac{1}{2}\left[a(t)x^2 + 2b(t)xy + m(t)y^2\right], \qquad (2.7.1)$$

where x and y are position coordinates and p_x and p_y are their conjugate momenta; $\Lambda(t), \Theta(t), a(t), b(t)$, and $m(t)$ are prescribed functions of time that are continuously differentiable functions, otherwise they are arbitrary.

Since the coefficient $b(t)$ is supposed to be different from zero we named this case a "coupled dynamical system." Note that quadratic conservation laws of the rheolinear "uncoupled dynamical system" (with $b(t) = 0$ and also $\Lambda(t) = \Theta(t) = 1$) have been studied in detail by numerous authors (see, for example, [51] and [97] and references cited therein).

Our method for finding conservation laws can be briefly described as follows. First, we introduce the canonical transformations

$$(x, y, p_x, p_y) \rightarrow (X, Y, P_X, P_Y) \tag{2.7.2}$$

for the Hamiltonian (2.7.1), by means of which we transform (2.7.1) into an absolute form-invariant Hamiltonian:

$$\bar{H} = \frac{1}{2}\Lambda(t)P_X^2 + \frac{1}{2}\Theta(t)P_Y^2 + \frac{1}{2}\left[a(t)X^2 + 2b(t)XY + m(t)Y^2\right]. \tag{2.7.3}$$

Note that to find a particular canonical transformation (2.7.2), which transforms the Hamiltonian (2.7.1) into the new Hamiltonian (2.7.3), is an essential part of our theory.

Second, the canonical differential equations based upon Hamiltonians (2.7.1) and (2.7.3) are

$$\begin{aligned}
\dot{x} &= \Lambda(t)p_x, & \dot{X} &= \Lambda(t)P_x, \\
\dot{y} &= \Theta(t)p_y, & \dot{Y} &= \Theta(t)P_y, \\
\dot{p}_x &= -a(t)x - b(t)y, & \dot{P}_x &= -a(t)X - b(t)Y, \\
\dot{p}_y &= -b(t)x - m(t)y, & \dot{P}_y &= -b(t)X - m(t)Y.
\end{aligned} \tag{2.7.4}$$

It is easy to verify that the eight differential equations (2.7.4) admit the conservation law of the form

$$I = xP_x - Xp_x + yP_y - Yp_y = const., \tag{2.7.5}$$

and for the special case in which

$$\Lambda(t) = \Theta(t), \quad a(t) = m(t), \quad b(t) \neq 0, \tag{2.7.6}$$

dynamical system (2.7.4) admits the angular momentum–type conservation law of the form

$$\bar{I} = xP_y - Xp_y + yP_x - Yp_x. \tag{2.7.7}$$

Thus, by supposing that we have succeeded in finding a concrete form of the canonical transformation (2.7.2) we are able to express conservation laws in terms of the original variables (x, y, p_x, p_y) by finding the inverse transformation (2.7.2).

Following the method just described, we introduce the generating function of the form

$$F = \frac{1}{2}\left[A(t)x^2 - 2B(t)xX + C(t)X^2\right] + \frac{1}{2}\left[\alpha(t)y^2 - 2\beta(t)yY + \gamma(t)Y^2\right]$$
$$-K(t)xY - D(t)yX - T(t)xy - S(t)XY, \tag{2.7.8}$$

where ten functions $A(t), ..., S(t)$ are unknown functions of time to be determined. This generating function specifies the equations of the canonical transformation (2.7.2) in the following way:

$$p_x = \frac{\partial F}{\partial x}, \quad p_y = \frac{\partial F}{\partial y}, \quad P_X = -\frac{\partial F}{\partial X}, \quad P_Y = -\frac{\partial F}{\partial Y}. \tag{2.7.9}$$

Substituting (2.7.8) into the last system of equations, we express the old variables in terms of the new ones:

$$x = \frac{1}{\Delta}\left[-\beta P_X + D P_Y - (\beta C + DS)X + (\beta S + \gamma D)Y\right],$$

$$y = \frac{1}{\Delta}\left[K P_X - B P_Y - (CK + BS)X - (KS + B\gamma)Y\right],$$

$$p_x = \frac{1}{\Delta}\{-(A\beta + TK)P_X + (AD + TB)P_Y$$
$$- [A(\beta C + DS) + T(CK + BS) + \Delta B]X$$
$$+ [A(\beta S + \gamma D) + T(KS + B\gamma) - \Delta K]Y\},$$

$$p_y = \frac{1}{\Delta}\{(T\beta + \alpha K)P_X - (TD + \alpha B)P_Y$$
$$+ [T(\beta C + DS) + \alpha(CK + BS) - \Delta D]X$$
$$- [T(\beta S + \gamma D) + \alpha(KS + B\gamma) + \Delta\beta]Y\}, \tag{2.7.10}$$

where

$$\Delta(t) = D(t)K(t) - B(t)\beta(t). \tag{2.7.11}$$

Expressing the new Hamiltonian $\bar{H} = H + \partial F/\partial t$ in terms of the new variables X, Y, P_X, and P_Y, where H is given by equation (2.7.1), we obtain

$$\bar{H} = \frac{1}{2\Delta^2}\left[V_1 P_X^2 + V_2 P_Y^2 + N_1 X^2 + N_2 Y^2\right.$$
$$-2N_3 XY - 2Q_1 P_X P_Y + 2Q_2 X P_X$$
$$\left.-2Q_3 X P_Y - 2Q_4 Y P_X + 2Q_5 Y P_Y\right]. \tag{2.7.12}$$

The functions $V_1, V_2, ..., Q_5$ are determined from the condition that the Hamiltonian $\bar{H}$ must be absolute form-invariant. Thus, by comparing the expressions (2.7.1) and (2.7.3), it is obvious that the following system of equations must be satisfied:

$$V_1 = \Lambda(t)\Delta^2, \quad V_2 = \Theta(t)\Delta^2,$$
$$N_1 = a(t)\Delta^2, \quad N_2 = m(t)\Delta^2,$$
$$N_3 = -b(t)\Delta^2, \quad Q_i = 0, \quad i = 1, ..., 5, \tag{2.7.13}$$

or, written explicitly in the same sequence, this system is

$$
\begin{aligned}
V_1 =\ & \beta^2(\dot{A} + \Lambda A + \Theta T^2 + a) + K^2(\dot{\alpha} + \Theta\alpha^2 + \Lambda T^2 + m) \\
& + 2K\beta(\dot{T} + \Lambda AT + \Theta T\alpha - b), \\
V_2 =\ & D^2(\dot{A} + \Lambda A^2 + \Theta T^2 + a) + B^2(\dot{\alpha} + \Theta\alpha^2 + \Lambda T^2 + m) \\
& + 2BD(\dot{T} + \Lambda AT + \Theta T\alpha - b), \\
N_1 =\ & C^2 V_1 + S^2 V_2 + 2CS Q_1 + \Delta^2(\dot{C} + \Lambda B^2 + \Theta D^2) \\
& + 2C\Delta\left[\dot{B}\beta - \dot{D}K + \Lambda B(A\beta + KT) - \Theta D(T\beta + K\alpha)\right] \\
& + 2S\Delta\left[\dot{B}D - \dot{D}B + \Lambda B(AD + BT) - \Theta D(DT + B\alpha)\right], \\
N_2 =\ & S^2 V_1 + \gamma^2 V_2 + 2CS Q_1 + \Delta^2(\dot{\gamma} + \Lambda K^2 + \Theta\beta^2) \\
& + 2S\Delta\left[\dot{\beta}K - \dot{K}\beta - \Lambda K(A\beta + KT) - \Theta\beta(T\beta + K\alpha)\right] \\
& + 2\gamma\Delta\left[\dot{\beta}B - \dot{K}D - \Lambda K(AD + BT) + \Theta\beta(DT + B\alpha)\right],
\end{aligned}
$$

$$
\begin{aligned}
N_3 =\ & SCV_1 + \gamma S V_2 + (C\gamma + S^2)Q_1 \\
& + S\Delta\left[\dot{B}\beta - \dot{D}K + \Lambda B(A\beta + KT) - \Theta D(T\beta + K\alpha)\right] \\
& + \gamma\Delta\left[\dot{B}D - \dot{D}B + \Lambda B(AD + BT) - \Theta D(TD + B\alpha)\right] \\
& + C\Delta\left[\dot{\beta}K - \dot{K}\beta - \Lambda K(A\beta + KT) + \Theta\beta(T\beta + K\alpha)\right] \\
& + S\Delta\left[\dot{\beta}B - \dot{K}D - \Lambda K(AD + BT) + \Theta\beta(TD + B\alpha)\right] \\
& + \Delta^2(\dot{S} - \Lambda BK - \Theta D\beta),
\end{aligned}
$$

$$
\begin{aligned}
Q_1 =\ & \beta D(\dot{A} + \Lambda A^2 + \Theta T^2 + a) + BK(\dot{\alpha} + \Theta\alpha^2 + \Lambda T^2 + m) \\
& + (DK + B\beta)(\dot{T} + \Lambda AT + \Theta T\alpha - b) = 0 \\
Q_2 =\ & CV_1 + SQ_1 \\
& + \Delta\left[\dot{B}\beta - \dot{D}K + \Lambda B(A\beta + KT) - \Theta D(T\beta + K\alpha)\right] = 0,
\end{aligned}
$$

$$
\begin{aligned}
Q_3 =\ & SV_2 + CQ_1 \\
& + \Delta\left[\dot{B}D - \dot{D}B + \Lambda B(AD + BT) - \Theta D(DT + B\alpha)\right] = 0, \\
Q_4 =\ & SV_1 + \gamma Q_1 \\
& + \Delta\left[\dot{\beta}K - \dot{K}\beta - \Lambda K(A\beta + KT) + \Theta\beta(T\beta + K\alpha)\right] = 0, \\
Q_5 =\ & \gamma V_2 + SQ_1 \\
& + \Delta\left[\dot{\beta}B - \dot{K}B - \Lambda K(AD + BT) + \Theta\beta(DT + B\alpha)\right] = 0.
\end{aligned}
$$

$$
(2.7.14)
$$

To analyze this system of ten differential equations with ten unknown functions $A, B, C, \alpha, \beta, \gamma, K, D, T$, and S we proceed as follows.

Since, according to equations (2.7.13), $V_1 = \Lambda\Delta^2, V_2 = \Theta\Delta^2$, and $Q_1 = 0$, we write the equations $(2.7.14)_6$–$(2.7.14)_{10}$ in the form

$$
\begin{aligned}
\dot{B}\beta - \dot{D}K + \Lambda B(A\beta + KT) - \Theta D(T\beta + K\alpha) + \Lambda C\Delta &= 0, \\
\dot{B}D - \dot{D}B + \Lambda B(AD + BT) - \Theta D(DT + B\alpha) + \Theta S\Delta &= 0, \\
\dot{\beta}K - \dot{K}\beta - \Lambda K(A\beta + KT) + \Theta\beta(T\beta + K\alpha) + \Lambda S\Delta &= 0, \\
\dot{\beta}B - \dot{K}B - \Lambda K(AD + BT) + \Theta\beta(DT + B\alpha) + \Theta\gamma\Delta &= 0. \quad (2.7.15)
\end{aligned}
$$

When solved for derivatives, this system can be written as

$$
\begin{aligned}
\dot{B} + \Lambda B(A - C) + \Theta(SK - DT) &= 0, \\
\dot{D} - \Lambda(BT + CD) + \Theta(D\alpha + S\beta) &= 0, \\
\dot{\beta} + \Lambda(DS - KT) + \Theta\beta(\alpha - \gamma) &= 0, \\
\dot{K} + \Lambda(AK + BS) - \Theta(T\beta + K\gamma) &= 0. \quad (2.7.16)
\end{aligned}
$$

where the explicit form of Δ given by (2.7.11) has been taken into account.

Substituting equations (2.7.15) into $(2.7.14)_5$ and supposing that $\Delta \neq 0$ we arrive at the equation

$$
\dot{S} - \Lambda(BK + CS) - \Theta(D\beta + S\gamma) + b = 0. \quad (2.7.17)
$$

Similarly, combining equations (2.7.15) with equations $(2.7.14)_3$ and $(2.7.14)_4$, we obtain, respectively,

$$
\dot{C} - \Lambda C^2 - \Theta S^2 + \Lambda B^2 + \Theta D^2 - a = 0, \quad (2.7.18)
$$

and

$$
\dot{\gamma} + \Lambda K^2 + \Theta\beta^2 - \Lambda S^2 - \Theta\gamma^2 - m = 0. \quad (2.7.19)
$$

By adding equation $(2.7.15)_1$ and $(2.7.15)_4$ and keeping in mind that $\Delta = KD - B\beta$, we also have

$$
\dot{\Delta} + \Delta(\Lambda A + \Theta\alpha) - \Delta(\Lambda C + \Theta\gamma) = 0. \quad (2.7.20)
$$

Finally, solving equations $(2.7.14)_1$, $(2.7.14)_2$, and $(2.7.14)_6$ for mutual terms $\dot{A} + \Lambda A^2 + \Theta T^2 + a, \dot{\alpha} + \Theta\alpha^2 + \Lambda T^2 + m$, and $\dot{T} + T(\Lambda A + \Theta\alpha) - b$, we obtain

$$
\begin{aligned}
\dot{A} + \Lambda A^2 + \Theta T^2 + a &= \Lambda B^2 + \Theta K^2, \\
\dot{\alpha} + \Theta\alpha^2 + \Lambda T^2 + m &= \Lambda K^2 + \Theta\beta^2, \\
\dot{T} + T(\Lambda A + \Theta\alpha) - b &= -\Lambda BD - \Theta K\beta. \quad (2.7.21)
\end{aligned}
$$

In order to reduce the number of unknown functions, we take

$$
C = -A, \quad \gamma = -\alpha, \quad S = -T, \quad K = D. \quad (2.7.22)
$$

For this case, four equations (2.7.16) are reduced to the following three equations:

$$\begin{aligned}
\dot{B} &= -2\Lambda AB + 2\Theta KT, \\
\dot{K} &= \Lambda(BT - AK) + \Theta(T\beta - K\alpha), \\
\dot{\beta} &= 2\Lambda KT - 2\Theta\alpha\beta,
\end{aligned} \tag{2.7.23}$$

while the six equations (2.7.17)–(2.7.19) and (2.7.21) are reduced to the following three equations:

$$\begin{aligned}
\dot{A} + \Lambda A^2 + \Theta T^2 + a &= \Lambda B^2 + \Theta K^2; \\
\dot{\alpha} + \Theta\alpha^2 + \Lambda T^2 + m &= \Lambda K^2 + \Theta\beta^2; \\
\dot{T} + T(\Lambda A + \Theta\alpha) - b &= -K(\Lambda B + \Theta\beta).
\end{aligned} \tag{2.7.24}$$

The remaining equation (2.7.20) is reduced to

$$\dot{\Delta} + 2\Delta(\Lambda A + \Theta\alpha) = 0, \tag{2.7.25}$$

where

$$\Delta = K^2 - B\beta. \tag{2.7.26}$$

Thus, ten differential equations (2.7.14) are reduced to six differential equations (2.7.23), (2.7.24) with six unknown functions A, α, β, K, B, and T. The seventh differential equation (2.7.25) is a direct consequence of the differential equations (2.7.24) and can be usefully employed as a characteristic compatibility condition.

The inverse transformation of (2.7.10) is now

$$X = \frac{1}{\Delta}\left[-(A\beta + KT)x + (\beta T + \alpha K)y + \beta p_x - K p_y\right],$$

$$\begin{aligned}
Y &= \frac{1}{\Delta}\left[(AK + BT)x - (KT + B\alpha)y - K p_x + B p_y\right], \\
P_X &= \frac{1}{\Delta}\left\{\left[B(K^2 - T^2) - \beta(A^2 + B^2) - 2AKT\right]x \right. \\
&\quad + \left[K(K^2 + T^2) + T(A\beta + B\alpha) + K(A\alpha - B\beta)\right]y \\
&\quad \left. + (A\beta + KT)p_x - (AK + BT)p_y\right\}, \\
P_Y &= \frac{1}{\Delta}\left\{\left[K(K^2 + T^2) + T(A\beta + B\alpha) + K(A\alpha - B\beta)\right]x \right. \\
&\quad + \left[\beta(K^2 - T^2) - B(\alpha^2 + \beta^2) - 2KT\alpha\right]y \\
&\quad \left. - (T\beta + K\alpha)p_x + (KT + B\alpha)p_y\right\}.
\end{aligned} \tag{2.7.27}$$

Therefore, the conservation law of the rheolinear dynamical system whose motion is described by the canonical equations of the Hamiltonian (2.7.1) which are given explicitly by the left column of the differential equations (2.7.4) is, in accordance with the equation (2.7.5) of the form

$$
\begin{aligned}
I &= xP_X - Xp_x + yP_Y - Yp_y \\
&= \frac{1}{\Delta} \left\{ -\beta p_x^2 - Bp_y^2 + 2Kp_xp_y \right. \\
&\quad + \left[B(K^2 - T^2) - \beta(A^2 + B^2) - 2AKT \right] x^2 \\
&\quad + \left[\beta(K^2 - T^2) - B(\alpha^2 + \beta^2) - 2KT\alpha \right] y^2 \\
&\quad + 2 \left[K(K^2 + T^2) + T(A\beta + B\alpha) + K(A\alpha - B\beta) \right] xy \\
&\quad + 2(A\beta + KT)xp_y - 2(T\beta + K\alpha)yp_x \\
&\quad \left. + 2(KT + B\alpha)yp_y \right\} = const.,
\end{aligned}
\tag{2.7.28}
$$

where the six functions $A(t), B(t), \alpha(t), \beta(t), K(t)$ and $T(t)$ are any solutions satisfying the six differential equations (2.7.23), (2.7.24). Therefore, since the auxiliary equations contain as many equations as unknowns, they will always have solutions in terms of the arbitrary, prescribed initial conditions.

The structure of the auxiliary differential equations (2.7.23), (2.7.24) shows that due to the nonlinearity of these equations, finding an exact or approximate closed form solution is not an easy task. Nevertheless, each solution (exact, approximate, or numerical) of this system will generate a conservation law (exact, approximate, or numerical) of our rheolinear dynamical system in the form of equation (2.7.28).

2.7.1 An Alternative Form of the Quadratic Conservation Law

The conservation law (2.7.28) can be expressed in an alternative form that, in some sense resembles the corresponding conservation law of a single-degree-of-freedom dynamical system studied in the proceeding section.

Instead of the functions $A(t), B(t), \alpha(t), \beta(t),$ and $T(t)$, let us introduce new functions of time $\rho_1(t), \rho_2(t), W(t), M(t), \Gamma(t),$ and $\Pi(t)$ by means of the following transformation formulae:

$$
\begin{aligned}
A &= -C = \frac{1}{\Lambda(t)} \frac{\dot{\rho}_1}{\rho_1}, \quad \alpha = -\gamma = \frac{1}{\Theta(t)} \frac{\dot{\rho}_2}{\rho_2}, \quad \beta = \frac{\Pi(t)}{\rho_2^2}, \\
T &= -S = \frac{W(t)}{\rho_1\rho_2}, \quad K = D = \frac{M(t)}{\rho_1\rho_2}, \quad B = \frac{\Gamma(t)}{\rho_1^2}.
\end{aligned}
\tag{2.7.29}
$$

Equation (2.7.25) is now reduced to the integrable form

$$
\frac{d\Delta}{\Delta} = -2 \left(\frac{d\rho_1}{\rho_1} + \frac{d\rho_2}{\rho_2} \right).
\tag{2.7.30}
$$

Integrating, one has

$$
\Delta = \frac{\Omega}{\rho_1^2\rho_2^2}, \quad \Omega = const.
\tag{2.7.31}
$$

At the same time we also have

$$\Delta = K^2 - B\beta = \frac{M^2(t) - \Gamma(t)\Pi(t)}{\rho_1^2 \rho_2^2}. \tag{2.7.32}$$

Therefore, we conclude that the functions $M(t), \Gamma(t)$, and $\Pi(t)$ satisfy the relation

$$M^2(t) - \Gamma(t)\Pi(t) = \Omega = const. \tag{2.7.33}$$

Substituting equations (2.7.29) into the auxiliary relations (2.7.23), (2.7.24) we obtain, respectively,

$$\begin{aligned}
\dot{\Gamma} &= 2\Theta MW \frac{1}{\rho_2^2}, \quad \dot{M} = W\left(\frac{\Lambda\Gamma}{\rho_1^2} + \frac{\Theta\Pi}{\rho_2^2}\right), \\
\dot{\Pi} &= 2\Lambda MW \frac{1}{\rho_2^2}, \quad \dot{W} = b\rho_1\rho_2 - \Lambda\frac{M\Gamma}{\rho_1^2} - \Theta\frac{M\Gamma}{\rho_2^2}, \\
\ddot{\rho}_1 + \Lambda a\rho_1 - \frac{\dot{\Lambda}}{\Lambda}\dot{\rho}_1 &= \frac{\Theta^2\Pi^2}{\rho_1^3} + \frac{\Lambda\Theta(M^2 - W^2)}{\rho_1\rho_2^2}, \\
\ddot{\rho}_2 + \Theta m\rho_2 - \frac{\dot{\Theta}}{\Theta}\dot{\rho}_2 &= \frac{\Theta^2\Pi^2}{\rho_2^3} + \frac{\Lambda\Theta(M^2 - W^2)}{\rho_1^2\rho_2}.
\end{aligned} \tag{2.7.34}$$

It is easy to see that relation (2.7.33) is a conservation law for the system of differential equations (2.7.34).

Finally, the quadratic conservation law (2.7.28) expressed in new variables becomes

$$\begin{aligned}
I = &-\left(\frac{\Pi}{\Omega}\right)\left[\left(\rho_1 p_x - \frac{\dot{\rho}_1}{\Lambda}x\right)^2 + (\Omega - W^2)\left(\frac{y}{\rho_2}\right)^2\right] \\
&-\left(\frac{\Gamma}{\Omega}\right)\left[\left(\rho_2 p_y - \frac{\dot{\rho}_2}{\Lambda}y\right)^2 + (\Omega - W^2)\left(\frac{x}{\rho_2}\right)^2\right] \\
&+2\left(\frac{M}{\Omega}\right)\left[(\Omega + W^2)\frac{x}{\rho_1}\frac{y}{\rho_2} + \rho_1\rho_2 p_x p_y xy - \dot{\rho}_1\rho_2 xp_y - \dot{\rho}_2\rho_1 yp_x\right] \\
&+2\left(\frac{MW}{\Omega}\right)\left[xp_x + yp_y - \frac{\dot{\rho}_1}{\Lambda\rho_1}x^2 - \frac{\dot{\rho}_2}{\Theta\rho_2}y^2\right] \\
&+2\left(\frac{W\Pi}{\Omega}\right)\left[\frac{\dot{\rho}_1}{\Lambda\rho_2}xy - \frac{\dot{\rho}_1}{\rho_2}yp_x\right] \\
&+2\left(\frac{W\Gamma}{\Omega}\right)\left[\frac{\dot{\rho}_2}{\Theta\rho_1}xy - \frac{\dot{\rho}_2}{\rho_1}xp_y\right] = const. \tag{2.7.35}
\end{aligned}$$

Unfortunately, it seems that the system of auxiliary equations (2.7.34) is of a more complicated structure than the corresponding system of auxiliary equations (2.7.23), (2.7.24). The main reason for this fact is that equations (2.7.34)$_{5,6}$ are of the second order.

2.7.2   Some Examples

(a) *Uncoupled oscillator.* As an illustrative example let us consider the uncoupled linear oscillator for which $b(t) = 0$. The canonical differential equations of motion are

$$
\begin{aligned}
\dot{x} &= \Lambda(t)p_x, & \dot{y} &= \Theta(t)p_y, \\
\dot{p}_x &= -a(t)x, & \dot{p}_y &= -m(t)y.
\end{aligned}
\tag{2.7.36}
$$

For this case we first take

$$
K(t) = T(t) = 0.
\tag{2.7.37}
$$

Hence, the auxiliary system (2.7.23) is reduced to two equations:

$$
\dot{B} = -2\Lambda AB, \quad \dot{\beta} = -2\Theta\alpha\beta,
\tag{2.7.38}
$$

and the set of equations (2.7.24) is reduced to two equations:

$$
\dot{A} + \Lambda A^2 + a = \Lambda B^2; \quad \dot{\alpha} + \Theta\alpha^2 + m = \Theta\beta^2.
\tag{2.7.39}
$$

Therefore, the quadratic conservation law of the rheolinear system (2.7.36) is, according to equation (2.7.28),

$$
I = \frac{1}{B}\left[(p_x - Ax)^2 + B^2x^2\right] + \frac{1}{\beta}\left[(p_y - \alpha y)^2 + \beta^2 y^2\right],
\tag{2.7.40}
$$

where $A(t), B(t), \alpha(t)$, and $\beta(t)$ are any solution of the auxiliary system (2.7.38), (2.7.39), subject to arbitrary initial conditions.

However, for this degenerate case, the conservation law (2.7.40) together with the auxiliary conditions can be split into two mutually independent conservation laws

$$
I_1 = \frac{1}{B}\left[(p_x - Ax)^2 + B^2x^2\right],
\tag{2.7.41}
$$

with the auxiliary conditions

$$
\dot{A} + \Lambda A^2 + a = \Lambda B^2, \quad \dot{B} = -2\Lambda AB,
\tag{2.7.42}
$$

and

$$
I_2 = \frac{1}{\beta}\left[(p_y - \alpha y)^2 + \beta^2 y^2\right],
\tag{2.7.43}
$$

with auxiliary conditions

$$
\dot{\alpha} + \Theta\alpha^2 + m = \Theta\beta^2, \quad \dot{\beta} = -2\Theta\alpha\beta.
\tag{2.7.44}
$$

In analogy with the one-degree-of-freedom system (see (2.5.23)), by taking $B = 1/\rho_1^2, \beta = 1/\rho_2^2$, where ρ_1 and ρ_2 are new unknown functions of time, we find from (2.7.42) and (2.7.44), that $A = \dot{\rho}_1/(\Lambda(t)\rho_1)$ and $\alpha = \dot{\rho}_2/(\Theta(t)\rho_2)$ and

$$
\ddot{\rho}_1 - \frac{\dot{\Lambda}}{\Lambda}\dot{\rho}_1 + \Lambda a\rho_1 = \frac{\Lambda^2}{\rho_1^3}, \quad \ddot{\rho}_2 - \frac{\dot{\Theta}}{\Theta}\dot{\rho}_2 + \Theta m\rho_2 = \frac{\Theta^2}{\rho_2^3}.
\tag{2.7.45}
$$

The conservation laws (2.7.41) and (2.7.43) become

$$I_1 = \left(\rho_1 p_x - \frac{\dot{\rho}_1}{\Lambda}\right)^2 + \left(\frac{x}{\rho_1}\right)^2, \quad I_2 = \left(\rho_2 p_y - \frac{\dot{\rho}_2}{\Theta}\right)^2 + \left(\frac{y}{\rho_2}\right)^2. \quad (2.7.46)$$

Note that for $\Lambda(t) = \Theta(t) = 1$, the auxiliary conditions (2.7.45) and conservation laws (2.7.46) are identical with the results of Günter and Leach [51].

As an alternative possibility (see [123]) for finding conservation laws of the uncoupled dynamical system (2.7.36), we take

$$B(t) = \beta(t) = T(t) = 0. \quad (2.7.47)$$

For this choice, the auxiliary conditions (2.7.23), (2.7.24) are reduced to

$$\dot{K} = -K(\Lambda A + \Theta\alpha), \quad \dot{A} + \Lambda A^2 + a = \Theta K^2, \quad \dot{\alpha} + \Theta\alpha^2 + m = \Lambda K^2, \quad (2.7.48)$$

and the conservation law of the system (2.7.36) is

$$I = \frac{1}{K}\left[p_x p_y + (K^2 + A\alpha)xy - Axp_y - \alpha y p_x\right], \quad (2.7.49)$$

where K, A, and α constitute a solution of the auxiliary equations (2.7.48) subject to the arbitrary initial conditions.

Similarly, as in the previous case, on taking $A = \dot{\rho}_1/(\Lambda\rho_1), \alpha = \dot{\rho}_2/(\Theta\rho_2)$, the first equation of the set (2.7.48) is reduced to the integrable form

$$\frac{dK}{K} = -\frac{d\rho_1}{\rho_1} - \frac{d\rho_2}{\rho_2}. \quad (2.7.50)$$

Integrating, we find

$$K = \frac{M}{\rho_1\rho_2}, \quad M = const. \neq 0. \quad (2.7.51)$$

The last two equations (2.7.48) are reduced to

$$\ddot{\rho}_1 - \frac{\dot{\Lambda}}{\Lambda}\dot{\rho}_1 + \Lambda a\rho_1 = \Lambda\Theta\frac{M^2}{\rho_1\rho_2^2}, \quad \ddot{\rho}_2 - \frac{\dot{\Theta}}{\Theta}\dot{\rho}_2 + \Theta m\rho_2 = \Lambda\Theta\frac{M^2}{\rho_1^2\rho_2}. \quad (2.7.52)$$

The conservation law (2.7.49) now becomes

$$I = \frac{1}{M}\left[\rho_1\rho_2 p_x p_y + \frac{\dot{\rho}_1}{\Lambda}\frac{\dot{\rho}_2}{\Theta}xy - \frac{\dot{\rho}_1}{\Lambda}\rho_2 x p_y\right.$$
$$\left. - \frac{\dot{\rho}_2}{\Theta}\rho_1 y p_x\right] + M\frac{x}{\rho_1}\frac{y}{\rho_2}, \quad (2.7.53)$$

where ρ_1, ρ_2 are any solutions of the system (2.7.52) for arbitrary initial conditions, and M is an arbitrary constant.

(b) *Semicoupled oscillator:* $\Lambda(t) = \Theta(t), a(t) = m(t)$. Let us consider the dynamical system whose Hamiltonian is of the form

$$H = \Lambda(t)\left(\frac{1}{2}p_x^2 + \frac{1}{2}p_y^2\right) + \frac{1}{2}ax^2 + bxy + \frac{1}{2}ay^2, \qquad (2.7.54)$$

with arbitrary $\Lambda(t), a(t),$ and $b(t)$.

For this case, if we take

$$A = \alpha, \quad B = \beta = 0, \quad S = -T, \quad D = -K, \quad C = -A, \quad \gamma = -A, \quad (2.7.55)$$

the auxiliary system of equations (2.7.23), (2.7.24) is reduced to

$$\dot{K} + 2\Lambda AK = 0, \quad \dot{T} + 2\Lambda AT - b = 0, \quad \dot{A} + \Lambda A^2 + a = \Lambda(K^2 - T^2), \qquad (2.7.56)$$

and for this case, the angular momentum–type conservation law (2.7.7) will be of the form

$$\bar{I} = \frac{1}{2K}\left[p_x^2 - p_y^2 - (T^2 - K^2 - A^2)x^2 + (T^2 - K^2 - A^2)y^2 \right. $$
$$\left. + 2A(yp_y - xp_x) + 2T(yp_x - xp_y)\right]. \qquad (2.7.57)$$

Taking $A = \dot{\rho}/(\Lambda\rho)$ as in the previous case, the first equation of the system (2.7.56) can be integrated to give $K = M/\rho^2$, where M is a constant of integration. The last two equations of (2.7.56) can be written as

$$\dot{T} + 2\left(\frac{\dot{\rho}}{\rho}\right)T - b = 0, \quad \ddot{\rho} - \frac{\dot{\Lambda}}{\Lambda}\frac{\dot{\rho}}{\rho} + \Lambda a\rho = \Lambda^2\left(\frac{M^2}{\rho^3} - T^2\rho\right). \qquad (2.7.58)$$

Thus, the conservation law (2.7.57) can be represented in the form

$$\bar{I} = \frac{1}{2M}\left\{\left(\rho p_x - \frac{\dot{\rho}x}{\Lambda}\right)^2 - \left(\rho p_y - \frac{y\dot{\rho}}{\Lambda}\right)^2 \right.$$
$$+ \left[T^2\left(y^2 - x^2\right) + T\left(yp_x - xp_y\right)\right]$$
$$\left. + M^2\left(\frac{x}{\rho}\right)^2 - M^2\left(\frac{y}{\rho}\right)^2\right\}. \qquad (2.7.59)$$

It is easy to see that for the special case $b(t) = 0$ the first auxiliary equation (2.7.58) is satisfied for $T = 0$. Note that for $T = 0$ and $\Lambda(t) = 1$, the second auxiliary equation (2.7.58) and the conservation law (2.7.59) become identical with the so-called Fradkin–Leach tensor [51] and [63].

(c) *A numerical example.* As indicated previously, the possibility of finding a quadratic conservation law of the rheolinear dynamical system with two degrees of freedom with the given Hamiltonian function in the form (2.7.1) strictly depends upon our ability to find any solution of the auxiliary differential equations

(2.7.23), (2.7.24). It is obvious, taking into account the nonlinear structure of these equations, that a solution cannot in general be found in a closed form in terms of known functions. Similar difficulty arises in finding approximate solutions in a closed analytical form due to the relatively large number of auxiliary equations. Besides that, even in the case of a single degree of freedom, the approximate analysis of the corresponding auxiliary equation like (2.6.30), as demonstrated by numerous authors, is a rather difficult mathematical task.

Here, we analyze a concrete dynamical rheolinear system employing purely numerical procedures.

Let us consider the rheolinear dynamical system defined as an initial value problem

$$
\begin{aligned}
\dot{x} &= 3(1+t^2)p_x, \quad x(0) = 1, \\
\dot{y} &= 2\left(1 + \frac{t}{2}\right)p_y, \quad y(0) = 3, \\
\dot{p}_x &= -35e^t x - 22(1+t^2)y, \quad p_x(0) = 1, \\
\dot{p}_y &= -22(1+t^2)x - 14(1+t^3)y, \quad p_y(0) = 2, \\
0 &\leq t \leq 6.
\end{aligned}
\tag{2.7.60}
$$

In accordance with the previously introduced notation, we have

$$
\begin{aligned}
\Lambda(t) &= 3(1+t^2), \\
\Theta(t) &= 2\left(1 + \frac{t}{2}\right), \\
a(t) &= 35e^t, \\
b(t) &= 22(1+t^2), \\
m(t) &= 14(1+t^3).
\end{aligned}
\tag{2.7.61}
$$

To obtain a solution of the auxiliary system (2.7.23), (2.7.24) we have to prescribe some initial conditions for the variables $A, \alpha, \beta, K, B,$ and T. For the problem in question, we assume

$$
A(0) = 0, \quad \alpha(0) = 0, \quad T(0) = 0,
\tag{2.7.62}
$$

and from (2.7.24) it follows that

$$
\begin{aligned}
\Lambda(0)B^2(0) + \Theta(0)K^2(0) &= a(0) = 35, \\
\Lambda(0)K^2(0) + \Theta(0)\beta^2(0) &= m(0) = 14, \\
K(0)\left[\Lambda(0)B(0) + \Theta(0)\beta(0)\right] &= b(0) = 22.
\end{aligned}
\tag{2.7.63}
$$

Solving this system using equations (2.7.61) for $t = 0$, we obtain

$$
B(0) = 3, \quad K(0) = 2, \quad \beta(0) = 1,
\tag{2.7.64}
$$

which, together with (2.7.62), comprise the complete set of the initial conditions for auxiliary variables.

The differential equations of the dynamical system (2.7.60) and six auxiliary differential equations (2.7.23), (2.7.24) are integrated numerically using the Runge–Kutta method. The step size of the dimensionless time has been taken to be $\Delta t = 10^{-6}$. The motion of the dynamical system, that is, $x(t), y(t), p_x(t)$, and $p_y(t)$, is depicted in Figure 2.7.1. The value of the conserved quantity I, given by equation (2.7.28), has been calculated in each step of integration and its closeness to the value along the exact trajectory $I = 19$ is assessed in Figure 2.7.1.

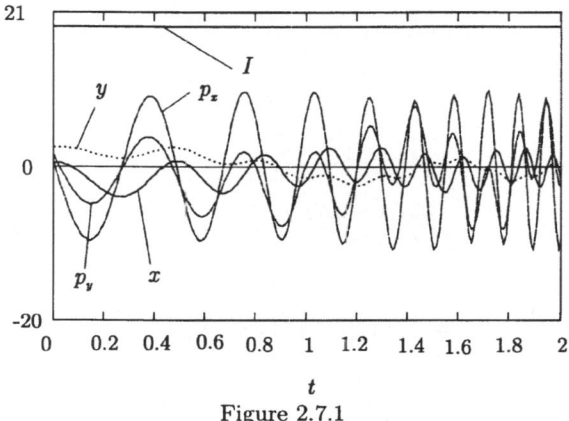

Figure 2.7.1

The values of the auxiliary variables B, K, A, β, and α are not presented in Figure 2.7.1.

It is of special interest to note that the conservation law obtained can be advantageously used as a reliable indicator of the quality of numerical solution of the rheolinear system (2.7.60). In Figure 2.7.2. the solution of the dynamical system is shown for values of the dimensionless time $t \geq 6$.

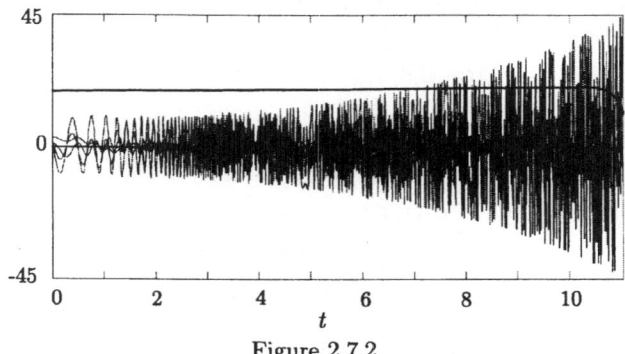

Figure 2.7.2

It is a well-known fact that every numerical solution of a system of ordinary differential equations inevitably brings some errors into results. If the range of the independent variable is large, the accumulation of error can be of such an order that the numerical solution becomes quite unreliable. In a comment of the adaptive step-size control for the Runge–Kutta method, Press et al observed that "sometimes accuracy may be demanded not directly in the solution itself, but in some related conserved quantity that can be monitored" [88, p. 554]. The value of the related conservation law has been calculated in each step and shown in Figure 2.7.2. The results shown clearly demonstrate that the solution of the dynamical system can be considered correct until, approximately, $t = 10$. After this period, the value of the conservation law starts to decrease rapidly, and the solution becomes inaccurate.

Chapter 3

Transformation Properties of the Lagrange–D'Alembert Variational Principle: Conservation Laws of Nonconservative Dynamical Systems

3.1 Introduction

In this chapter we shall demonstrate that the Lagrange–D'Alembert differential variational principle can be used for the study of conservation laws of conservative and purely nonconservative dynamical systems. The basic idea of this approach is to consider the transformation properties of the Lagrange–D'Alembert principle with respect to the infinitesimal transformation of the generalized coordinates and time. It is of interest to note that for the Lagrangian and Hamiltonian dynamical systems (i.e., for the systems that are completely described by the Lagrangian or Hamiltonian functions and in which the nonconservative forces are absent, $Q_i = 0$) the way of obtaining the conservation laws is identical with the famous theory of Emmy Noether, which is based upon the transformation properties of the Hamiltonian action integral $\int^t L du$. However, the approach based upon the Lagrange–D'Alembert differential variational principle admits the possibility to include into consideration purely nonconservative dynamical systems for which $Q_i \neq 0$.

3.2   Simultaneous and Nonsimultaneous Virtual Displacements (Variations), Infinitesimal Transformations

For the sake of completeness we will briefly describe the forms of the infinitesimal transformations (variations, virtual displacements) used in the proceeding text. Let the position of dynamical system with n degrees of freedom be specified by the set of generalized coordinates

$$q_1, ..., q_n, \tag{3.2.1}$$

which are supposed to be continuous functions of time.

   The symbol δ will denote the *simultaneous variation:* a representative point A that is on the actual path at time t and is correlated to an infinitesimally close point B occupied at the *same time* t in the varied path by the relation

$$\bar{q}_i(t) = q_i(t) + \delta q_i, \tag{3.2.2}$$

where $\bar{q}_i(t)$ and $q_i(t)$ are the coordinates of the point B and A, respectively (see Figure 3.2.1).

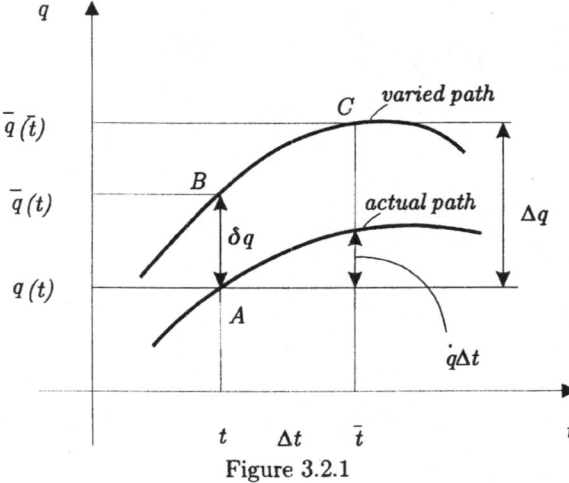

Figure 3.2.1

Note that the infinitesimal quantities δq_i are identical to the components of virtual displacement vector introduced in section 1.2.

   At the same time, in order to consider a much broader class of infinitesimal transformations, we introduce a new kind of variation by supposing that the time t suffers an infinitesimal deformation Δt, such that

$$\bar{t} = t + \Delta t. \tag{3.2.3}$$

If the motion on the actual path is given by $q_i(t)$, we will determine the corresponding infinitesimally close motion on the varied path, which is taking place in accordance with the holonomic constraints acting on the system by $\bar{q}(\bar{t})$. Let us define the *nonsimultaneous (generalized) variation* by the equation

$$\Delta q_i = \bar{q}_i(\bar{t}) - q_i(t) = \bar{q}_i(t + \Delta t) - q_i(t). \tag{3.2.4}$$

Developing term $\bar{q}(t + \Delta t)$ into a Taylor series and retaining the first-order terms only, one has $\bar{q}_i(t + \Delta t) \approx \bar{q}_i(t) + \dot{\bar{q}}_i(t) \Delta t$, and hence (3.2.4) can be written in the form

$$\Delta q_i = \delta q_i + \dot{q}_i \Delta t, \tag{3.2.5}$$

where we have employed (3.2.2). The geometrical interpretation of (3.2.2), (3.2.3), (3.2.4), and (3.2.5) is provided in Figure 3.2.1.

We note that the relation (3.2.5) can serve as a useful pattern for finding the nonsimultaneous variations of any scalar vector or tensor functional $L(t, \mathbf{q}, \dot{\mathbf{q}})$, where $\mathbf{q} = \{q_1, ..., q_n\}$ and $\dot{\mathbf{q}} = \{\dot{q}_1, ..., \dot{q}_n\}$. In fact, it is easy to show that

$$\Delta L(t, \mathbf{q}, \dot{\mathbf{q}}) = \delta L(t, \mathbf{q}, \dot{\mathbf{q}}) + \dot{L}(t, \mathbf{q}, \dot{\mathbf{q}}) \Delta t, \tag{3.2.6}$$

where

$$
\begin{aligned}
\Delta L(t, t, \mathbf{q}, \dot{\mathbf{q}}) &= L\left(\bar{t}, \bar{\mathbf{q}}(\bar{t}), \frac{d\bar{\mathbf{q}}(\bar{t})}{d\bar{t}}\right) - L\left(t, \mathbf{q}(t), \frac{d\mathbf{q}(t)}{dt}\right) \\
&= \frac{\partial L}{\partial q_i} \Delta q_i + \frac{\partial L}{\partial \dot{q}} \Delta \dot{q}_i + \frac{\partial L}{\partial t} \Delta t.
\end{aligned}
\tag{3.2.7}
$$

Also,

$$
\begin{aligned}
\delta L(t, \mathbf{q}, \dot{\mathbf{q}}) &= L\left(t, \bar{\mathbf{q}}(t), \frac{d\bar{\mathbf{q}}(t)}{dt}\right) - L\left(t, \mathbf{q}(t), \frac{d\mathbf{q}(t)}{dt}\right) \\
&= \frac{\partial L}{\partial q_i} \delta q_i + \frac{\partial L}{\partial \dot{q}} \delta \dot{q}_i
\end{aligned}
\tag{3.2.8}
$$

and

$$\dot{L}(t, \mathbf{q}, \dot{\mathbf{q}}) = \frac{\partial L}{\partial t} + \frac{\partial L}{\partial q_i} \dot{q}_i + \frac{\partial L}{\partial \dot{q}_i} \ddot{q}_i. \tag{3.2.9}$$

We note that from (3.2.6) it follows that

$$\Delta \dot{q}_i = \delta \dot{q}_i + \ddot{q}_i \Delta t. \tag{3.2.10}$$

Differentiating (3.2.5) with respect to time we find

$$(\Delta q_i)^{\cdot} = (\delta q_i)^{\cdot} + \ddot{q}_i \Delta t + \dot{q}_i (\Delta t)^{\cdot}. \tag{3.2.11}$$

Combining the last two equations we find

$$\Delta \dot{q}_i - (\Delta q_i)^{\cdot} = \delta \dot{q}_i - (\delta q_i)^{\cdot} - \dot{q}_i (\Delta t)^{\cdot}. \tag{3.2.12}$$

As usual in variational calculus and also accepting the *commutative rules* (1.3.18), that is, $\delta \dot{q}_i = (\delta q_i)^{\cdot}$, we see that

$$\Delta \dot{q}_i - (\Delta q_i)^{\cdot} = -\dot{q}_i (\Delta t)^{\cdot}, \tag{3.2.13}$$

and this equation indicates that the symbols of nonsimultaneous variations are not commutative, namely,

$$\Delta d(\cdot) - d\Delta(\cdot) \neq 0. \tag{3.2.14}$$

Note for completeness that the formula (3.2.6) can also be applied to the "action-type" functionals that are defined in the form of a definite integral,

$$\Delta \int_{t_0}^{t_1} L(t, \mathbf{q}, \dot{\mathbf{q}}) dt = \delta \int_{t_0}^{t_1} L(t, \mathbf{q}, \dot{\mathbf{q}}) dt + L(t, \mathbf{q}, \dot{\mathbf{q}}) \Delta t |_{t_0}^{t_1}, \tag{3.2.15}$$

where

$$
\begin{aligned}
\Delta \int_{t_0}^{t_1} L(t, \mathbf{q}, \dot{\mathbf{q}}) dt &= \int_{\bar{t}_0}^{\bar{t}_1} L\left(\bar{t}, \bar{\mathbf{q}}(\bar{t}), \frac{d\bar{\mathbf{q}}(\bar{t})}{d\bar{t}}\right) d\bar{t} \\
&\quad - \int_{t_0}^{t_1} L\left(t, \mathbf{q}(t), \frac{d\mathbf{q}(t)}{dt}\right) dt
\end{aligned}
\tag{3.2.16}
$$

and $\bar{t}_0 = t_0 + \Delta t(t_0), \bar{t}_1 = t_1 + \Delta t(t_1)$. Also,

$$
\begin{aligned}
\delta \int_{t_0}^{t_1} L(t, \mathbf{q}, \dot{\mathbf{q}}) dt &= \int_{t_0}^{t_1} \left[L\left(t, \bar{\mathbf{q}}(t), \frac{d\bar{\mathbf{q}}(t)}{dt}\right) - L\left(t, \mathbf{q}(t), \frac{d\mathbf{q}(t)}{dt}\right) \right] dt \\
&= \int_{t_0}^{t_1} \left(\frac{\partial L}{\partial q_i} \delta q_i + \frac{\partial L}{\partial \dot{q}_i} \delta \dot{q}_i \right) dt.
\end{aligned}
\tag{3.2.17}
$$

Consequently

$$\delta \int_{t_0}^{t_1} L(t, \mathbf{q}, \dot{\mathbf{q}}) dt = \int_{t_0}^{t_1} \delta L(t, \mathbf{q}, \dot{\mathbf{q}}) dt. \tag{3.2.18}$$

At this point we are going to interpret equations (3.2.2) and (3.2.5) as the *infinitesimal transformations* of generalized coordinates $q_i(t)$ and time t. Let us suppose that the infinitesimal transformations are of the form

$$
\begin{aligned}
\bar{q}_i(\bar{t}) &= q_i(t) + \varepsilon F_i(t, \mathbf{q}(t), \dot{\mathbf{q}}(t)), \quad i = 1, ..., n, \\
\bar{t} &= t + \varepsilon f(t, \mathbf{q}(t), \dot{\mathbf{q}}(t)),
\end{aligned}
\tag{3.2.19}
$$

where the functions $F_i(t, \mathbf{q}(t), \dot{\mathbf{q}}(t)) = F_i(t, q_1, ..., q_n, \dot{q}_1, ..., \dot{q}_n), i = 1, ..., n$, and $f(t, \mathbf{q}(t), \dot{\mathbf{q}}(t)) = f(, q_1, ..., q_n, \dot{q}_1, ..., \dot{q}_n)$ are also called generators of the infinitesimal transformation of space and time, respectively. The parameter ε is a small constant positive number. Comparing (3.2.19) with (3.2.4) and (3.2.3) we have

$$\Delta q_i = \varepsilon F_i(t, \mathbf{q}(t), \dot{\mathbf{q}}(t)), \quad \Delta t = \varepsilon f(t, \mathbf{q}(t), \dot{\mathbf{q}}(t)). \tag{3.2.20}$$

The infinitesimal transformations of generalized coordinates and time (3.2.19) constitute the general infinitesimal transformation in which the space and time generators F_i and f are supposed to be functions of generalized coordinates q_i, generalized velocities $\dot{q}_i$, and time t. From (3.2.5) and (3.2.13) with the notation (3.2.20), we find

$$
\begin{aligned}
\delta q_i &= \varepsilon \left[F_i(t, \mathbf{q}(t), \dot{\mathbf{q}}(t)) - \dot{q}_i f(t, \mathbf{q}(t), \dot{\mathbf{q}}(t)) \right], \\
\Delta \dot{q}_i &= \varepsilon \left[\dot{F}_i(t, \mathbf{q}(t), \dot{\mathbf{q}}(t)) - \dot{q}_i \dot{f}(t, \mathbf{q}(t), \dot{\mathbf{q}}(t)) \right].
\end{aligned} \tag{3.2.21}
$$

Finally, note that in the case when the time t is not varied (i.e., $\delta t = 0$) the operators δ and Δ coincide, that is,

$$
\bar{q}_i(t) = q_i(t) + \delta q_i = q_i + \varepsilon F_i(t, \mathbf{q}(t), \dot{\mathbf{q}}(t)), \quad \bar{t} = t. \tag{3.2.22}
$$

By using the transformation rules (3.2.21) we shall transform the Lagrange–D'Alembert principle into a form that reveals possibilities to obtain conservation laws of conservative and nonconservative dynamical systems.

3.3 A Transformation of the Lagrange–D'Alembert Principle

Let us consider a dynamical system of N particles subject to holonomic constraints. We shall assume that the given impressed forces $\mathbf{F}_1, ..., \mathbf{F}_N$ act at some points of the system. The virtual displacements of these points are denoted by $\delta \mathbf{r}_1, ..., \delta \mathbf{r}_N$. According to the Lagrange–D'Alembert principle of virtual work, we have (see (1.3.4))

$$
\sum_{i=1}^{N} (m_i \ddot{\mathbf{r}}_i - \mathbf{F}_i) \cdot \delta \mathbf{r}_i = 0. \tag{3.3.1}
$$

Introducing the generalized coordinates (3.2.1) in such a way that every position vector can be expressed as a function of these coordinates and time $\mathbf{r}_i = \mathbf{r}_i(t, q_1, ..., q_n)$ and applying the procedure described in section 1.3, we find (see equation (1.3.33)) that

$$
\left(\frac{d}{dt} \frac{\partial L}{\partial \dot{q}_i} - \frac{\partial L}{\partial q_i} - Q_i \right) \delta q_i = 0, \quad i = 1, ..., n, \tag{3.3.2}
$$

where $L = L(t, \mathbf{q}(t), \dot{\mathbf{q}}(t))$ is the Lagrangian function and $Q_i = Q_i(t, \mathbf{q}(t), \dot{\mathbf{q}}(t))$ are the components of the nonconservative (nonpotential) generalized forces, which are supposed to be arbitrary functions of time t, generalized coordinates q_i, and generalized velocities $\dot{q}_i$.

We write equation (3.3.2) in the form

$$
\frac{d}{dt} \left(\frac{\partial L}{\partial \dot{q}_i} \delta q_i \right) - \frac{\partial L}{\partial q_i} \delta \dot{q}_i - \frac{\partial L}{\partial q_i} \delta q_i - Q_i \delta q_i = 0, \tag{3.3.3}
$$

where we have used commutativity rule $\delta\dot{q}_i = (d/dt)\,\delta q_i$. Taking into account (3.2.5) and (3.2.10), that is, $\delta q_i = \Delta q_i - \dot{q}_i\Delta t$ and $\delta\dot{q}_i = \Delta q_i - \ddot{q}_i\Delta t$, the equation (3.3.2) becomes

$$\frac{d}{dt}\left[\frac{\partial L}{\partial \dot{q}_i}(\Delta q_i - \dot{q}_i\Delta t)\right] - \left(\frac{\partial L}{\partial \dot{q}_i}\Delta\dot{q}_i + \frac{\partial L}{\partial q_i}\Delta q_i + \frac{\partial L}{\partial t}\Delta t\right)$$

$$+ \left(\frac{\partial L}{\partial \dot{q}_i}\ddot{q}_i + \frac{\partial L}{\partial q_i}\dot{q}_i + \frac{\partial L}{\partial t}\right)\Delta t - Q_i\left(\Delta q_i - \dot{q}_i\Delta t\right) = 0. \qquad (3.3.4)$$

Denoting by

$$\begin{aligned}\Delta L &= \frac{\partial L}{\partial \dot{q}_i}\Delta\dot{q}_i + \frac{\partial L}{\partial q_i}\Delta q_i + \frac{\partial L}{\partial t}\Delta t,\\[1mm]\dot{L} &= \frac{\partial L}{\partial \dot{q}_i}\ddot{q}_i + \frac{\partial L}{\partial q_i}\dot{q}_i + \frac{\partial L}{\partial t},\end{aligned} \qquad (3.3.5)$$

we write equation (3.3.4) in the form

$$\frac{d}{dt}\left[\frac{\partial L}{\partial \dot{q}_i}\left(\Delta q_i - \dot{q}_i\Delta t\right) + L\Delta t\right] - \Delta L - L(\Delta t)^{\cdot} - Q_i\left(\Delta q_i - \dot{q}_i\Delta t\right) = 0. \quad (3.3.6)$$

By adding and subtracting the total time derivative of an arbitrary function $\varepsilon \dot{P}(t, \mathbf{q}(t), \dot{\mathbf{q}}(t))$, we can write (3.3.6) in the form

$$\frac{d}{dt}\left[\frac{\partial L}{\partial \dot{q}_i}\left(\Delta q_i - \dot{q}_i\Delta t\right) + L\Delta t - \varepsilon P\right]$$

$$- \left[\Delta L + L(\Delta t)^{\cdot} + Q_i\left(\Delta q_i - \dot{q}_i\Delta t\right) - \varepsilon\dot{P}\right] = 0, \qquad (3.3.7)$$

which is the transformation of the Lagrange–D'Alembert principle we have been seeking. As in the Noetherian theory based upon the integral variational principles, we are referring to the function $P(t, \mathbf{q}(t), \dot{\mathbf{q}}(t))$ as a *gauge-variant function*.

3.4   The Conditions for the Existence of a Conserved Quantity of the Given Dynamical System

From the equation (3.3.7), which represents the transformed form of the Lagrange-D'Alembert principle, it is obvious that if the relation

$$\Delta L + L(\Delta t)^{\cdot} + Q_i\left(\Delta q_i - \dot{q}_i\Delta t\right) - \varepsilon\dot{P} = 0 \qquad (3.4.1)$$

is satisfied, the dynamical system admits a conservation law of the form

$$\frac{\partial L}{\partial \dot{q}_i}\left(\Delta q_i - \dot{q}_i\Delta t\right) + L\Delta t - \varepsilon P = C = const. \qquad (3.4.2)$$

Introducing the generators F_i and f of the infinitesimal transformations defined by equations (3.2.19), we can write (3.4.1), (3.4.2) in the form

$$\frac{\partial L}{\partial q_i} F_i + \frac{\partial L}{\partial \dot{q}_i} \left(\dot{F}_i - \dot{q}_i \dot{f} \right) + \frac{\partial L}{\partial t} f + L\dot{f} + Q_i \left(F_i - \dot{q}_i f \right) - \dot{P} = 0 \qquad (3.4.3)$$

and

$$\frac{\partial L}{\partial \dot{q}_i} \left(F_i - \dot{q}_i f \right) + Lf - P = const. \qquad (3.4.4)$$

Note that the equation (3.4.3) is usually referred to as the generalized *basic Noether identity*.

Thus, we can state the following.

For every infinitesimal transformation of the generalized coordinates and time of the form (3.2.18) *and for every gauge function* $P(t, \mathbf{q}(t), \dot{\mathbf{q}}(t))$ *that satisfy the scalar equation* (3.4.3)*, there exists a conservation quantity of the form* (3.4.4).

Note that this statement for the case when the dynamical system is fully specified by the Lagrangian function $L(t, \mathbf{q}(t), \dot{\mathbf{q}}(t))$ and $Q_i = 0$ can be considered identical with the famous theorem of Emmy Noether [78], whose results have been based upon the invariant properties of the Hamilton action integral $\int^t L dt$ with respect to the infinitesimal transformations of generalized coordinates and time, with the generators depending only upon the generalized coordinates. Equations (3.4.3) and (3.3.4) were first derived in [37] starting from a generalized form of Hamilton's variational principle. The invariant form of the Lagrange–D'Alembert principle presented here have was first derived in [112] (see also [122], where (3.4.3) and (3.4.4) were obtained by studying invariant properties of Gauss and Jourdain differential variational principles). It is also of interest to note that the generalized Noetherian approach to the study of conservation laws has been extended to more complex dynamical systems whose structures demand introduction of quasi coordinates [36]. Djukic was the first to introduce the gauge functions in his study of Noether's theorem (see [35]).

The relations (3.4.3) and (3.4.4) can be easily expressed in the Hamiltonian canonical variables q_i and p_i.

Recalling the relations introduced in section 1.8,

$$\begin{aligned} p_i &= \frac{\partial L}{\partial \dot{q}_i}, \quad \frac{\partial L}{\partial \dot{q}_i} \dot{q}_i - L = H(t, q_1, ..., q_n, p_1, ..., p_n), \\ \frac{\partial L}{\partial q_i} &= -\frac{\partial H}{\partial q_i}, \quad \frac{\partial L}{\partial t} = -\frac{\partial H}{\partial t}, \end{aligned} \qquad (3.4.5)$$

we write (3.4.3) and (3.4.4) in canonical form:

$$-\frac{\partial H}{\partial q_i} F_i + p_i \dot{F}_i - H\dot{f} - \frac{\partial H}{\partial t} f + Q_i \left(F_i - \frac{\partial H}{\partial p_i} \right) + \dot{P}(t, \mathbf{q}(t), \mathbf{p}(t)) = 0 \quad (3.4.6)$$

and

$$I = p_i F_i - Hf + P = const. \qquad (3.4.7)$$

Naturally, we suppose that the generators of the infinitesimal transformations F_i, f, the gauge function P, and the generalized forces Q_i are functions of time t and the canonical variables q_i and $p_i, i = 1, ..., n$.

Thus, we can formulate a similar result as stated above.

If the space and time generators $F_i(t, \mathbf{q}(t), \mathbf{p}(t))$, $f(t, \mathbf{q}(t), P(t))$ and the gauge function $P(t, \mathbf{q}(t), \mathbf{p}(t))$ satisfy Noether's identity (3.4.6) identically along the dynamical trajectory of the dynamical system

$$\dot{q}_i = \frac{\partial H}{\partial p_i}, \quad \dot{p}_i = -\frac{\partial H}{\partial q_i} + Q_i(t, \mathbf{q}(t), P(t)), \quad i = 1, ..., n, \tag{3.4.8}$$

then there exists a conservation law (3.4.7).

It should be noted that the term "along the trajectory" means that we are able to express the acceleration vector $\ddot{q}_i$ in terms of $t, \dot{q}_i$, and q_i by using the Euler–Lagrangian equations of motion and to substitute this into Noether's identity (3.4.3). Similarly, for the dynamical problems expressed in the canonical variables, we are generally able to express $\dot{q}_i$ and $\dot{p}_i$ by means of (3.4.8) in terms of q_i, p_i, and t and to substitute this into Noether's identity in canonical form (3.4.6).

As has been shown (see, for example, [122] and the references cited therein), the basic Noether identity (3.4.3) can be considered from the various points of views. Frequently, this identity can be transformed into a system of partial differential equations. If any solution of this system is available, a conservation law of the dynamical system follows immediately.

3.5   The Generalized Killing Equations

As noted at the end of the previous section, the basic Noether identity can be decomposed into a system of partial differential equations. These equations are frequently linear partial differential equations of the first order with respect to generators F_i, f and gauge function P.

For example, let us suppose that the set of $n + 2$ functions F_i, f, and P depends only upon time t and generalized coordinates q_i:

$$F_i = F_i(t, \mathbf{q}(t)), \quad f = f(t, \mathbf{q}(t)), \quad P = P(t, \mathbf{q}(t)), \quad i = 1, ..., n, \tag{3.5.1}$$

where $\mathbf{q} = \{q_1, ..., q_n\}$ and the generalized forces are absent, that is, $Q_i = 0$. Entering with this into the basic Noether identity (3.4.3), we find

$$L\left(t, \mathbf{q}(t), \dot{\mathbf{q}}(t)\right)\left(\frac{\partial f}{\partial t} + \frac{\partial f}{\partial q_i}\dot{q}_i\right) + \frac{\partial L}{\partial t}f + \frac{\partial L}{\partial q_i}F_i$$
$$+\frac{\partial L}{\partial \dot{q}_i}\left(\frac{\partial F_i}{\partial t} + \frac{\partial F_i}{\partial q_j}\dot{q}_j - \frac{\partial f}{\partial t}\dot{q}_i - \frac{\partial f}{\partial q_j}\dot{q}_i\dot{q}_j\right) - \frac{\partial P}{\partial t} - \frac{\partial P}{\partial q_i}\dot{q}_i = 0. \tag{3.5.2}$$

Specifying the form of the Lagrangian function, this equation can be transformed into a system of partial differential equations by equating to zero terms of the corresponding degree of $\dot{\mathbf{q}}(t)$.

For example, let us consider a rheonomic dynamical system with n degrees of freedom whose Lagrangian function is of the form

$$L = T - \Pi = \frac{1}{2} g\left(t, \mathbf{q}\right) + a_i\left(t, \mathbf{q}\right) \dot{q}_i + \frac{1}{2} g_{ij}\left(t, \mathbf{q}\right) \dot{q}_i \dot{q}_j - \Pi\left(t, \mathbf{q}\right). \qquad (3.5.3)$$

Entering with this into (3.5.2) and grouping the terms according to degree of $\dot{\mathbf{q}}$, we have

$$\frac{1}{2} \frac{\partial}{\partial t}\left(gf\right) - \frac{\partial}{\partial t}\left(\Pi f\right) + \frac{1}{2} \frac{\partial g}{\partial q_i} F_i - \frac{\partial \Pi}{\partial q_i} F_i + a_i \frac{\partial F_i}{\partial t} - \frac{\partial P}{\partial t}$$

$$+ \dot{q}_s \left(\frac{\partial a_s}{\partial q_i} F_i + a_i \frac{\partial F_i}{\partial q_s} + g_{is} \frac{\partial F_i}{\partial t} + \frac{1}{2} g \frac{\partial f}{\partial q_s} - \Pi \frac{\partial f}{\partial q_s} + \frac{\partial a_s}{\partial t} f - \frac{\partial P}{\partial q_s}\right)$$

$$+ \frac{1}{2} \dot{q}_s \dot{q}_r \left(\frac{\partial g_{sr}}{\partial q_i} F_i + g_{is} \frac{\partial F_i}{\partial q_s} - g_{rs} \frac{\partial f}{\partial t} + \frac{\partial g_{rs}}{\partial t} f\right)$$

$$- \frac{1}{6} \dot{q}_s \dot{q}_r \dot{q}_p \left(g_{sr} \frac{\partial f}{\partial q_p} + g_{sp} \frac{\partial f}{\partial q_r} + g_{rp} \frac{\partial f}{\partial q_s}\right) = 0 \qquad (3.5.4)$$

for $i, s, r, p = 1, ..., n$. Note that the expression in parentheses with coefficients $\dot{q}_s \dot{q}_r$ and $\dot{q}_s \dot{q}_r \dot{q}_p$ is written in symmetric form.

Equating each factor of the various powers of $\dot{\mathbf{q}}$ to zero, we obtain the following system of partial differential equations of the first order, which we refer to as the *generalized Killing equations*:

$$\frac{1}{2} \frac{\partial}{\partial t}\left(gf\right) - \frac{\partial}{\partial t}\left(\Pi f\right) + \frac{1}{2} F_i \frac{\partial g}{\partial q_i} - \frac{\partial \Pi}{\partial q_i} F_i + a_i \frac{\partial F_i}{\partial t} - \frac{\partial P}{\partial t} = 0,$$

$$\frac{\partial a_s}{\partial q_i} F_i + a_i \frac{\partial F_i}{\partial q_s} + g_{is} \frac{\partial F_i}{\partial t} + \frac{1}{2} g \frac{\partial f}{\partial q_s} - \Pi \frac{\partial f}{\partial q_s} + \frac{\partial a_s}{\partial t} f - \frac{\partial P}{\partial q_s} = 0,$$

$$\frac{\partial g_{sr}}{\partial q_i} F_i + g_{is} \frac{\partial F_i}{\partial q_r} + g_{ir} \frac{\partial F_i}{\partial q_s} - g_{rs} \frac{\partial f}{\partial t} + \frac{\partial g_{rs}}{\partial t} f = 0,$$

$$g_{sr} \frac{\partial f}{\partial q_p} + g_{sp} \frac{\partial f}{\partial q_r} + g_{rp} \frac{\partial f}{\partial q_s} = 0. \qquad (3.5.5)$$

These equations were first derived for the case $P = 0$ in [111], and the case $P \neq 0$ was reported in [35]. The reason why we call this system the generalized Killing equations lies in the fact that this system for the case of free, that is, inertial motion of a dynamical system can be reduced to the Killing equations well known in Riemannian geometry and general relativity.

To show this, we suppose that the following conditions are satisfied:

(a) The Lagrangian function does not depend upon time: $\partial L / \partial t = 0$.

(b) The time is not varied, i.e., $f = 0$.

(c) The space generators F_i do not depend upon time t, i.e., $\partial F / \partial t = 0$.

(d) The gauge function P is equal to zero: $P = 0$.

(e) The potential energy $\Pi\left(t, \mathbf{q}\right) = 0$. Therefore, according to (3.5.3) the Lagrangian function becomes $L = \left(\frac{1}{2}\right) g_{ij} \dot{q}_i \dot{q}_j$. Thus, the motion is inertial.

With these suppositions, the equations $(3.5.5)_{1,2,4}$ are identically equal to zero, and the equation $(3.5.5)_3$ becomes

$$\frac{\partial g_{sr}}{\partial q_i} F_i + g_{is} \frac{\partial F_i}{\partial q_r} + g_{ir} \frac{\partial F_i}{\partial q_s} = 0, \quad i, s, r = 1, ..., n, \qquad (3.5.6)$$

which are classical Killing equations. Because $g_{ij} = g_{ji}$, they constitute a system of $n(n+1)/2$ independent equations, and in the general case they do not have a solution. However, if for some particular dynamical system the solution exists, according to (3.4.4) the dynamical system has a linear conservation law of the form

$$g_{ij} F_i \dot{q}_j = I = const. \qquad (3.5.7)$$

Note that for more complicated dynamical systems and more complex forms of the generators of infinitesimal transformations and the gauge function, we are faced with a more complicated system of partial differential equations that stem from the basic Noether identity (3.4.3).

However, very frequently, we are able to find the conservation laws directly from the basic Noether identity, without reducing it to the system of generalized Killing equations.

For example, we are easily able to find two important conservation laws: the Jacobi conservation law and the cyclic integral of the scleronomic dynamical system, which are derived in section 1.4.

Example 3.5.1. *Jacobi conservation law.* Let us consider the scleronomic dynamical system whose Lagrangian function is $L = L(\mathbf{q}(t), \dot{\mathbf{q}}(t))$, where $\mathbf{q} = \{q_1, ..., q_n\}$ and the nonconservative forces Q_i are absent.

It is easy to verify that for

$$f = A = const., \quad F_i = 0, \quad P = 0, \qquad (3.5.8)$$

the basic Noether identity (3.4.3) is identically satisfied. Therefore, from (3.4.4) it follows that

$$L - \frac{\partial L}{\partial \dot{q}_i} \dot{q}_i = E = const., \qquad (3.5.9)$$

which is identical to the Jacobi integral (1.4.45). From (3.5.8) it follows that the infinitesimal transformation which leaves the Lagrange–D'Alembert principle invariant is of the form

$$\bar{q}_i = q_i, \quad \bar{t} = t + \varepsilon A, \qquad (3.5.10)$$

which is the time translation mentioned in section 1.4 (see equation (1.4.48)).

Similarly, considering the equivalent dynamical system expressed in canonical variables whose Hamiltonian function is $H = H(\mathbf{q}, \mathbf{p})$ and $Q_i = 0$, for F_i, f, and P given by (3.5.8), we see that the basic Noether identity (3.4.6) is identically satisfied and from (3.4.7) we obtain that

$$H(\mathbf{q}, \mathbf{p}) = const., \qquad (3.5.11)$$

which is identical with the previously obtained result (1.8.17) obtained by a different reasoning.

Example 3.5.2. *Conservation of the generalized momentum.* If the Lagrangian function L does not depend upon a specific generalized coordinate, say q_1, but it depends on $\dot{q}_1$, then if $Q_i = 0$, the basic Noether identity (3.4.3) will be identically satisfied for the infinitesimal transformations of the form

$$\bar{q}_1 = q_1 + \varepsilon A, \quad \bar{q}_\alpha = q_\alpha, \quad \bar{t} = t, \quad \alpha = 2, ..., n, \tag{3.5.12}$$

whose generators are obviously

$$F_1 = A = const., \quad F_\alpha = 0, \quad f = 0. \tag{3.5.13}$$

Again, the basic Noether identity will be identically satisfied for $P = const.$ and from (3.4.4) it follows that the conserved quantity that corresponds to the infinitesimal transformation (3.5.12) is

$$\frac{\partial L}{\partial \dot{q}_1} = const., \tag{3.5.14}$$

which is the well-known integral of momentum or cyclic integral discussed in section 1.4.

3.6 The Basic Noether Identity and Integrating Factors of Equations of Motion

The basic Noether identity (3.4.3) is of a very complex structure and can be connected with almost all vital and important parts of analytical mechanics.

In this section we demonstrate that this identity is intimately related to the integrating factors of the differential equations of motion of holonomic nonconservative dynamical systems.

Rewriting (3.4.3) in the form

$$\frac{\partial L}{\partial \dot{q}_i} \dot{F}_i + \frac{\partial L}{\partial q_i} F_i + \left(L - \frac{\partial L}{\partial \dot{q}_i} \dot{q}_i \right) \dot{f} + \frac{\partial L}{\partial t} f + Q_i \left(F_i - \dot{q}_i f \right) - \dot{P} = 0, \tag{3.6.1}$$

and transforming the first and third terms by using Leibniz's rule $u\dot{v} = (uv)^{\cdot} - \dot{u}v$, we have

$$\frac{d}{dt} \left[\frac{\partial L}{\partial \dot{q}_i} F_i + \left(L - \frac{\partial L}{\partial \dot{q}_i} \dot{q}_i \right) f - P \right] + F_i \left(\frac{\partial L}{\partial q_i} - \frac{d}{dt} \frac{\partial L}{\partial \dot{q}_i} \right)$$
$$- \frac{d}{dt} \left(L - \frac{\partial L}{\partial \dot{q}_i} \dot{q}_i \right) f + \frac{\partial L}{\partial t} f + Q_i \left(F_i - \dot{q}_i f \right) = 0. \tag{3.6.2}$$

It is easy to demonstrate by differentiation with respect to time that the following identity holds:

$$\frac{d}{dt} \left(L - \frac{\partial L}{\partial \dot{q}_i} \dot{q}_i \right) = \frac{\partial L}{\partial t} + \dot{q}_i \left(\frac{\partial L}{\partial q_i} - \frac{d}{dt} \frac{\partial L}{\partial \dot{q}_i} \right). \tag{3.6.3}$$

Substituting this into (3.6.2) one has

$$\left(\frac{\partial L}{\partial q_i} - \frac{d}{dt}\frac{\partial L}{\partial \dot{q}_i} + Q_i\right)(F_i - \dot{q}_i f) = \frac{d}{dt}\left[P - \frac{\partial L}{\partial \dot{q}_i}F_i - \left(L - \frac{\partial L}{\partial \dot{q}_i}\dot{q}_i\right)f\right]. \quad (3.6.4)$$

Note that the expression in square brackets on the right-hand side is, according to (3.4.4), a constant for every solution F_i, f, and P of the basic Noether identity (3.6.1). Thus it is evident that the expression $(F_i - \dot{q}_i f)$ can be interpreted as integrating factors of the Euler–Lagrangian equations

$$\frac{d}{dt}\frac{\partial L}{\partial \dot{q}_i} - \frac{\partial L}{\partial q_i} - Q_i = 0, \quad i = 1, ..., n. \quad (3.6.5)$$

More precisely, the terms $(F_i - \dot{q}_i f) \neq 0$ are generally proportional to the integrating factors, since the Euler–Lagrangian equations (3.6.5) can contain the differential equations of motion (DEM) multiplied by some scalar factors $\theta_{(i)}(t, \mathbf{q}(t), \dot{\mathbf{q}}(t))$. Written symbolically, we can express (3.6.5) in the form

$$(\text{DEM})_i\, \theta_{(i)}(t, \mathbf{q}(t), \dot{\mathbf{q}}(t)) = 0, \quad (3.6.6)$$

where the summation convention with respect to indices i does *not* hold. Therefore, the actual integrating factors are

$$(\text{Integrating factor})_{(i)} = \theta_{(i)}(t, \mathbf{q}(t), \dot{\mathbf{q}}(t))(F_i - \dot{q}_i f), \quad (3.6.7)$$

where index i is not summed.

Note that a rather exhaustive account of the theory of integrating factors of nonconservative dynamical systems and related conservation laws are published in [38] from a different point of view.

3.7 Quadratic Conservation Laws of Euler's Equation

Let us consider the famous Euler's differential equation

$$\ddot{x} + \frac{a}{t}\dot{x} + \frac{b}{t^2}x = 0, \quad (3.7.1)$$

where $q = x$ is a generalized coordinate and a and b are real constant parameters. This equation can be derived from the Euler–Lagrangian equation whose Lagrangian function is

$$L = \frac{1}{2}\dot{x}^2 t^a - \frac{1}{2}bx^2 t^{a-2}. \quad (3.7.2)$$

Note that the scalar factor $\theta(t, x, \dot{x})$ mentioned at the end of the last section is

$$\theta(t) = t^a, \quad (3.7.3)$$

since (see (3.6.6))

$$\frac{d}{dt}\frac{\partial L}{\partial \dot{x}} - \frac{\partial L}{\partial x} = t^a \left(\ddot{x} + \frac{a}{t}\dot{x} + \frac{b}{t^2}x \right) = 0. \qquad (3.7.4)$$

In order to study quadratic conservation laws of this equation, with respect to generalized velocity $\dot{x}$, we shall suppose that the time generators f, the space generator F, and the gauge function P are, respectively, of the form

$$t = At^m, \quad F = Kt^p x, \quad P = Rt^s x^2, \qquad (3.7.5)$$

where A, K, R, m, p, and s are unknown constants to be determined.

The basic Noether identity and the corresponding conservation law in our notation are (see (3.4.3), (3.4.4))

$$\frac{\partial L}{\partial x}F + \frac{\partial L}{\partial \dot{x}}\dot{F} + \left(L - \frac{\partial L}{\partial \dot{x}}\dot{x} \right)\dot{f} + \frac{\partial L}{\partial t} - \dot{P} = 0 \qquad (3.7.6)$$

and

$$\frac{\partial L}{\partial \dot{x}}F + \left(L - \frac{\partial L}{\partial \dot{x}}\dot{x} \right) - P = I = const. \qquad (3.7.7)$$

Substituting (3.7.5) into (3.7.6) and equating to zero terms with $x^2, x\dot{x}$, and $\dot{x}$, we obtain, respectively,

$$-bKt^{a+p-2} - \frac{1}{2}Amt^{a+m-3} - Rst^{s-1} - \frac{1}{2}Ab(a-2)t^{a+m-3} = 0,$$
$$Kpt^{a+p-1} - 2Rt^s = 0,$$
$$Kt^{a+p} - \frac{1}{2}Amt^{a+m-1} + \frac{1}{2}aAt^{a+m-1} = 0. \qquad (3.7.8)$$

These equations will be compatible with respect to the exponents for

$$p = m - 1, \quad s = a + m - 2. \qquad (3.7.9)$$

Thus, equations (3.7.8) become

$$-bK - \frac{1}{2}Amb - \frac{1}{2}Ab(a-2) - Rs = 0,$$
$$Kp - 2R = 0,$$
$$K - \frac{1}{2}Am + \frac{1}{2}aA = 0. \qquad (3.7.10)$$

From $(3.7.10)_{2,3}$ and (3.7.9) we obtain

$$K = \frac{1}{2}A(m-a), \quad R = \frac{1}{4}A(m-1)(m-a), \qquad (3.7.11)$$

and $(3.7.10)_1$ becomes

$$(m - 1) [4b + (m - a)(a + m - 2)] = 0. \tag{3.7.12}$$

This cubic equation has three roots that are given as

$$m_1 = 1, \quad m_2 = 1 + \sqrt{(a - 1)^2 - 4b}, \quad m_3 = 1 - \sqrt{(a - 1)^2 - 4b}. \tag{3.7.13}$$

A simple inspection shows that the constant A can be taken to be $A = 1$ since f, F, and K are depending linearly on this constant, which is evident from $(3.7.5)$ and $(3.7.11)$.

The generators of transformations and the gauge function are

$$f_{(i)} = t^{m_{(i)}}, \quad F_{(i)} = \frac{1}{2} [m_{(i)} - a] t^{m_{(i)} - 1} x,$$

$$P_{(i)} = \frac{1}{4} [m_{(i)} - 1][m_{(i)} - a] t^{a + m_{(i)} - 2} x^2, \quad i = 1, 2, 3. \tag{3.7.14}$$

Multiplying both sides of $(3.7.7)$ with -1, for ease, and substituting $(3.7.14)$ we find for each given parameters a and b the following three quadratic conservation laws:

$$\frac{1}{2} \dot{x}^2 t^{a + m_{(i)}} - \frac{1}{2} [m_{(i)} - a] t^{a + m_{(i)} - 1} x \dot{x}$$

$$+ \frac{x^2}{4} \{ 2b + [m_{(i)} - 1][m_{(i)} - a] \} t^{a + m_{(i)} - 2} = = -I_{(i)} = const.,$$

$$i = 1, 2, 3. \tag{3.7.15}$$

For the arbitrary a and b and $m_{(1)} = 1$, the gauge function $P_{(1)} = 0$, and we have the following conservation law:

$$\frac{1}{2} \dot{x}^2 t^{a+1} - \frac{1}{2} (1 - a) t^a x \dot{x} + \frac{1}{2} b x^2 t^{a-1} = -I_{(1)} = const., \tag{3.7.16}$$

while the other two conservation laws, for $m_{(\alpha)}, \alpha = 2, 3$ given by $(3.7.13)$, become

$$\frac{1}{2} \dot{x}^2 t^{a + m_{(\alpha)}} - \frac{1}{2} [m_{(\alpha)} - a] t^{a + m_{(\alpha)} - 1} x \dot{x}$$

$$+ \frac{x^2}{4} \{ 2b + [m_{(\alpha)} - 1][m_{(\alpha)} - a] \} t^{a + m_{(\alpha)} - 2}$$

$$= -I_{(\alpha)} = const., \quad \alpha = 2, 3, \tag{3.7.17}$$

where a and b are also arbitrary.

It is now easy to verify that for each given $\theta_{(i)}, f_{(i)}$, and $F_{(i)}$ we can find the corresponding integrating factor of the differential equation of motion.

For the sake of simplicity let us find the integrating factor of the differential equation $(3.7.1)$ that generates the conservation law $(3.7.16)$. For this case $m_{(1)} = 1, f_{(1)} = t, F_{(1)} = \frac{1}{2}(1 - a) x$, and since $\theta = t^a$ (see $(3.7.3)$) we have on the basis of (3.6.7) that the integrating factor of (3.7.1) is in accordance to (3.6.7):

$$\text{(Integrating factor)}_{(1)} = \theta\left(F_{(1)} - f_{(1)}\dot{x}\right) = \frac{1}{2}(1-a)t^a x - \dot{x}t^{a+1}. \quad (3.7.18)$$

Therefore, it is easy to verify that the following relation holds:

$$
\begin{aligned}
\left(\ddot{x} + \frac{a}{t}\dot{x} + \frac{b}{t^2}x\right) & \left[\frac{1}{2}(1-a)t^a x - \dot{x}t^{a+1}\right] \\
& \equiv \frac{d}{dt}\left[\frac{1}{2}\dot{x}^2 t^{a+1} - \frac{1}{2}(1-a)t^a x\dot{x} + \frac{1}{2}bx^2 t^{a-1}\right] \\
& = \frac{d}{dt}\left(-I_{(1)}\right) = 0. \quad (3.7.19)
\end{aligned}
$$

As a concrete example, let us consider the case

$$\ddot{x} + \frac{6}{t}\dot{x} + \frac{4}{t^2}x = 0, \quad (3.7.20)$$

for which $a = 6$ and $b = 4$. From (3.7.13) it follows that the exponent m has the values $m_{(1)} = 1, m_{(2)} = 4, m_{(3)} = -2$. From (3.7.16) and (3.7.17) it follows that the differential equation (3.7.20) possesses the following three quadratic conservation laws:

$$
\begin{aligned}
\frac{1}{2}\dot{x}^2 t^7 + \frac{5}{2}x\dot{x}t^6 + 2x^2 t^5 &= -I_{(1)} = const., \\
\frac{1}{2}\dot{x}^2 t^{10} + x\dot{x}t^9 + \frac{1}{2}x^2 t^8 &= -I_{(2)} = const., \\
\frac{1}{2}\dot{x}^2 t^4 + 4x\dot{x}t^3 + 8x^2 t^2 &= -I_{(3)} = const., \quad (3.7.21)
\end{aligned}
$$

It should be noted that all three conservation laws (3.7.16), (3.7.17) and consequently three conservation laws (3.7.21) are *not mutually independent*. In fact, we can consider any two of the conservation laws (3.7.21) as mutually independent, and the third one should be a consequence of the first two arbitrarily selected.

To demonstrate this, we note that the differential equation (3.7.20) has two linear conservation laws of the form

$$C_1 = 4tx + t^2\dot{x}, \quad C_2 = t^4 x + t^5 \dot{x}, \quad (3.7.22)$$

which can be considered as the general solution of (3.7.20). From (3.7.22) it follows that

$$-I_{(1)} = \frac{1}{2}C_1 C_2, \quad -I_{(2)} = \frac{1}{2}C_2^2, \quad -I_{(3)} = \frac{1}{2}C_1^2, \quad (3.7.23)$$

which confirm our statement.

3.8   Quadratic Conservation Laws of the Scleronomic Duffing Oscillator

In this section we consider the possibility of finding the conservation laws of the Duffing oscillator whose differential equation of motion is

$$\ddot{x} + k\dot{x} + \omega^2 x + \lambda x^n = 0, \qquad (3.8.1)$$

where $k, \omega, \lambda,$ and n are given constant parameters.

One of the basic suppositions for treating this problem is that we will assume that the velocity can be represented as a field depending on time t and position x, namely,[11]

$$\dot{x} = \phi(t, x). \qquad (3.8.2)$$

Despite the fact that the dynamical system (3.8.1) can be completely derived from a Lagrangian function, we will treat it as a purely nonconservative system whose Lagrangian function is $L = \frac{1}{2}\dot{x}^2 - \frac{1}{2}\omega^2 x^2 - \frac{\lambda}{n+1}x^{n+1}$, and the nonconservative force is $Q = -k\dot{x}$.

If we select the generators of the infinitesimal transformations in the form $F = -1, f = 0$, and the gauge function $P = \phi(t, x)$, Noether's identity (3.4.3) becomes

$$\frac{\partial L}{\partial x}F + QF + \dot{\phi} = 0 \qquad (3.8.3)$$

or, in explicit form,

$$\frac{\partial \phi}{\partial t} + \phi\frac{\partial \phi}{\partial x} + k\phi + \omega^2 x + \lambda x^n = 0. \qquad (3.8.4)$$

In the next chapter we will call this quasi-linear differential equation the *basic field equation*. Note that the equation (3.4.4) in this case reduces to an identity $(-\dot{x} + \dot{x} = 0)$, from which it follows that the constant appearing in (3.4.4) is equal to zero. This fact means that we can consider the basic supposition (3.8.2) as a conservation law of the dynamical system (3.8.1). Note also that by combining (3.8.1) with (3.8.2) we can immediately derive partial differential equation (3.8.4). However, we have demonstrated here that the method based upon the theory of this chapter has a rather operative possibility. To find an incomplete solution of (3.8.4) (which in fact represents a conservation law of the dynamical system (3.8.1)), we select the solution in the form

$$\dot{x} = \phi(t, x) = Ax + [If(t) + W(x)]^{1/2}, \qquad (3.8.5)$$

where A and I are constants and $f(t)$ and $W(x)$ are unknown functions. Entering with this into (3.8.4) we find

$$I\left(\dot{f} + 2kf + 2Af\right) + (AxW' + 2AW + 2kW)$$
$$+\sqrt{R}\left[2x\left(A^2 + kA + \omega^2\right) + (W' + 2\lambda x^n)\right] = 0, \qquad (3.8.6)$$

---

[11]Note that this supposition will be widely used in the next chapter of this book.

where $\sqrt{R} = (If + W)^{1/2}$ and prime denotes the differentiation with respect to x. By equating to zero the terms in round brackets, we arrive at the following system of generalized Killing's equations:

$$\dot{f} + 2kf + 2Af = 0, \quad AxW' + 2kW + 2AW = 0,$$
$$A^2 + kA + \omega^2 = 0, \quad W' + 2\lambda x^n = 0. \tag{3.8.7}$$

Integrating $(3.8.7)_4$ we obtain

$$W(x) = -\frac{2\lambda}{n+1} x^{n+1}. \tag{3.8.8}$$

Substituting this into $(3.8.7)_2$ we find the following relation between the coefficients k, A, and n:

$$k = -A \frac{n+3}{2}. \tag{3.8.9}$$

Now, from $(3.8.7)_3$ it follows that

$$A = \pm \sqrt{\frac{2}{n+1}} \omega. \tag{3.8.10}$$

Taking the minus sign, that is, selecting the damping factor to be positive, we see that the coefficient k, the natural frequency ω, and the nonlinear exponent n are not independent:

$$k = \frac{n+3}{2} \left(\frac{2}{n+1} \right)^{1/2} \omega. \tag{3.8.11}$$

Finally, integrating $(3.8.7)_1$ one finds

$$f(t) = e^{\left[-\sqrt{2(n+1)}\omega t \right]}. \tag{3.8.12}$$

Therefore, from (3.8.8), (3.8.10)–(3.8.12), and (3.8.5) it follows that the nonlinear, damped Duffing oscillator

$$\ddot{x} + \frac{n+3}{2} \sqrt{\frac{2}{n+1}} \omega \dot{x} + \omega^2 x + \lambda x^n = 0 \tag{3.8.13}$$

has the conservation law

$$\dot{x} = -\sqrt{\frac{2}{n+1}} \omega x + \sqrt{Ie^{-\sqrt{2(n+1)}\omega t} - \frac{2\lambda}{n+1} x^{n+1}}, \tag{3.8.14}$$

or

$$\left[\left(\dot{x} + \sqrt{\frac{2}{n+1}} \omega x \right)^2 + \frac{2\lambda}{n+1} x^{n+1} \right] e^{\sqrt{2(n+1)}\omega t} = I = const. \tag{3.8.15}$$

For example, let $n = 7$, for which $k = (5/2)\,\omega$. Thus, the differential equation

$$\ddot{x} + \frac{5}{2}\omega\dot{x} + \omega^2 x + \lambda x^7 = 0 \tag{3.8.16}$$

possesses, according to (3.8.15), the quadratic conservation law of the form

$$\left[\left(\dot{x} + \frac{1}{2}\omega x\right)^2 + \frac{\lambda}{4}x^8\right]e^{4\lambda t} = I = const. \tag{3.8.17}$$

As a next concrete example, let us consider the case of a conservative dynamical system for which $k = 0$. From (3.8.11) it follows that $n = -3$. Therefore, we conclude that the differential equation

$$\ddot{x} + \omega^2 x + \frac{\lambda}{x^3} = 0 \tag{3.8.18}$$

has quadratic conservation law

$$\left[(\dot{x} - i\omega x)^2 - \frac{\lambda}{x^2}\right]e^{2i\omega t} = C_1 = const., \tag{3.8.19}$$

where $i = \sqrt{-1}$. Since the differential equation (3.8.18) is invariant with respect to transformation $x = -x, t = -t$, we have another conservation law of (3.8.18) in the form

$$\left[(\dot{x} + i\omega x)^2 - \frac{\lambda}{x^2}\right]e^{-2i\omega t} = C_2 = const. \tag{3.8.20}$$

Two first integrals (3.8.19), (3.8.20) are independent, and we can use them to find the general solution of (3.8.18). Adding (3.8.19) and (3.8.20) and multiplying the result by x^2, we have

$$2\left(x\dot{x}\right)^2 - 2\omega^2 x^4 - 2\lambda - x^2\left(C_1 e^{2i\omega t} + C_2 e^{-2i\omega t}\right) = 0. \tag{3.8.21}$$

By subtracting (3.8.19) and (3.8.20) we have

$$x\dot{x} = \frac{C_1}{4i\omega}e^{2i\omega t} - \frac{C_2}{4i\omega}e^{-2i\omega t}. \tag{3.8.22}$$

Integrating we find

$$x^2 = \frac{C_1}{4\omega^2}e^{2i\omega t} - \frac{C_2}{4\omega^2}e^{-2i\omega t} + D, \tag{3.8.23}$$

where D is a constant of integration which depends on C_1 and C_2. To find the constant D, we substitute (3.8.22) into (3.8.23) to obtain

$$D = \sqrt{\frac{1}{4}\frac{C_1 C_2}{\omega^2} - \frac{\lambda}{\omega^2}}. \tag{3.8.24}$$

Employing $e^{iX} = \cos X + i \sin X$ we can write (3.8.23) in the form

$$x^2 = A \cos 2\omega t + B \sin 2\omega t + \sqrt{\frac{1}{4} \frac{C_1 C_2}{\omega^2} - \frac{\lambda}{\omega^2}}, \qquad (3.8.25)$$

where

$$A = -\frac{C_1 + C_2}{4\omega^2}, \quad B = i\frac{-C_1 + C_2}{4\omega^2}. \qquad (3.8.26)$$

Finally, since $A^2 + B^2 = \frac{1}{4\omega^2} C_1 C_2$, we can write the general solution (3.8.25) of the differential equation (3.8.18) in the form

$$x^2 = \sqrt{A^2 + B^2 - \frac{\lambda}{\omega^2}} + A \cos 2\omega t + B \sin 2\omega t, \qquad (3.8.27)$$

which is, for $\lambda = -1$ identical with the solution of the equation (2.5.40) given by (2.5.41).

It should be noted that the scleronomic differential equation (3.8.18) admits the total energy conservation law of the form

$$\frac{\dot{x}^2}{2} + \omega^2 \frac{x^2}{2} - \frac{\lambda}{2} \frac{1}{x^2} = E = const. \qquad (3.8.28)$$

Naturally, all three conservation laws (3.8.19), (3.8.20), and (3.8.28) are not mutually independent. They are connected by the relation

$$C_1 C_2 = (2E)^2, \qquad (3.8.29)$$

which is easy to verify by direct calculation.

3.9 Conservation Laws of the Arbitrary Degree of a Purely Dissipative Dynamical System

Let us consider a purely nonconservative dissipative dynamical system whose differential equation is of the form

$$\ddot{x} = X(t, x)\dot{x}. \qquad (3.9.1)$$

Treating this system as purely nonconservative, we have

$$L = \frac{\dot{x}^2}{2}, \quad Q(t, x, \dot{x}) = X(t, x)\dot{x}. \qquad (3.9.2)$$

As a trial form of the space and time generators of the infinitesimal transformations, we take

$$F = B\dot{x}^b, \quad f = AS(t, x)\dot{x}^a, \qquad (3.9.3)$$

where A, B, a, and b are unknown constants and $S(t, x)$ is an unknown function of time t and the position x. We also suppose that the gauge function P is equal to zero.

By noting that the time derivatives of the generators F and f along the dynamical system trajectory are

$$\dot{F} = Bb\dot{x}^b X(t, x), \quad \dot{f} = A\left(\frac{\partial S}{\partial t} + \frac{\partial S}{\partial x} + aSX\right)\dot{x}^a, \qquad (3.9.4)$$

the Noether identity (3.4.3)

$$\frac{\partial L}{\partial \dot{x}}\dot{F} + \left(L - \frac{\partial L}{\partial \dot{x}}\dot{x}\right)\dot{f} + Q(F - f\dot{x}) = 0 \qquad (3.9.5)$$

becomes

$$B(b+1)X\dot{x}^{b+1} - \frac{1}{2}A\frac{\partial S}{\partial x}\dot{x}^{a+3} - A\left[\frac{1}{2}\frac{\partial S}{\partial t} + \left(\frac{a}{2}+1\right)SX\right]\dot{x}^{a+2} = 0. \quad (3.9.6)$$

Taking

$$a = b - 2, \qquad (3.9.7)$$

the Noether identity becomes

$$\left[BX(b+1)X - \frac{1}{2}A\frac{\partial S}{\partial x}\right]\dot{x}^{b+1} - A\dot{x}^b\left[\frac{\partial S}{\partial t} + SX\left(\frac{a}{2}+1\right)\right] = 0. \qquad (3.9.8)$$

Since the terms in brackets are independent of the velocity $\dot{x}$, we arrive at the following system of partial differential equations, which represents the generalized Killing system:

$$\frac{1}{2}\frac{A}{B(b+1)}\frac{\partial S}{\partial x} = X(t, x), \quad \frac{\partial S}{\partial t} + \frac{A}{B}\frac{b}{2(b+1)}S\frac{\partial S}{\partial x} = 0. \qquad (3.9.9)$$

The parameters A and B are free, and we select, for example,

$$A = -2, \quad B = \frac{1}{b+1}. \qquad (3.9.10)$$

Thus, we obtain

$$X(t, x) = -\frac{\partial S}{\partial x} \qquad (3.9.11)$$

and

$$\frac{\partial S}{\partial t} - bS\frac{\partial S}{\partial x} = 0. \qquad (3.9.12)$$

The generators of the infinitesimal transformations are

$$F = \frac{1}{b+1}\dot{x}^b, \quad f = -2S(t, x)\dot{x}^{b-2}. \qquad (3.9.13)$$

Referring to equation (3.4.4), we conclude that the dynamical system

$$\ddot{x} = -\frac{\partial S}{\partial x}(t, x)\dot{x} \tag{3.9.14}$$

has the conservation law of the form

$$\frac{\dot{x}^{b+1}}{b+1} + S(t, y)\dot{x}^b = I = const., \tag{3.9.15}$$

where $S(t, x)$ is *any solution* of the partial differential equation (3.9.12) and b is an arbitrary constant parameter $b \neq -1$.

If we introduce a new function $U(t, x)$ by the relation

$$S(t, x) = \frac{1}{b}U(t, x), \tag{3.9.16}$$

the equation (3.9.12) is reduced to a standard form,

$$\frac{\partial U}{\partial t} - U\frac{\partial U}{\partial x} = 0. \tag{3.9.17}$$

A rather broad class of solutions of this equation is known to be

$$x + tU = R(U), \tag{3.9.18}$$

where $R(U)$ is an arbitrary function of U.

For example, if $R(U) = U^2/2$, from the previous equation we have

$$U(t, x) = t + \left(t^2 + 2x\right)^{1/2}. \tag{3.9.19}$$

Thus,

$$S(t, x) = \frac{1}{b}\left[t + \left(t^2 + 2x\right)^{1/2}\right]. \tag{3.9.20}$$

Therefore, from (3.9.14) and (3.9.15) it follows that the differential equation

$$\ddot{x} = -\frac{1}{b}\frac{1}{\sqrt{(t^2 + 2x)}}\dot{x} \tag{3.9.21}$$

has a conservation law of the arbitrary degree (in our case $b + 1$) with respect to $\dot{x}$:

$$\frac{\dot{x}^{b+1}}{b+1} + \frac{1}{b}\left[t + \sqrt{(t^2 + 2x)}\right]\dot{x}^b = I = const. \tag{3.9.22}$$

3.10   Polynomial Conservation Laws of the Generalized Emden–Fowler Equation

In this section we shall consider the existence of conservation laws of the generalized Emden–Fowler equation

$$\ddot{x} + \frac{b}{t}\dot{x} + \lambda x^{\alpha} t^{k} = 0, \tag{3.10.1}$$

where b, λ, α, and k are constants. Note that this equation appears in many branches of physics and engineering, for example, in stellar dynamics, quantum mechanics, and fluid dynamics. The closed form solution of the Emden–Fowler equation is rather scarce and the search for conservation laws, which can in many respects shed light on the physical mechanism of the dynamical problem, has its full vindication. In the following presentation we pursue [129].

Our interest will be focused upon the problem of finding conservation laws of the fourth degree with respect to generalized velocity $\dot{x}$, which we shall refer to as the polynomial conservation laws. Recently, more interest in polynomial conservation laws of the fourth degree has been aroused in theoretical physics and quantum mechanics (see, for example, [2] and [3]).

The differential equation (3.10.1) can be derived as an Euler–Lagrangian equation for the Lagrangian L given by

$$L = \frac{1}{2}\dot{x}^{2}t^{b} - \frac{\lambda}{\alpha+1}x^{\alpha+1}t^{k+b}. \tag{3.10.2}$$

The crucial step in searching for conservation laws of any kind is the selection of the generators of the infinitesimal transformations $F\left(t, x, \dot{x}\right), f\left(t, x, \dot{x}\right)$ and the gauge function $P\left(t, x, \dot{x}\right)$. Naturally, the structure of these functions has to be selected in such a way that the conservation law of the given degree can be obtained. According to our experience, the generators of the infinitesimal transformations must contain as a basic germ the expression of the "total energy" of the dynamical system, namely $E\left(t, x, \dot{x}\right) = \left(\partial L/\partial \dot{x}\right)\dot{x} - L$, that can be multiplied by some function that depends upon the dynamical variables.

Let the generator of the infinitesimal space transformation be equal to zero, that is, $F = 0$, and the generator of the infinitesimal time transformation be given as the total energy of the dynamical system multiplied by the factor t^{m}, that is,

$$f\left(t, x, \dot{x}\right) = \left[\frac{\partial L}{\partial \dot{x}}\dot{x} - L\right]t^{m} = \frac{1}{2}\dot{x}^{2}t^{b+m} + \frac{\lambda}{\alpha+1}t^{k+b+m}. \tag{3.10.3}$$

Let the form of the gauge function be given by

$$P = f_{0}\left(t, x\right) + f_{1}\left(t, x\right)\dot{x} + f_{2}\left(t, x\right)\dot{x}^{2}, \tag{3.10.4}$$

where $f_{0}, ..., f_{2}$ are unknown functions of the position and time to be determined in the course of analysis.

The total time derivative of f and P along the dynamical trajectory are

$$
\begin{aligned}
\dot{f}(t, x, \dot{x}) &= \frac{m-b}{2}\dot{x}^2 t^{b+m-1} + \frac{\lambda(k+b+m)}{\alpha+1} x^{\alpha+1} t^{k+b+m}, \\
\dot{P}(t, x, \dot{x}) &= \frac{\partial f_0}{\partial t} - \lambda f_1 x^\alpha t^k + \dot{x}\left(\frac{\partial f_0}{\partial x} + \frac{\partial f_1}{\partial t} - \frac{b}{t}f_1 - 2\lambda f_2 x^\alpha t^k\right) \\
&\quad + \dot{x}^2\left(\frac{\partial f_1}{\partial x} + \frac{\partial f_2}{\partial t} - 2\frac{b}{t}f_2 - 3\lambda f_3 x^\alpha t^k\right) \\
&\quad + \dot{x}^3\left(\frac{\partial f_2}{\partial x} + \frac{\partial f_3}{\partial t} - 3\frac{b}{t}f_3\right) + \frac{\partial f_3}{\partial x}\dot{x}^3. \qquad (3.10.5)
\end{aligned}
$$

The basic Noether identity (3.4.3) becomes

$$
\begin{aligned}
&\left[\frac{\partial f_0}{\partial t} - \lambda f_1 x^\alpha t^k + \frac{\lambda^2(2k+2b+m)}{(\alpha+1)^2} x^{2\alpha+2} t^{2k+2b+m-1}\right] \\
&+ \dot{x}\left[\frac{\partial f_0}{\partial x} + \frac{\partial f_1}{\partial t} - \frac{b}{t}f_1 - 2\lambda f_2 x^\alpha t^k\right] \\
&+ \dot{x}^2\left[\frac{\partial f_1}{\partial x} + \frac{\partial f_2}{\partial t} - 2\frac{b}{t}f_2 - 3\lambda f_3 x^\alpha t^k + \frac{\lambda(k+m)}{\alpha+1} x^{\alpha+1} t^{k+2b+m-1}\right] \\
&+ \dot{x}^3\left[\frac{\partial f_2}{\partial x} + \frac{\partial f_3}{\partial t} - 3\frac{b}{t}f_3\right] + \dot{x}^4\left[\frac{\partial f_3}{\partial x} + \frac{1}{4}(m-2b)t^{2b+m-1}\right] = 0.
\end{aligned}
$$
$$(3.10.6)$$

Since the expressions in brackets do not depend on $\dot{x}$, equating the different powers of $\dot{x}$ to zero we get the following set of first-order partial differential equations

$$
\begin{aligned}
\frac{\partial f_3}{\partial x} + \frac{1}{4}(m-2b)t^{2b+m-1} &= 0, \\
\frac{\partial f_2}{\partial x} + \frac{\partial f_3}{\partial t} - 3\frac{b}{t}f_3 &= 0, \\
\frac{\partial f_1}{\partial x} + \frac{\partial f_2}{\partial t} - 2\frac{b}{t}f_2 - 3\lambda f_3 x^\alpha t^k + \frac{\lambda(k+m)}{\alpha+1} x^{\alpha+1} t^{k+2b+m-1} &= 0, \\
\frac{\partial f_0}{\partial x} + \frac{\partial f_1}{\partial t} - \frac{b}{t}f_1 - 2\lambda f_2 x^\alpha t^k &= 0, \\
\frac{\partial f_0}{\partial t} - \lambda f_1 x^\alpha t^k + \frac{\lambda^2(2k+2b+m)}{(\alpha+1)^2} x^{2\alpha+2} t^{2k+2b+m-1} &= 0.
\end{aligned}
$$
$$(3.10.7)$$

Integrating $(3.10.7)_1$ with respect to x and neglecting an arbitrary function of t, we find

$$
f_3(t, x) = -\frac{1}{4}(m-2b)xt^{2b+m-1}. \qquad (3.10.8)
$$

Substituting this into $(3.10.7)_2$ and integrating, we obtain

$$f_2\left(t,x\right) = -\frac{1}{8}\left(m-2b\right)\left(b-m+1\right)x^2 t^{2b+m-2}. \qquad (3.10.9)$$

Repeating the procedure, we find from $(3.10.7)_3$ and $(3.10.7)_4$ that

$$
\begin{aligned}
f_1\left(t,x\right) &= -\frac{1}{24}\left(m-2b\right)\left(b-m+1\right)\left(2-m\right)x^3 t^{2b+m-3} \\
&\quad -\lambda\frac{\left(3\alpha m - 6\alpha b + 7m + 4k - 6b\right)}{4\left(\alpha+1\right)\left(\alpha+2\right)}, \\
f_0\left(t,x\right) &= \frac{1}{96}\left(m-2b\right)\left(b-m+1\right)\left(2-m\right)\left(b+m-3\right)x^4 t^{2b+m-4} \\
&\quad +\lambda\frac{x^{\alpha+3}t^{k+2b+m-2}}{4}\left[\frac{\left(2b-m\right)\left(b-m+1\right)}{\alpha+3}\right. \\
&\quad \left. +\frac{\left(3\alpha m - 6b\alpha + 4k - 6b\right)\left(k+b+m-1\right)}{\left(\alpha+1\right)\left(\alpha+2\right)\left(g\alpha+3\right)}\right].
\end{aligned}
\qquad (3.10.10)
$$

Finally, equation $(3.10.7)_5$ becomes

$$
\begin{aligned}
&\frac{x^4 t^{2b+m-5}}{96}\left[\left(m-2b\right)\left(b-m+1\right)\left(2-m\right)\left(b+m-3\right)\left(2b+m-4\right)\right] \\
&+\lambda\frac{x^{\alpha+3}t^{k+2b+m-3}}{4}\left(k+2b+m-2\right)\left[\frac{\left(2b-m\right)\left(b-m+1\right)}{\alpha+3}\right. \\
&\quad \left. +\frac{\left(3\alpha m - 6b\alpha + 7m + 4k - 6b\right)\left(k+b+m-1\right)}{\left(\alpha+1\right)\left(\alpha+2\right)\left(\alpha+3\right)}\right] \\
&+\lambda\frac{x^{\alpha+3}t^{k+2b+m-3}}{24}\left(m-2b\right)\left(b-m+1\right)\left(2-m\right) \\
&+\lambda^2 x^{2\alpha+2}t^{2k+2b+m-1}\left[\frac{\left(3\alpha m - 6b\alpha + 7m + 4k - 6b\right)}{4\left(\alpha+1\right)\left(\alpha+2\right)}\right. \\
&\quad \left. +\frac{2k+2b+m}{\left(\alpha+1\right)^2}\right] = 0.
\end{aligned}
\qquad (3.10.11)
$$

A simple inspection shows that this equation will be identically satisfied if the following algebraic system is satisfied:

$$(m - 2b)(b - m + 1)(2 - m)(b + m - 3)(2b + m - 4) = 0,$$

$$(k + 2b + m - 2)\left[\frac{(2b - m)(b - m + 1)}{\alpha + 3}\right.$$

$$\left. + \frac{(3\alpha m - 6b\alpha + 7m + 4k - 6b)(k + b + m - 1)}{(\alpha + 1)(\alpha + 2)(\alpha + 3)}\right]$$

$$+ \frac{1}{6}(m - 2b)(b - m + 1)(2 - m) = 0;$$

$$\frac{3\alpha m - 6b\alpha + 7m + 4k - 6b}{4(\alpha + 1)(\alpha + 2)} + \frac{2k + 2b + m}{(\alpha + 1)^2} = 0.$$

$$(3.10.12)$$

Therefore, the problem of finding first integrals is reduced to an algebraic problem with three equations which depend upon four constants $k, \alpha, b,$ and n. Note that the equation $(3.10.12)_3$ can be written in the form

$$\left(\alpha + \frac{5}{3}\right)[m(\alpha + 3) - 2b(\alpha - 1) + 4k] = 0. \qquad (3.10.13)$$

Since the gauge function $P(t, x, \dot{x})$ given by (3.10.4) is fully determined, taking into account the time generator $f(t, x, \dot{x})$ of the infinitesimal transformation in the form (3.10.3), the conservation law of the fourth degree of the dynamical system (3.10.1) with respect to $\dot{x}$ given by the general expression (3.4.4) becomes

$$I = \frac{1}{4}\dot{x}^4 t^{2b+m} - \frac{1}{4}(m - 2b)x\dot{x}^3 t^{2b+m-1}$$

$$+ \left[-\frac{1}{8}(m - 2b)(b - m + 1)x^2 t^{2b+m-2} + \lambda\frac{x^{\alpha+1}}{\alpha + 1}t^{k+2b+m}\right]\dot{x}^2$$

$$- \left[\frac{1}{24}(m - 2b)(b - m + 1)(2 - m)x^3 t^{2b+m-3}\right.$$

$$\left. + \lambda\frac{(3\alpha m - 6\alpha b + 7m + 4k - 6b)}{4(\alpha + 1)(\alpha + 2)}x^{\alpha+2}t^{k+2b+m-1}\right]\dot{x}$$

$$+ \frac{1}{96}(m - 2b)(b - m + 1)(2 - m)(b + m - 3)x^4 t^{2b+m-4}$$

$$+ \frac{1}{4}\lambda x^{\alpha+3}t^{k+2b+m-2}\left[\frac{(2b - m)(b - m + 1)}{\alpha + 3}\right.$$

$$\left. + \frac{(3\alpha - 6\alpha b + 7m + 4k - 6b)(k + b + m - 1)}{(\alpha + 1)(\alpha + 2)(\alpha + 3)}\right]$$

$$+ \lambda^2\frac{x^{2\alpha+2}}{(\alpha + 1)^2}t^{2k+2b+m} = const. \qquad (3.10.14)$$

Example 3.10.1. As a typical example, let us consider the case

$$m = 3 - b, \tag{3.10.15}$$

which satisfies equation $(3.10.12)_1$. Substituting this into $(3.10.12)_3$ we obtain

$$9\alpha^2 + 42\alpha + 45 - b\left(9\alpha^2 + 18\alpha + 5\right) + 12k\left(\alpha + \frac{5}{3}\right) = 0 \tag{3.10.16}$$

or

$$\left(\alpha + \frac{5}{3}\right)\left[3\left(\alpha + 3\right) - 3b\left(\alpha + \frac{1}{3}\right) - 4k\right] = 0. \tag{3.10.17}$$

Discarding the case $\alpha = -5/3$, we have

$$k = \frac{b}{4}\left(3\alpha + 1\right) - \frac{3}{4}\left(\alpha + 3\right). \tag{3.10.18}$$

Substituting (3.10.15) and (3.10.18) into $(3.10.12)_2$, we find after a laborious algebraic calculation

$$(b-1)^3\left\{\frac{1}{4}\left(3\alpha + 5\right)\left[\frac{6}{\alpha + 3} - \frac{3}{2}\frac{3\alpha + 1}{(\alpha + 1)(\alpha + 3)}\right] - 1\right\} = 0. \tag{3.10.19}$$

Supposing that $b \neq 1$, we find $\alpha = -7$. Thus, $k = 3 - 5b$. Substituting this and $m = 3 - b$ into (3.10.1) and (3.10.14), we find that the differential equation

$$\ddot{x} + \frac{b}{t}\dot{x} + \frac{\lambda}{x^7}t^{3-5b} = 0 \tag{3.10.20}$$

has a conservation law of the form

$$
\begin{aligned}
I \;=\;& \frac{1}{4}\dot{x}^4 t^{b+3} + \frac{3}{4}\left(b - 1\right)x\dot{x}^3 t^{b+2} + \frac{3}{4}x^2\dot{x}^2 t^{b+1} - \frac{1}{6}\lambda x^{-6}\dot{x}^2 t^{-4b+6} \\
&+ \frac{1}{4}\left(b - 1\right)^3 x^3\dot{x}t^b - \frac{1}{4}\lambda\left(b - 1\right)x^{-5}\dot{x} + \frac{1}{36}x^{-12}t^{-9b+9} \\
&- \lambda\frac{(b-1)^2}{16}x^{-4}t^{4-4b} = const.
\end{aligned}
\tag{3.10.21}
$$

For the case $b = 0$, it follows that the differential equation

$$\ddot{x} + \lambda\frac{1}{x^7 t^2} = 0 \tag{3.10.22}$$

has a conservation law

$$
\begin{aligned}
I \;=\;& \frac{1}{4}\dot{x}^4 t^3 - \frac{3}{4}x\dot{x}^3 t^2 + \left(\frac{3}{4}x^2 t - \frac{1}{6}\lambda x^{-6}t^6\right)\dot{x}^2 - \frac{1}{4}\left(x^3 - \lambda x^{-5}t^5\right)\dot{x} \\
&+ \frac{1}{36}\lambda^2 x^{-12}t^9 - \frac{1}{16}x^{-4}t^4 = const.,
\end{aligned}
\tag{3.10.23}
$$

which is the same conservation law as obtained by Airault [2] by some undefined method.

Note that for $b = 1$ the polynomial conservation law (3.10.21) reduces to a quadratic, namely,

$$I = \frac{1}{4}\dot{x}^4 t^4 - \frac{1}{6}\lambda x^{-6}\dot{x}^2 t^2 + \frac{1}{36}\lambda^2 x^{-12} = \left(\frac{1}{2}\dot{x}^2 t^2 - \frac{1}{6}\lambda x^{-6}\right)^2 = const.,$$

(3.10.24)

which means that the differential equation

$$\ddot{x} + \frac{\dot{x}}{t} + \lambda\frac{1}{x^7 t^2} = 0$$

(3.10.25)

has a conservation law of the form

$$\frac{1}{2}\dot{x}^2 t^2 - \frac{\lambda}{6x^6} = const.$$

(3.10.26)

Consider now the case $\alpha = -5/3$ and $m = 3 - b$, which means that the equation $(3.10.12)_2$ becomes

$$14b^3 - 3b^2\,(3k + 20) - 3b\,(27k^2 + 102k + 82)$$
$$- \left(54k^3 + 243k^2 + 333k + 140\right) = 0.$$

(3.10.27)

This is an algebraic relation between parameters b and k. As an example, some numerical values for b, k, and $m = 3 - b$ are given in Table 3.10.1.

Table 3.10.1

No.	b	k	$m = 3 - b$
1	0	$-5/6$	3
2	0	$-7/3$	3
3	0	$-4/3$	3
4	$-5/7$	0	$26/7$
5	-2	0	5
6	7	0	10
7	$-3/4$	$-5/6$	$15/4$
8	$9/2$	$-5/6$	$15/2$
9	$3/2$	$-7/3$	$3/2$
10	$9/7$	$-7/3$	$12/7$
11	$3/7$	$-4/3$	$18/7$
12	3	$-4/3$	0

For example, selecting the data from the seventh case, we find, by using (3.10.14), that the differential equation

$$\ddot{x} - \frac{3}{4}\frac{\dot{x}}{t} + \lambda\frac{1}{x^{5/3} t^{5/6}} = 0$$

(3.10.28)

has a conservation law of the fourth order

$$
\begin{aligned}
I \;=\;& \frac{1}{4}\dot{x}^4 t^{9/4} - \frac{21}{16}x\dot{x}^3 t^{5/4} + \left(\frac{147}{64}x^2 t^{1/4} - \frac{3}{2}\lambda x^{-2/3}t^{17/12} \right)\dot{x}^2 \\
& - \left(\frac{343}{256}x^3 t^{-3/4} - \frac{21}{16}x^{1/3}t^{5/12} \right)\dot{x} \\
& + \frac{147}{64}\lambda x^{4/3}t^{-7/12} + \frac{9}{4}\lambda^2 x^{-4/3}t^{7/12} \;=\; const.
\end{aligned}
\tag{3.10.29}
$$

Similarly, for the case $b = 3$, $k = -4/3$, and $m = 0$ given in the 12th case of the table, it follows that the differential equation

$$
\ddot{x} + 3\frac{\dot{x}}{t} + \lambda\frac{1}{x^{5/3}t^{4/3}} = 0
\tag{3.10.30}
$$

has a fourth-degree conservation law of the form

$$
\begin{aligned}
I \;=\;& \frac{1}{4}\dot{x}^4 t^6 + \frac{3}{2}x\dot{x}^3 t^5 + 3\left(x^2 t^4 - \frac{1}{2}\lambda x^{-2/3}t^{14/3} \right)\dot{x}^2 \\
& + \left(2x^3 t^3 + \frac{15}{2}\lambda x^{1/3}t^{11/3} \right)\dot{x} \\
& + \frac{3}{4}\lambda x^{4/3}t^{8/3} + \frac{9}{4}\lambda^2 x^{-4/3}t^{10/3} \;=\; const.
\end{aligned}
\tag{3.10.31}
$$

Example 3.10.2. As another example, let us consider the case for $m = 2$. It is evident that the algebraic equation $(3.10.12)_1$ is identically satisfied. Then, equation $(3.10.12)_3$ reduces to

$$
k = \frac{b}{2}(\alpha - 1) - \frac{1}{2}(\alpha + 3),
\tag{3.10.32}
$$

where the common multiplicative factor $\alpha + 5/3$ has been dropped. Substituting $(3.10.32)$ into $(3.10.12)_2$, we see that this equation is identically satisfied for the arbitrary values of α and b (except for $\alpha = -1, -2, -3$ and $b = 1$). For example, taking $\alpha = 3$ and $b = 4$, it follows that $k = 1$.

Therefore, the differential equation

$$
\ddot{x} + 4\frac{\dot{x}}{t} + \lambda x^3 t = 0
\tag{3.10.33}
$$

has the conservation law of the form

$$
\begin{aligned}
I \;=\;& \frac{1}{4}\dot{x}^4 t^{10} + \frac{3}{2}x\dot{x}^3 t^9 + \left(\frac{9}{4}x^2 t^8 + \frac{1}{4}\lambda x^4 t^{11} \right)\dot{x}^2 \\
& + \frac{3}{4}\lambda x^5 \dot{x}t^{10} + \frac{1}{16}\lambda^2 x^8 t^{12} \;=\; const.
\end{aligned}
\tag{3.10.34}
$$

Chapter 4

A Field Method Suitable for Application in Conservative and Nonconservative Mechanics

4.1 Introduction

As demonstrated in the last several paragraphs, the Hamilton–Jacobi method can be advantageously used in many practical situations as an *exact* method for solving the canonical differential equations of motion. In addition, a variety of approximate methods can be built up, based upon this method, for solving nonlinear problems for which an exact, complete solution of the Hamilton-Jacobi nonlinear partial differential equation is not available. An exhaustive review of applications of the Hamilton–Jacobi method is presented in the monographs of Kevorkian and Kole [60] and Neyfeh [76].

As indicated previously, the method of Hamilton and Jacobi can be employed only with those dynamical systems described by the Lagrangian or Hamiltonian function, and purely nonconservative (non-Hamiltonian) systems remain outside of the areas treated by this method.

In this chapter we shall discuss a field method suitable for finding the motion of conservative or *purely nonconservative dynamical systems,* which is conceptually different than the method of Hamilton and Jacobi. The main characteristic of the field method presented here is that we are dealing with a single quasi-linear partial differential equation whose complete solution leads to the general solution of corresponding differential equations of motion. It is well known that finding a complete solution of a quasi-linear partial differential equation is much more manageable in comparison with the nonlinear partial differential equations of the Hamilton–Jacobi type.

4.2 The Field Concept and Its Partial Differential Equation

In this section we demonstrate a method for solving the system of ordinary differential equations of motion of a rheonomic dynamical system

$$\dot{x}_1 = X_1(t, x_1, ..., x_n),$$

$$\dots\dots\dots\dots$$

$$\dots\dots\dots\dots$$

$$\dot{x}_n = X_n(t, x_1, ..., x_n). \tag{4.2.1}$$

One of the central points in this study is the supposition that we are able to consider one of the variables entering in the system (4.2.1), say x_1, as a field function depending upon the time t, and the rest of the variables, $x_2, ..., x_n$, that is,

$$x_1 = U(t, x_2, ..., x_n). \tag{4.2.2}$$

This supposition was introduced in a series of papers (see [113]–[117]) and applied to various problems of Hamiltonian and non-Hamiltonian (nonconservative) dynamics and vibration theory (see also [122]).

By differentiating (4.2.2) with respect to time t and using the last $(n-1)$ equations (4.2.1), we write the first differential equation of (4.2.1) in the form

$$\frac{\partial U}{\partial t} + \frac{\partial U}{\partial x_i} X_i(t, U, x_2, ..., x_n) - X_1(t, U, x_2, ..., x_n) = 0, \quad i = 2, 3, ..., n. \tag{4.2.3}$$

We shall call this quasi-linear partial differential equation of the first order the *basic field equation*.

Instead of integrating the system of ordinary differential equations (4.2.1) directly, we shall demonstrate that we can find a general solution of (4.2.1) from a *complete solution* of the basic equation (4.2.3). The complete solution of (4.2.3) is of the form

$$x_1 = U(t, x_2, ..., x_n, C_1, ..., C_n), \tag{4.2.4}$$

which, in addition to the variables $t, x_2, ..., x_n$, contains n arbitrary constants $C_1, ..., C_n$ and satisfies identically the basic field equation (4.2.3) for all admissible values of parameters $t, x_2, ..., x_n, C_1, ..., C_n$.

We shall now demonstrate that if a complete solution (4.2.4) of the basic field equation (4.2.3) is available, then the solution of the system (4.2.1) follows immediately without any additional integration. However, there are several ways to obtain the solution of the dynamical system (4.2.1) by means of (4.2.4), and we will demonstrate these ways briefly.

4.2.1   The Bundle of Conservation Laws

As demonstrated in [103] every complete solution of a quasi-linear partial differential equation can be expressed in the form

$$\theta_0 \left(t, x_1, ..., x_n\right) + D_1 \theta_1 \left(t, x_1, ..., x_n\right) + \cdots$$
$$+ D_{n-1} \theta_{n-1} \left(t, x_1, ..., x_n\right) + D_n = 0, \tag{4.2.5}$$

where

$$
\begin{aligned}
\theta_0 \left(t, x_1, ..., x_n\right) &= K_0 = const., \\
\theta_1 \left(t, x_1, ..., x_n\right) &= K_1 = const., \\
&\dotsi \\
\theta_{n-1} \left(t, x_1, ..., x_n\right) &= K_{n-1} = const.
\end{aligned}
\tag{4.2.6}
$$

are a complete set of the conservation laws of the dynamical system (4.2.1), and n constants $D_1, ..., D_n$ are constant parameters depending upon the constants $C_1, ..., C_n$ figuring in the complete solution (4.2.4). Therefore, every complete solution of the form (4.2.4) can be precomposed in the form (4.2.5) from which we find the general solution of the dynamical system (4.2.1) in the form (4.2.6). It is clear from (4.2.5) and its equivalent (4.2.4) that these two expressions represent a *bundle of conservation laws* of the dynamical system (4.2.1), that is, a scalar equation that contains a complete set of conservation laws (first integrals) fastened together by means of the arbitrary constant parameters D_i, which stand to mark these conservation laws.

Note also that we can recover all n conservation laws of (4.2.1) by giving particular values to $n - 1$ constants C_i in (4.2.4) in their relevant domain and allowing one of them to be arbitrary. For example, we find n conservation laws of (4.2.1) from (4.2.4) in the following way:

$$
\begin{aligned}
x_1 &= U_1 \left(t, x_2, ..., x_n, C_1\right), & C_2 = C_3 = \cdots = C_n = 0, \\
x_1 &= U_2 \left(t, x_2, ..., x_n, C_2\right), & C_{12} = C_3 = \cdots = C_n = 0, \\
&\dotsi \\
x_1 &= U_n \left(t, x_2, ..., x_n, C_n\right), & C_1 = C_2 = \cdots = C_{n-1} = 0.
\end{aligned}
\tag{4.2.7}
$$

4.2.2   The Initial Value Problems

Let the variables of the dynamical system (4.2.1) be specified at the time instant $t = 0$ as

$$x_\alpha \left(0\right) = a_\alpha, \quad \alpha = 1, ..., n, \tag{4.2.8}$$

where a_α are given constants.

By substituting (4.2.8) into (4.2.4) and expressing one constant, say C_1 in terms of a_α and $C_i, \alpha = 1, ..., n, i = 2, ..., n$, we obtain

$$x_1 = u \left(t, x_2, ..., x_n, a_1, ..., a_n, C_2, ..., C_n\right). \tag{4.2.9}$$

We shall refer to this form of a complete solution of the basic field equation (4.2.3) as the *conditioned form solution*.

We now prove the following statement that is the central result of this chapter.

The initial value problem (4.2.1) *and* (4.2.8) *has the solution given by* (4.2.9) *and* $n - 1$ *algebraic equations*

$$\frac{\partial u}{\partial C_i} = 0, \quad i = 2, ..., n \tag{4.2.10}$$

under the condition that the following determinant is nowhere zero in the relevant domain of x_i *and* C_i, *that is,*

$$\det\left(\frac{\partial^2 u}{\partial C_i \partial x_j}\right) \neq 0, \quad i, j = 2, ..., n. \tag{4.2.11}$$

To prove this statement we assume that (4.2.10) holds. By differentiating (4.2.10) with respect to time, we find

$$\frac{\partial^2 u}{\partial C_i \partial t} + \frac{\partial^2 u}{\partial C_i \partial x_j}\dot{x}_j = 0, \quad i, j = 2, ..., n. \tag{4.2.12}$$

Substituting (4.2.9) into (4.2.3) we obtain an identity. Making the partial derivatives of this identity with respect to C_i one has

$$\frac{\partial^2 u}{\partial C_i \partial t} + \frac{\partial^2 u}{\partial C_i \partial x_j}X_j + \left(\frac{\partial u}{\partial x_j}\frac{\partial X_j}{\partial u} - \frac{\partial X_1}{\partial u}\right)\frac{\partial u}{\partial C_i} = 0, \quad i, j = 2, ..., n, \tag{4.2.13}$$

Since $\frac{\partial u}{\partial C_i} = 0$ by assumption, we find by combining the last two equations

$$\frac{\partial^2 u}{\partial C_i \partial x_j}(\dot{x}_j - X_j) = 0, \quad i, j = 2, ..., n, \tag{4.2.14}$$

and since (4.2.11) is also satisfied, we conclude that the last $n - 1$ equations (4.2.1) are satisfied for the arbitrary values of the constant parameters C_i. To show that the first differential equation (4.2.1) is satisfied, we calculate the total time derivative of (4.2.9) by using the last $n - 1$ differential equations (4.2.1): $\dot{x}_1 = \partial u/\partial t + X_i \partial u/\partial x_i$. Substituting this into (4.2.3) we conclude that the first equation of the dynamical system (4.2.1) is satisfied, which completes the proof of our statement.

At this point the following comments are of interest.

(a) From the foregoing discussion it is obvious that the theory presented here can be equally applied to both non-Hamiltonian (nonconservative) and Hamiltonian dynamical systems for which we can form corresponding Hamilton–Jacobi partial differential equation.

(b) However, in contrast to the Hamilton–Jacobi partial differential equation, which is always nonlinear, the basic field equation (4.2.3) is quasi-linear and its analysis for finding complete (or incomplete) solutions is, as a rule, considerably simpler in comparison with the Hamilton–Jacobi method.

(c) The main point of the method presented is in the fact that one of the dynamical variables (generalized coordinate, say x_1) is interpreted as the basic field. Thus, the corresponding field equation is more intimately connected with the dynamical problem than the Hamiltonian principal function $S = S(t, x_1, ..., x_n)$, which is not by itself a constituent of the dynamical problem

(d) For the case of linear rheonomic (time-dependent) dynamical systems, a complete solution of the corresponding basic field equation can be sought in the form

$$x_1 = f_1(t) + \sum_{i=2}^{n} f_i(t) x_i, \qquad (4.2.15)$$

where the unknown functions f_i are functions of time.

To illustrate the foregoing theory we turn to a simple example of the harmonic oscillator whose differential equations of motion are

$$\begin{aligned} \dot{x}_1 &= x_2, \quad \dot{x}_2 = -\omega^2 x_1, \\ x_1(0) &= a_1, \quad x_2(0) = a_2. \end{aligned} \qquad (4.2.16)$$

Taking the coordinate x_1 as the basic field

$$x_1 = U(t, x_2), \qquad (4.2.17)$$

the basic field equation reads

$$\frac{\partial U}{\partial t} - U \frac{\partial U}{\partial x_2} \omega^2 - x_2 = 0. \qquad (4.2.18)$$

Since the dynamical problem (4.2.16) is linear, we seek a complete solution in accordance with (4.2.15) in the form

$$x_1 = U(t, x_2) = f_1(t) + f_2(t) x_2. \qquad (4.2.19)$$

Entering with this into (4.2.18) and equating to zero terms with x_2 and free terms, we obtain the following system

$$\begin{aligned} \dot{f}_2 - \omega^2 f_2^2 - 1 &= 0, \\ \dot{f}_1 - \omega^2 f_1 f_2 &= 0. \end{aligned} \qquad (4.2.20)$$

Integrating, we find

$$f_1 = \frac{C_2}{\cos(\omega t + C_1)}, \quad f_2 = \frac{1}{\omega} \tan(\omega t + C_1), \qquad (4.2.21)$$

where C_1 and C_2 are constants of integration. A complete solution of (4.2.18) reads

$$x_1 = U(t, x_2, C_1, C_2) = \frac{x_2}{\omega} \tan(\omega t + C_1) + \frac{C_2}{\cos(\omega t + C_1)}. \qquad (4.2.22)$$

To obtain the *bundle of conservation laws* given by equation (4.2.5), we multiply (4.2.22) by $\omega \cos(\omega t + C_1)$, develop $\sin(\omega t + C_1)$ and $\cos(\omega t + C_1)$, and group corresponding terms, so that

$$x_1 \cos \omega t - \frac{x_2}{\omega} \sin \omega t + D_1 \left(x_1 \sin \omega t + \frac{x_2}{\omega} \cos \omega t \right) + D_2 = 0, \qquad (4.2.23)$$

where $D_1 = -\tan C_1, D_2 = -C_2 / \cos C_1$. According to (4.2.6), two conservation laws of dynamical system (4.2.16) are of the form

$$\begin{aligned}
x_1 \cos \omega t - \frac{x_2}{\omega} \cos \omega t &= K_0 = const., \\
x_1 \sin \omega t + \frac{x_2}{\omega} \cos \omega t &= K_1 = const.,
\end{aligned} \qquad (4.2.24)$$

where K_0 and K_1 are constants that can be determined from the given initial conditions (4.2.16).

As shown in (4.2.7), we can find two conservation laws from (4.2.22) by putting $C_1 \neq 0, C_2 = 0$ and $C_1 = 0, C_2 \neq 0$. Thus we have, respectively,

$$\begin{aligned}
x_1 &= \frac{x_2}{\omega} \tan(\omega t + C_1), \qquad \text{i.e., } C_1 = -\omega t + \arctan\left(\frac{\omega x_1}{x_2}\right), \\
x_1 &= \frac{x_2}{\omega} \tan \omega t + \frac{C_2}{\cos \omega t}, \qquad \text{i.e., } C_2 = x_1 \cos \omega t - \frac{x_2}{\omega} \sin \omega t. \quad (4.2.25)
\end{aligned}$$

Finally, to find the solution of the dynamical system (4.2.16) for the given initial conditions $x_1(0) = a_1, x_2(0) = a_2$, we enter with this into (4.2.22) and express C_2 in terms of C_1, that is, $C_2 = a_1 \cos C_1 - (a_2/\omega) \sin C_1$. The *conditioned form* of the complete solution becomes

$$x_1 = u(t, x_2, a_1, a_2, C_1) = \frac{x_2}{\omega} \tan(\omega t + C_1) + \frac{a_1 \cos C_1 - \left(\frac{a_2}{\omega}\right) \sin C_1}{\cos(\omega t + C_1)}. \qquad (4.2.26)$$

In accordance with (4.2.10) we easily obtain that the equation $\partial u / \partial C_1 = 0$ under the condition (4.2.11), that is, $\partial^2 u / \partial x_2 \partial C_1 \neq 0$, gives

$$\frac{x_2}{\omega} = -a_1 \sin \omega t + \frac{a_2}{\omega} \cos \omega t. \qquad (4.2.27)$$

Substituting this into (4.2.26), the parameter C_1 completely disappears and, after some elementary calculation, we obtain

$$x_1 = a_1 \cos \omega t + \frac{a_2}{\omega} \sin \omega t. \qquad (4.2.28)$$

4.3   A Non-Hamiltonian Rheonomic System

To demonstrate that the field method presented in this chapter can be applied to the dynamical systems that do not have Lagrangian or Hamiltonian structure, we consider a dynamical rheonomic system whose physical manifestations are described by the following system of three differential equations of the first order [56]:

$$
\begin{aligned}
\dot{x} &= y - z, \\
\dot{y} &= x + y + t, \\
\dot{z} &= x + z + t.
\end{aligned}
\tag{4.3.1}
$$

Taking as the field function $x = U(t, y, z)$, we arrive at the basic equation of the form

$$
\frac{\partial U}{\partial t} + \frac{\partial U}{\partial y}(U + y + t) + \frac{\partial U}{\partial z}(U + z + t) - y + z = 0.
\tag{4.3.2}
$$

Since the problem is linear, we seek a complete solution in the form suggested by (4.2.15), namely,

$$
x = U(t, y, z) = f_1(t)\, y + f_2(t)\, z + f_3(t).
\tag{4.3.3}
$$

Entering with this into (4.3.2) and grouping terms with y, z and free terms, we obtain the following system of equations:

$$
\begin{aligned}
\dot{f}_1 + f_1(f_1 + f_2 + 1) - 1 &= 0, \\
\dot{f}_2 + f_2(f_1 + f_2 + 1) - 1 &= 0, \\
\dot{f}_3 + f_3(f_1 + f_2) + (f_1 + f_2)\, t &= 0.
\end{aligned}
\tag{4.3.4}
$$

By adding the first two equations we find that

$$
\frac{d}{dt}(f_1 + f_2) + (f_1 + f_2)^2 + (f_1 + f_2) = 0.
\tag{4.3.5}
$$

Integrating, we have

$$
f_1 + f_2 = \frac{e^{-t}}{C_2 - e^{-t}},
\tag{4.3.6}
$$

where C_2 is a constant of integration. Entering with this into $(4.3.4)_1$ and integrating, we have

$$
f_1 = \frac{C_1 e^{-t} + C_2 - t e^{-t}}{C_2 - e^{-t}},
\tag{4.3.7}
$$

where C_1 is a constant. From (4.3.6), (4.3.7) it follows that

$$
f_2 = \frac{e^{-t} - C_1 e^{-t} + t e^{-t} - C_2}{C_2 - e^{-t}}.
\tag{4.3.8}
$$

Substituting (4.3.6) into $(4.3.4)_3$ and integrating, we have

$$
f_3 = \frac{e^{-t}(t + 1)}{C_2 - e^{-t}} + \frac{C_3}{C_2 - e^{-t}},
\tag{4.3.9}
$$

where C_3 is a constant.

Therefore, a complete solution of (4.3.2) is

$$x = U(t, y, z, C_1, C_2, C_3) = \left(\frac{C_1 e^{-t} - C_2 - te^{-t}}{C_2 - e^{-t}} \right) y$$
$$+ \left(\frac{e^{-t} - C_1 e^{-t} + te^{-t} - C_2}{C_2 - e^{-t}} \right) z + \frac{e^{-t}(t+1)}{C_2 - e^{-t}} + \frac{C_3}{C_2 - e^{-t}}.$$

(4.3.10)

Let us write this expression in the form of the *bundle of conservation laws* suggested by (4.2.5). Multiplying (4.3.10) by $c_2 - e^{-t}$ and grouping free terms and the terms multiplied by C_1 and C_2, we find

$$- [x - yt + z(t+1) + t + 1] e^{-t} + C_1 (y - z) e^{-t}$$
$$+ C_2 (-x + y - z) + C_3 = 0, ,$$

(4.3.11)

whence we obtain the following complete set of the conservation laws of dynamical system (4.3.1):

$$[x - yt + z(t+1) + t + 1] e^{-t} = K_1 = const.,$$
$$(y - z) e^{-t} = K_2 = const.,$$
$$-x + y - z = K_3 = const.,$$

(4.3.12)

where K_1, K_2, and K_3 are arbitrary constants that can be determined from the given initial conditions, which completes the calculation of the solution of (4.3.1).

4.4 Some Examples with Many-Degrees-of-Freedom Dynamical Systems

To illustrate the foregoing theory we consider in this section a couple of problems with two degrees of freedom.

4.4.1 Projectile Motion with Linear Air Resistance

Consider the motion of a heavy particle of the unit mass moving in a vertical plane with linear air friction depending on the velocity. If x and y denote the horizontal and vertical axes, respectively, the differential equations of motion are

$$\ddot{x} = -k\dot{x}; \quad \ddot{y} = -k\dot{y} - g,$$

(4.4.1)

where k and g are given constants. Let the initial conditions be

$$x(0) = 0, \quad y(0) = 0, \quad \dot{x}(0) = v_0 \cos \alpha, \quad \dot{y}(0) = v_0 \sin \alpha,$$

(4.4.2)

where v_0 is the initial velocity of the particle and α is the initial angle of inclination.

Introducing new variables

$$x = x_1, \quad y = x_2, \quad \dot{x} = x_3, \quad \dot{y} = x_4, \tag{4.4.3}$$

we arrive at the following system:

$$
\begin{aligned}
\dot{x}_1 &= x_3, \\
\dot{x}_2 &= x_4, \\
\dot{x}_3 &= -kx_3, \\
\dot{x}_4 &= -kx_4 - g.
\end{aligned}
\tag{4.4.4}
$$

Let us suppose that the variable x_1 can be interpreted as a field depending on t and the rest of the variables x_2, x_3, and x_4, namely,

$$x_1 = U(t, x_2, x_3, x_4). \tag{4.4.5}$$

Differentiating this with respect to time and using (4.4.4), we arrive at the field equation

$$\frac{\partial U}{\partial t} + \frac{\partial U}{\partial x_2} x_4 - kx_3 \frac{\partial U}{\partial x_3} - (kx_4 + g) \frac{\partial U}{\partial x_4} - x_3 = 0. \tag{4.4.6}$$

In accordance with remark (d) given in section 4.2, we seek a complete solution in the form

$$x_1 = U = f_2(t) x_2 + f_3(t) x_3 + f_4(t) x_4 + f_5(t). \tag{4.4.7}$$

Entering with this into (4.4.6) we find

$$x_2 \dot{f}_2 + x_3 \left(\dot{f}_3 - kf_3 - 1 \right) + x_4 \left(\dot{f}_4 + f_2 - kf_4 \right) + \dot{f}_5 - gf_4 = 0. \tag{4.4.8}$$

This expression will be satisfied identically for the arbitrary values of $x_i, i = 2, 3, 4$, if

$$
\begin{aligned}
\dot{f}_2 &= 0, \\
\dot{f}_3 - kf_3 - 1 &= 0, \\
\dot{f}_4 + f_2 - kf_4 &= 0, \\
\dot{f}_5 - gf_4 &= 0,
\end{aligned}
\tag{4.4.9}
$$

whence

$$
\begin{aligned}
f_2 &= C_2, \quad f_3 = -\frac{1}{k} + \frac{1}{k} C_3 e^{kt}, \quad f_4 = \frac{1}{k} C_2 + \frac{1}{k} C_4 e^{kt}, \\
f_5 &= \frac{1}{k} C_2 gt + \frac{g}{k^2} e^{kt} + C_5,
\end{aligned}
\tag{4.4.10}
$$

where $C_2, ..., C_5$ are constants of integration. Therefore, the complete solution of (4.4.6) is found to be (after grouping terms with respect to constants)

$$-x_1 + \frac{1}{k}x_3 + C_2\left(x_2 + \frac{1}{k}x_4 + \frac{1}{k}gt\right) + C_3\left(\frac{1}{k}x_3 e^{\kappa t}\right)$$
$$+C_4\left(\frac{1}{k}x_4 e^{kt} + \frac{1}{k^2}g e^{kt}\right) + C_5 = 0. \tag{4.4.11}$$

This form of a complete solution is equivalent to (4.2.5). On the basis of (4.2.5) and (4.2.6), we have the following complete system of conservation laws:

$$-x_1 + \frac{1}{k}x_3 = K_1 = const., \tag{4.4.12}$$

$$x_2 + \frac{1}{k}x_4 + \frac{1}{k}gt = K_2 = const.,$$

$$\frac{1}{k}x_3 e^{\kappa t} = K_3 = const.,$$

$$\frac{1}{k}x_4 e^{kt} + \frac{1}{k^2}g e^{kt} = K_4 = const.,$$

where $K_i, i = 1, ..., 4$, are constants that can be determined from the given initial conditions (4.4.2). After simple calculation we finally find the motion of the particle

$$x_1 = \frac{v_0\cos\alpha}{k}\left(1 - e^{-kt}\right), \quad x_2 = -\frac{gt}{k} + \left(\frac{g}{k^2} + \frac{v_0\sin\alpha}{k}\right)\left(1 - e^{-kt}\right),$$

$$x_3 = e^{-kt}v_0\cos\alpha, \quad x_4 = -\frac{g}{k} + \left(\frac{g}{k} + v_0\sin\alpha\right)e^{-kt}, \tag{4.4.13}$$

which completes the calculation of motion of the system.

To demonstrate how to find the conditioned form solution given by (4.2.10) we substitute the initial conditions (4.4.2) into (4.4.11) and express the constant C_5 in terms of C_2, C_3, and C_4:

$$C_5 = \left(\frac{1}{k} - \frac{C_3}{k}\right)v_0\cos\alpha - \left(\frac{C_2}{k} + \frac{C_4}{k}\right)v_0\sin\alpha - \frac{C_4 g}{k^2}. \tag{4.4.14}$$

Substituting this into (4.4.11) and separating terms with the constants C_2, C_3 and C_4, we obtain

$$x_1 = \frac{1}{k}v_0\cos\alpha - \frac{1}{k}x_3 + C_2\left(x_2 + \frac{1}{k}x_4 + \frac{1}{k}gt - \frac{1}{k}v_0\sin\alpha\right)$$
$$+C_3\left(\frac{1}{k}e^{kt}x_3 - \frac{1}{k}v_0\cos\alpha\right)$$
$$+C_4\left(\frac{1}{k}x_4 e^{kt} + \frac{g}{k^2}e^{kt} - \frac{v_0\sin\alpha}{k} - \frac{g}{k^2}\right). \tag{4.4.15}$$

This expression is equivalent to the conditioned form solution of the basic equation given by (4.2.10). It is quite clear that the equations (4.2.11), namely,

$$\frac{\partial x_1}{\partial C_2} = 0, \quad \frac{\partial x_1}{\partial C_2} = 0, \quad \frac{\partial x_1}{\partial C_3} = 0, \tag{4.4.16}$$

will generate the expressions for $x_2(t), x_3(t)$, and $x_4(t)$ given by (4.4.13), and entering with this into (4.4.15) we obtain $x_1(t)$. It is also clear that all constants C_2, C_3, C_4, and C_5 remain undetermined.

Remark. It is interesting to note that the system of original differential equations (4.4.1) can be derived from the Lagrangian of the form

$$L = \left[\frac{1}{2}\left(\dot{x}^2 + \dot{y}^2\right) - gy \right] e^{kt}. \tag{4.4.17}$$

Thus, the corresponding Hamiltonian reads

$$H = \frac{1}{2}p_x^2 e^{-kt} + \frac{1}{2}p_y^2 e^{-kt} + gy e^{kt}. \tag{4.4.18}$$

The canonical differential equations are

$$\dot{x} = p_x e^{-kt}, \quad \dot{y} = p_y e^{-kt}, \quad \dot{p}_x = 0, \quad \dot{p}_y = ge^{kt}. \tag{4.4.19}$$

Therefore, the general solution of this system can be found by means of the Hamilton–Jacobi method. The Hamilton–Jacobi partial differential equation is

$$\frac{\partial S}{\partial t} + \frac{1}{2}\left(\frac{\partial S}{\partial x}\right)^2 e^{-kt} + \frac{1}{2}\left(\frac{\partial S}{\partial y}\right)^2 e^{-kt} + gy e^{kt} = 0. \tag{4.4.20}$$

However, to find a complete solution of this nonlinear equation is much more difficult in comparison with the quasi-linear equation (4.4.6). Note also that despite the fact that the dynamical system is linear, we have no way of knowing in what form the complete solution of (4.4.20) should be sought. Let us seek a complete solution of (4.4.20) in the form

$$S(t, x, y) = \Psi(t) + \frac{1}{2}A(t)x^2 e^{kt} + f(t)y e^{kt} + \frac{1}{2}Ky^2 e^{kt}, \tag{4.4.21}$$

where $\Psi(t), A(t)$, and $f(t)$ are unknown functions of time and K is an adjustable constant. Substituting (4.4.21) into (4.4.20) and equating to zero free terms and terms of various powers of x and y, we arrive at the following system:

$$\begin{aligned}
\dot{\Psi} + \frac{1}{2}f^2 e^{kt} &= 0, \\
\dot{A} + kA + A^2 &= 0, \\
\dot{f} + kf + KF + g &= 0, \\
Kk + K &= 0. \tag{4.4.22}
\end{aligned}$$

Integrating, we find a complete solution to (4.4.20) in the form

$$S\left(t,x,y,C_1,C_2\right) = -\frac{1}{2}\frac{C_2^2}{k}e^{kt} + C_2 g\left(\frac{t}{k} - \frac{1}{k^2}\right)e^{kt}$$

$$+\frac{1}{2}\frac{kC_1}{1-C_1 e^{-kt}}x^2 + (C_2 - gt)e^{kt}y + \frac{1}{2}k^2 y^2 e^{kt}$$

$$+ \text{(terms not containing } x, y, C_1, C_2), \qquad (4.4.23)$$

where C_1 and C_2 are constants. It is easy to verify that the first group of equations of the Jacobi theorem

$$\frac{\partial S}{\partial C_1} = B_1, \quad \frac{\partial S}{\partial C_2} = B_2, \qquad (4.4.24)$$

where B_1, B_2 are constants, generates the equations of motion of the particle, having coordinates $x(t)$ and $y(t)$, namely,

$$x = \sqrt{\frac{2B_1}{k}}\left(1 - C_1 e^{-kt}\right), \quad y = \frac{C_2}{k} - g\left(\frac{t}{k} - \frac{1}{k^2}\right) + B_2 e^{-kt}. \qquad (4.4.25)$$

Finding the constants C_1, C_2, B_1, and B_2, these expressions become identical with $x_1(t)$ and $x_2(t)$ given by (4.4.13).

From the remark just stated, the following are demonstrated.

(a) The field method described in the previous section can be applied to nonconservative dynamical systems notwithstanding if they are describable by the Hamiltonian or not.

(b) For linear dynamical systems, application of the field method is, as a rule, more simple in comparison with the method of Hamilton and Jacobi.

4.4.2 Application of the Field Method to Nonholonomic Dynamical Systems

As another example of application of the field method, let us consider the one discussed in section 1.5, where a rectangular plate can move on the inclined plane (the so-called Chaplygin sled problem); see Figure 1.5.1.

Keeping the same notation, the differential equations of the problem are given by the system (1.5.25):

$$\begin{aligned}
\ddot{x}_c - \bar{g} &= -\lambda \tan\varphi, \quad (\bar{g} = g\sin\alpha = const.), \\
\ddot{y}_c &= \lambda, \\
\ddot{\varphi} &= 0, \qquad (4.4.26)
\end{aligned}$$

where λ is an undetermined Lagrangian multiplier and the nonholonomic constraint is prescribed in the form

$$\dot{y}_c - \dot{x}_c \tan\varphi = 0. \qquad (4.4.27)$$

Differentiating (4.4.27) with respect to time and substituting the result into the first two equations of (4.4.26), we find that

$$\lambda = \ddot{x}_c \dot{\varphi} + \bar{g} \sin\varphi \cos\varphi. \tag{4.4.28}$$

Integrating the equation $(4.4.26)_3$ and using the initial conditions (1.5.34), the first two equations of the system (4.4.26) become

$$\begin{aligned}
\ddot{x}_c &= -\dot{x}_c \omega \tan\omega t + \bar{g}\cos^2\omega t, \\
\ddot{y}_c &= \dot{x}_c \omega + \bar{g}\sin\omega t \cos\omega t.
\end{aligned} \tag{4.4.29}$$

To solve this system of simultaneous equations, we employ the field method. Let us denote by

$$x_c = x_1, \quad \dot{x}_c = x_2, \quad y_c = x_3, \quad \dot{y}_c = x_4. \tag{4.4.30}$$

Therefore, we have the following completely nonconservative system of differential equations of the first order:

$$\begin{aligned}
\dot{x}_1 &= x_2, \\
\dot{x}_2 &= -x_2 \omega \tan\omega t + \bar{g}\cos^2\omega t, \\
\dot{x}_3 &= x_4, \\
\dot{x}_4 &= x_2 \omega + \bar{g}\sin\omega t \cos\omega t.
\end{aligned} \tag{4.4.31}$$

Let us suppose that the coordinate x_1 can be represented as a field depending upon time and the rest of the coordinates x_2, x_3, and x_4:

$$x_1 = U(t, x_2, x_3, x_4). \tag{4.4.32}$$

The basic field equation becomes

$$\frac{\partial U}{\partial t} + \frac{\partial U}{\partial x_2}\left(-x_2\omega\tan\omega t + \bar{g}\cos^2\omega t\right) + \frac{\partial U}{\partial x_3}x_4$$
$$+\frac{\partial U}{\partial x_4}\left(x_2\omega + \bar{g}\sin\omega t\cos\omega t\right) - x_2 = 0. \tag{4.4.33}$$

Since the problem is linear we can on the basis of (4.2.16) suppose that the complete solution can be presented in the form

$$x_1 = f_1(t) + f_2(t)x_2 + f_3(t)x_3 + f_4(t)x_4. \tag{4.4.34}$$

Entering with this into (4.4.33) we have

$$\dot{f}_1 + f_2\bar{g}\cos^2\omega t + f_4\bar{g}\sin\omega t\cos\omega t + x_2\left(\dot{f}_2 - f_2\omega\tan\omega t + f_4\omega - 1\right)$$
$$+x_3\dot{f}_3 + x_4\left(\dot{f}_4 + f_3\right) = 0. \tag{4.4.35}$$

This expression will be satisfied for the arbitrary x_i, $= 2, 3, 4$ if

$$
\begin{aligned}
\dot{f}_1 + f_2 \bar{g} \cos^2 \omega t + f_4 \bar{g} \sin \omega t \cos \omega t &= 0, \\
\dot{f}_2 - f_2 \omega \tan \omega t + f_4 \omega - 1 &= 0, \\
\dot{f}_3 &= 0, \\
\dot{f}_4 + f_3 &= 0.
\end{aligned}
\tag{4.4.36}
$$

Integrating, we find

$$
\begin{aligned}
f_1 &= -\frac{\bar{g} C_3}{\omega^2} \left(\frac{\omega t}{2} + \frac{1}{2} \sin \omega t \cos \omega t \right) - \frac{\bar{g}}{2\omega^2} \sin^2 \omega t \\
&\quad - \frac{\bar{g} C_2}{\omega} \sin \omega t + C_1, \\
f_2 &= \frac{C_3}{\omega} (\omega t \tan \omega t + 1) - C_4 \tan \omega t + \frac{1}{\omega} \tan \omega t + \frac{C_2}{\cos \omega t}, \\
f_3 &= C_3, \\
f_4 &= C_4 - C_3 t,
\end{aligned}
\tag{4.4.37}
$$

where $C_1, ..., C_4$ are arbitrary constants.

Therefore, the complete solution of the field equation (4.4.33) is found to be

$$
\begin{aligned}
x_1 &= -\frac{\bar{g}}{2\omega^2} + \frac{x_2}{\omega} \tan \omega t + C_2 \left(\frac{x_2}{\cos \omega t} - \frac{\bar{g}}{\omega} \sin \omega t \right) \\
&\quad + C_3 \left(-\frac{\bar{g} t}{2\omega} - \frac{\bar{g}}{2\omega^2} \sin \omega t \cos \omega t - x_4 t + x_2 t \tan \omega t + \frac{x_2}{\omega} + x_3 \right) \\
&\quad + C_4 (x_4 - x_2 \tan \omega t) + C_1.
\end{aligned}
\tag{4.4.38}
$$

Since, according to (1.5.34) and (4.4.30), the initial conditions are

$$
x_1(0) = x_2(0) = x_3(0) = x_4(0) = 0,
\tag{4.4.39}
$$

we find, entering with this into (4.4.38), that $C_1 = 0$. Hence the expression (4.4.38) is at the same time the conditioned solution.

Applying the rule (4.2.11) we find that the equations

$$
\frac{\partial x_1}{\partial C_2} = 0, \quad \frac{\partial x_1}{\partial C_3} = 0, \quad \frac{\partial x_1}{\partial C_4} = 0,
\tag{4.4.40}
$$

will generate the following expressions

$$
\begin{aligned}
x_2 - \frac{\bar{g}}{\omega} \sin \omega t \cos \omega t &= 0, \\
x_3 + \frac{x_2}{\omega} + x_2 t \tan \omega t - x_4 t - \frac{\bar{g}}{2\omega^2} - \frac{\bar{g} t}{2\omega} &= 0, \\
x_4 - x_2 \tan \omega t &= 0,
\end{aligned}
\tag{4.4.41}
$$

and from the rest of (4.4.38) we obtain

$$
x_1 = -\frac{\bar{g}}{2\omega^2} \sin^2 \omega t + \frac{x_2}{\omega} \tan \omega t,
\tag{4.4.42}
$$

which comprise the solution of the problem. By solving these equations with respect to $x_i, i = 1, ..., 4$ we finally have

$$x_1 = \frac{\bar{g}}{2\omega^2} \sin^2 \omega t, \qquad\qquad x_2 = \frac{\bar{g}}{\omega} \sin \omega t \cos \omega t,$$

$$x_3 = \frac{\bar{g}}{2\omega^2} \left(\omega t - \frac{1}{2} \sin 2\omega t \right), \quad x_4 = \frac{\bar{g}}{\omega} \sin^2 \omega t. \qquad (4.4.43)$$

It is of interest to note that Mei Fenxiang (see [42]–[47], [48]), in his numerous papers devoted to the study of modern nonholonomic dynamical problems, adopted the field method presented here as a basic tool in his analysis and demonstrated that this method has important advantages in comparison to other methods of integration.

4.5   Nonlinear Analysis

In this section we demonstrate that the field method presented in this chapter can be advantageously applied to the study of motion and conservation laws of nonlinear dynamical problems. However, in contrast to linear time-dependent problems for which we can suggest a rather general procedure, the complete solutions of nonlinear problems with one degree of freedom is frequently difficult and strongly individual. To explain these facts, we turn first to some nonlinear examples in which a complete solution of the corresponding basic field equations can be obtained.

(a) Let us consider the nonlinear problem [125]

$$\dot{x}_1 = x_2, \quad \dot{x}_2 = \frac{a\,(tx_2 - x_1)^2}{t^3}, \qquad (4.5.1)$$

where a is a constant. Taking as the basic field the coordinate x_2, namely,

$$x_2 = F\,(t, x_1), \qquad (4.5.2)$$

the basic field equation becomes

$$\frac{\partial F}{\partial t} + F \frac{\partial F}{\partial x_1} - \frac{a\,(tF - x_1)^2}{t^3} = 0. \qquad (4.5.3)$$

It is easily seen that $F = x_1/t$ is a particular solution of (4.5.3). Therefore, it seems natural to try with a complete solution in the form

$$F\,(t, x_1) = X + f\,(X)\,S\,(t), \quad X = \frac{x_1}{t}, \qquad (4.5.4)$$

where $f\,(X)$ and $S\,(t)$ are unknown functions. Substituting (4.5.4) into the basic equation (4.5.3), we obtain two separated groups of terms, each of which should be equated to an arbitrary constant M:

$$\frac{\dot{S}\,(t)}{S^2\,(t)} + \frac{1}{S\,(t)} = -\frac{df\,(X)}{dX} + af\,(X) = M = const. \qquad (4.5.5)$$

Integrating separately, we find

$$S(t) = \frac{1}{Gt + M}, \quad f(X) = \frac{M}{a} + De^{aX}, \tag{4.5.6}$$

where G and D are new arbitrary constants. Thus the complete solution of (4.5.3) becomes

$$x_2 = X + \frac{\frac{M}{a} + De^{aX}}{Gt + M} = X + \frac{C_2 + e^{aX}}{C_1 t + C_2}, \tag{4.5.7}$$

where we have introduced two independent constants $C_1 = G/D$ and $C_2 = M/(aD)$. Writing (4.5.7) in the form of bundle of the conservation laws

$$-\frac{e^{aX}}{(x_2 - X)t} + C_2 \frac{(x_2 - X - 1)}{x_2 - X} + C_1 = 0, \tag{4.5.8}$$

we find that in accordance with (4.2.5) and (4.2.6), the expressions

$$-\frac{e^{aX}}{(x_2 - X)t} = K_1 = const.,$$

$$\frac{(x_2 - X - 1)}{x_2 - X} = K_2 = const. \tag{4.5.9}$$

represent two independent conservation laws of the dynamical system (4.5.1).

(b) Let us consider the nonlinear initial value problem

$$\ddot{x} - \frac{\dot{x}^2}{x} = 0, \quad x(0) = a, \quad \dot{x}(0) = b. \tag{4.5.10}$$

The basic differential equation is

$$\frac{\partial U}{\partial t} + U \frac{\partial U}{\partial x} - \frac{U^2}{x} = 0, \tag{4.5.11}$$

where

$$p = \dot{x} = U(t, x). \tag{4.5.12}$$

Let us suppose that the variables x and t are separated, that is,

$$U = F(x) f(t). \tag{4.5.13}$$

Therefore, one finds

$$\frac{\dot{f}}{f^2} = -F' + \frac{F}{x} = -C = const. \tag{4.5.14}$$

Integrating, we find $f = 1/\left(Ct - C_1^*\right), F = Cx \ln x + C_2^* x$, where C_1^* and C_2^* are constants of integration. Hence

$$U = \frac{Cx \ln x + C_2^*}{Ct - C_1^*} = \frac{x \ln x + C_2 x}{t - C_1}, \tag{4.5.15}$$

where $C_1 = C_1^*/C, C_2 = C_2^*/C$. Applying the initial conditions and expressing C_2 in terms of C_1, a, and b, we find

$$U = \frac{x \left[\ln\left(x/a\right) - bC_1/a\right]}{t - C_1}, \tag{4.5.16}$$

which is a conditioned form solution. The equation $\partial U/\partial C_1 = 0$ gives

$$\ln\left(\frac{x}{a}\right) - b\frac{t}{a} = 0. \tag{4.5.17}$$

Substituting this expression into (4.5.16), the parameter C_1 disappears and one has the momentum

$$p = b\frac{x}{a} = be^{bt/a}. \tag{4.5.18}$$

(c) Incomplete solutions of basic equation. It is important to note that *incomplete solutions* of the basic field equation for the case of dynamical systems with one degree of freedom represent the conservation laws of that dynamical system and can be important in dynamical analysis of linear and nonlinear dynamical systems [125]. The rather general representation of an incomplete solution of the basic field equation can be anticipated from the form of the Hamiltonian function. Here we briefly describe this representation. Let us suppose that the dynamical system has the Hamiltonian of the form

$$H = \frac{1}{2}p^2\theta\left(t\right) + \sum_{i=1}^{N} \Pi_i\left(x\right) \lambda_i\left(t\right), \tag{4.5.19}$$

where $\theta\left(t\right), \lambda_i\left(t\right), \Pi_i\left(x\right)$, and N are specified. Naturally, the Hamiltonian (4.5.19) is not a constant of motion since the problem is not conservative. Calculating p from (4.5.19), we have

$$p = \left[2\frac{H}{\theta} - \frac{1}{\theta}\sum_{i=1}^{N} \Pi_i\left(x\right) \lambda_i\left(t\right)\right]^{1/2}, \quad 2H/\theta \neq const. \tag{4.5.20}$$

We shall demonstrate that a rather broad class of dynamical systems which possess Hamiltonians of the form (4.5.19) also possess the incomplete solutions of the basic field equation in the form

$$p = \Phi\left(t, x\right) = A\left(t\right)x + C\left(t\right)\left[I - \sum_{i=1}^{N} \Pi_i\left(x\right) \lambda_i\left(t\right)\right]^{1/2}, \tag{4.5.21}$$

where I is a constant of motion and $A(t), B(t)$, and $C(t)$ are unknown functions of time that are to be determined in the course of analysis. Since (4.5.21) contains only one constant I, it is clear that this expression represents an incomplete solution of the corresponding field equation.

Let us apply the foregoing consideration to the case of a dynamical system whose differential equation is given by

$$\ddot{x} + 2k(t)\dot{x} + q(t)x = \lambda f(t)x^n, \quad \lambda = const. \tag{4.5.22}$$

Note that Ranganathan [90] considered a class of dynamical problems which, when reduced to a single degree of freedom, are of the type (4.5.22).

Taking as the basic field $\dot{x} = \Phi(t, x)$, the basic field equation becomes

$$\frac{\partial \Phi}{\partial t} + \Phi \frac{\partial \Phi}{\partial x} + 2k\Phi + qx = \lambda f(t)x^n. \tag{4.5.23}$$

As suggested by (4.5.21), we suppose that an incomplete solution (a conservation law) of this equation can be selected in the form

$$\dot{x} = \Phi(t, x) = A(t)x + C(t)\left[I - B(t)x^2 + \frac{\lambda D(t)x^{n+1}}{n+1}\right]^{1/2}, \tag{4.5.24}$$

where $A(t)$, $B(t)$, $C(t)$, and $D(t)$ are adjustable functions and I is a constant. Substituting (4.5.24) into (4.5.23), we obtain

$$x\left(\dot{A} + A^2 + 2kA + q - BC^2\right) + \frac{1}{2}\lambda x^n\left(C^2 D - 2f\right)$$

$$+ \frac{1}{\Lambda}\left\{2IC\left(\frac{\dot{C}}{C} + A + 2k\right) - BCx^2\left(2\frac{\dot{C}}{C} + \frac{\dot{B}}{B} + 4A + 4k\right)\right.$$

$$\left. + \lambda x^{n+1}CD\left[2\frac{\dot{C}}{C} + \frac{\dot{D}}{D} + (n+3)A + 4k\right]\right\} = 0, \tag{4.5.25}$$

where

$$\Lambda = 2\left[I - Bx^2 + \frac{\lambda D x^{n+1}}{n+1}\right]^{1/2}. \tag{4.5.26}$$

Equating to zero terms with various powers of x, we arrive at the following system of differential equations

$$\dot{A} + A^2 + 2kA + q - BC^2 = 0,$$

$$C^2 D - 2f = 0,$$

$$\frac{\dot{C}}{C} + A + 2k = 0,$$

$$2\frac{\dot{C}}{C} + \frac{\dot{B}}{B} + 4A + 4k = 0,$$

$$2\frac{\dot{C}}{C} + \frac{\dot{D}}{D} + (n+3)A + 4k = 0. \tag{4.5.27}$$

To analyze this system, we introduce a new function $w(t)$ by the relation

$$A(t) = \frac{\dot{w}}{w} - k(t).$$ (4.5.28)

Entering this into $(4.5.27)_3$ and integrating, we find

$$C(t) = \frac{\bar{C}}{w} e^{\left[-\int^t k(u)du\right]}, \quad \bar{C} = const.$$ (4.5.29)

Combining $(4.5.27)_3$ and $(4.5.28)$ with $(4.5.27)_5$ we obtain $\dot{D}/D + (n+1)(\dot{w}/w - k) = 0$. Integrating, we find

$$D(t) = \frac{\bar{D}}{w^{n+1}} e^{\left[(n+1)\int^t k(u)du\right]}, \quad \bar{D} = const.$$ (4.5.30)

Since the auxiliary equations (4.5.27) are not restricted by any initial or boundary conditions, we are free to introduce any particular values for the integration constants. Thus, by taking $\bar{C} = 1$ and $\bar{D} = 2$, the relation $(4.5.27)_2$ gives

$$f(t) = \frac{1}{w^{n+3}} e^{\left[(n+1)\int^t k(u)du\right]}.$$ (4.5.31)

Similarly, combining $(4.5.27)_3$ and $(4.5.27)_5$ with $(4.5.27)_4$, one has after integration

$$B(t) = \frac{1}{w^2} e^{\left[2\int^t k(u)du\right]},$$ (4.5.32)

where we selected the integration constant $\bar{B} = 1$.

Substituting (4.5.28), (4.5.29), and (4.5.32) into $(4.5.27)_1$ we have the auxiliary equation in the form

$$\ddot{w} + \left(q(t) - k^2(t) - \dot{k}(t)\right)w - \frac{1}{w^3} = 0.$$ (4.5.33)

By substituting $A(t), B(t), C(t), D(t)$, and $f(t)$ given here in (4.5.24), we can formulate the following theorem.

The dynamical system

$$\ddot{x} + 2k(t)\dot{x} + q(t)x = \lambda f(t) x^n$$ (4.5.34)

has a quadratic conservation law in the form

$$\dot{x} = \left(\frac{\dot{w}}{w} - k\right)x + \frac{1}{w} e^{\left[-\int^t k(u)du\right]} \left\{ I - \left(\frac{x}{w}\right)^2 e^{\left[2\int^t k(u)du\right]} \right.$$
$$\left. + \frac{2\lambda}{n+1} \left(\frac{x}{w}\right)^{n+1} e^{\left[(n+1)\int^t k(u)du\right]} \right\}^{1/2},$$ (4.5.35)

if $f(t)$ is given by (4.5.31) and $w(t)$ is any solution of the auxiliary equation (4.5.33). This conservation law can be written in a more traditional form as

$$I = \left\{ [\dot{x}w - x(\dot{w} - kw)]^2 + \left(\frac{x}{w}\right)^2 \right\} e^{[2\int^t k(u)du]}$$
$$-\frac{2\lambda}{n+1}\left(\frac{x}{w}\right)^{n+1} e^{[(n+1)\int^t k(u)du]} = const. \qquad (4.5.36)$$

As an illustration, several examples are in order.

Example 4.5.1. Let us consider the dynamical system

$$\ddot{x} + \frac{\dot{x}}{2t} + \frac{a}{2t}x = \lambda x^n f(t), \qquad (4.5.37)$$

where a and λ are given constants and $f(t)$ is a function that will be specified later. For this case, we have $q = a/2t, k = 1/4t$, and the auxiliary equation (4.5.33) becomes

$$\ddot{w} + \left(\frac{a}{2t} + \frac{3}{16t^2}\right)w - \frac{1}{w^3} = 0. \qquad (4.5.38)$$

A particular solution of this equation is found to be $w = (2t/a)^{1/4}$. Therefore, according to (4.5.31), $f(t)$ must be of the form

$$f(t) = \left(\frac{a}{2}\right)^{(n+3)/4}\frac{1}{t}. \qquad (4.5.39)$$

The quadratic conservation law (4.5.36) of the dynamical system (4.5.37) for $f(t)$ given by (4.5.39) is

$$I = \dot{x}^2 t + \left(\frac{a}{2}\right)x^2 - 2\lambda\left(\frac{a}{2}\right)^{(n+3)/4}\frac{x^{n+1}}{n+1}. \qquad (4.5.40)$$

Example 4.5.2. Let a dynamical system be described by the differential equation

$$\ddot{x} + \frac{b^2}{(a^2 + t^2)^2}x = \lambda x^n f(t). \qquad (4.5.41)$$

Note that the linear part of this equation, that is, $\lambda = 0$, describes the equilibrium configuration of an elastic rod with variable cross section (see [56, p. 447]).

Since $k(t) = 0$ and $q(t) = b^2/(a^2 + t^2)^2$, the auxiliary equation is

$$\ddot{w} + \left[\frac{b^2}{(a^2 + t^2)^2}\right]w - \frac{1}{w^3} = 0, \qquad (4.5.42)$$

whose particular solution is $w = \left(a^2 + t^2\right)^{1/2} / \left(a^2 + b^2\right)^{1/4}$. Equation (4.5.31) gives

$$f(t) = \frac{\left(a^2 + b^2\right)^{(n+3)/4}}{\left(a^2 + t^2\right)^{(n+3)/2}}. \tag{4.5.43}$$

Therefore, according to (4.5.36) we find that the dynamical system (4.5.41) with $f(t)$ given by (4.5.43) has a quadratic conservation law of the form

$$\begin{aligned}
I &= \left[\dot{x}\left(a^2 + t^2\right)^{1/2} - xt\left(a^2 + b^2\right)^{-1/2}\right]^2 \\
&\quad + \frac{x^2\left(a^2 + b^2\right)}{2\left(a^2 + t^2\right)} - \lambda\frac{x^{n+1}}{n+1}\frac{\left(a^2 + b^2\right)^{(n+3)/4}}{\left(a^2 + t^2\right)^{(n+1)/2}}.
\end{aligned} \tag{4.5.44}$$

Note that the case $n = -3$ and $a = 0$ were considered in [90]. For this case, the differential equation (4.5.41) becomes

$$\ddot{x} + \frac{b^2}{t^4}x = \frac{\lambda}{x^3}, \tag{4.5.45}$$

and has the conservation law

$$I = (t\dot{x} - x)^2 + \frac{b^2 x^2}{t^2} + \lambda\frac{t^2}{x^2} = const., \tag{4.5.46}$$

which is identical to the result reported by Ranganathan [90].

   Example 4.5.3. Let us consider a dynamical system whose differential equation is of the form

$$\ddot{x} + \dot{x}\tan t + x\cos^2 t = \lambda x^n \cos^2 t. \tag{4.5.47}$$

The linear case $\lambda = 0$ was considered in [56]. For this case, we have $k(t) = (1/2)\tan t$, $q(t) = \cos^2 t$, and the auxiliary equation is

$$\ddot{w} + \left(\cos^2 t - \frac{1}{4}\tan^2 t - \frac{1}{2\cos^2 t}\right)w - \frac{1}{w^3} = 0. \tag{4.5.48}$$

A particular solution of this equation is $w = (\cos t)^{-1/2}$. Since according to (4.5.31), $f(t) = \cos^2 t$, we have the conservation law of (4.5.47) in the form

$$I = \frac{\dot{x}^2}{\cos^2 t} + x^2 - 2\lambda\frac{x^{n+1}}{n+1}. \tag{4.5.49}$$

4.6 Conservation Laws and Reduction to Quadratures of the Generalized Time-Dependent Duffing Equation

In this section we attempt to answer the following question: Under what conditions will the time-dependent differential equation

$$\ddot{x} + Q_1 f_1(t)x + Q_2 f_2(t)x^2 + Q_3 f_3(t)x^3 + Q_4 f_4(t) = 0 \tag{4.6.1}$$

possess, the quadratic conservation laws and, simultaneously the solution to (4.6.1) be obtained by quadratures? Here $Q_1, ..., Q_4$ are constants and $f_i(t)$ are supposed to be power-type functions of time t. Note that for $Q_2 = 0$, the equation (4.6.1) falls into the class of Duffing equation widely used in physics and engineering, while for $Q_1 = Q_3 = Q_4 = 0$ and $Q_1 = Q_2 = Q_4 = 0$ the corresponding equations can be classified as the Emden–Fowler equations. Thus, we will conditionally call (4.6.1) the generalized Duffing equation. We shall demonstrate that for a large class of dynamical systems for which the functions $f_i(t)$ are of the power form, that is, $f_i(t) = t^{k_i}, i = 1, ..., 4$, the differential equation (4.6.1) has conservation laws with respect to velocity $\dot{x}$ which by application of the Hamilton–Jacobi method can be reduced to quadratures.

By supposing that the linear momentum (i.e., velocity) can be represented as a field depending on time and position x, namely, $\dot{x} = \phi(t, x)$, we write the basic field equation in the form

$$\frac{\partial \phi}{\partial t} + \phi \frac{\partial \phi}{\partial x} + Q_1 f_1(t) x + Q_2 f_2(t) x^2 + Q_3 f_3(t) x^3 + Q_4 f_4(t) = 0. \quad (4.6.2)$$

As demonstrated in the previous section regarding finding a quadratic conservation law of (4.6.1), we can find an incomplete solution of equation (4.6.2).

Thus, we seek an incomplete solution in the form

$$\begin{aligned}
\dot{x} = \phi(t, x) &= P(t) + A(t) x \\
&+ C(t) \left[I + K(t) + M(t) x + B(t) x^2 \right. \\
&\left. + D(t) x^3 + E(t) x^4 \right]^{1/2},
\end{aligned} \quad (4.6.3)$$

where $P(t), A(t), ..., E(t)$ are unknown functions of time and I is an arbitrary constant. Substituting (4.6.3) into (4.6.2), we have

$$\begin{aligned}
&\dot{P} + AP + \frac{1}{2} C^2 M + Q_4 f_4(t) + x \left[\dot{A} + A^2 + BC^2 + Q_1 f_1(t) \right] \\
&+ x^2 \left[\frac{3}{2} C^2 D + q_2 f_2(t) \right] + x^3 \left[2EC^2 + Q_3 f_3(t) \right] \\
&+ \frac{1}{2\Lambda} \left[2I \left(\dot{C} + AC \right) + \left(C\dot{K} + 2\dot{C}K + 2ACK + CMP \right) \right. \\
&\qquad + x \left(C\dot{M} + 2\dot{C}M + 3ACM + 2BCP \right) \\
&\qquad + x^2 \left(\dot{B}C + 2\dot{C}B + 4ABC + 3CDP \right) \\
&\qquad + x^3 \left(C\dot{D} + 2\dot{C}D + 5ACD + 4CEP \right) \\
&\qquad \left. + x^4 \left(C\dot{E} + 2\dot{C}E + 6ACE \right) \right] = 0, \quad (4.6.4)
\end{aligned}$$

where

$$\Lambda = \left(I + K + Mx + Bx^2 + Dx^3 + Ex^4 \right)^{1/2}. \quad (4.6.5)$$

Equating coefficients of equal powers of x, we arrive at the following system of auxiliary differential equations of the first order:

$$\begin{aligned}
\dot{P} + AP + \frac{1}{2}C^2M + Q_4 f_4(t) &= 0, \\
\dot{A} + A^2 + BC^2 + Q_1 f_1(t) &= 0, \\
\frac{3}{2}C^2D + Q_2 f_2(t) &= 0, \\
2EC^2 + Q_3 f_3(t) &= 0, \\
\dot{C} + AC &= 0, \\
C\dot{K} + 2\dot{C}K + 2ACK + CMP &= 0, \\
C\dot{M} + 2\dot{C}M + 3ACM + 2BCP &= 0, \\
\dot{B}C + 2\dot{C}B + 4ABC + 3CDP &= 0, \\
C\dot{D} + 2\dot{C}D + 5ACD + 4CEP &= 0, \\
C\dot{E} + 2\dot{C}E + 6ACE &= 0.
\end{aligned} \qquad (4.6.6)$$

Therefore, we have obtained eight differential equations with eight unknown functions $P, A, C, K, M, B, D,$ and E, and two algebraic equations, $(4.6.6)_{3,4}$. It is clear that every solution of this system will generate a conservation law of the dynamical system (4.6.1). However, it should be noted that there exists some difference between the constant I introduced in (4.6.3) and the integration constants obtained in the process of integration of the auxiliary system (4.6.6). The constant I can be specified from the given initial conditions of the dynamical equation (4.6.1), while the integration constants stemming from the integration of the auxiliary system must be selected arbitrarily since the system (4.6.6) is not confined by any initial or boundary conditions. It is also clear that the auxiliary system cannot be integrated generally for the arbitrary functions $f_i(t)$ and coefficients $Q_i, i = 1, ..., 4$.

We shall seek a solution of the auxiliary system in the form of the power functions of time. Namely, we suppose that the form of the unknown functions introduced in (4.6.3) are

$$\begin{aligned}
A(t) &= A_0 t^{-a}, & B(t) &= B_0 t^{-b}, & C(t) &= C_0 t^{-c}, \\
K(t) &= K_0 t^{-k}, & M(t) &= M_0 t^{-m}, & P(t) &= P_0 t^{-p}, \\
D(t) &= D_0 t^{-d}, & E(t) &= E_0 t^{-e},
\end{aligned} \qquad (4.6.7)$$

where $M_0, m, ..., E_0$ and e are constants. Substituting $A(t)$ and $C(t)$ into $(4.6.6)_5$, we find

$$-ct^{-1} + A_0 t^{-a} = 0. \qquad (4.6.8)$$

This equation will be identically satisfied for $c = A_0$ and $a = 1$. Thus

$$A = A_0 t^{-1}, \quad C = C_0 t^{-A_0}. \qquad (4.6.9)$$

Repeating the same procedure, we find from equation $(4.6.6)_6$ that $e = 4A_0$ and therefore

$$E\left(t\right) = E_0 t^{-4A_0}. \tag{4.6.10}$$

From equation $(4.6.6)_6$, we find that

$$K_0 k = P_0 M_0, \qquad p - k + m = 1, \tag{4.6.11}$$

so that by using $(4.6.6)_7$ one has

$$M_0\left(A_0 - m\right) + 2B_0 P_0 = 0, \quad b - m + p = 1. \tag{4.6.12}$$

From $(4.6.6)_8$ we have

$$B_0\left(2A_0 - b\right) + 3D_0 P_0 = 0, \quad -b + d + p = 1, \tag{4.6.13}$$

so that $(4.6.6)_9$ implies

$$D_0\left(3A_0 - d\right) + 4E_0 P_0 = 0, \quad e - d + p = 1. \tag{4.6.14}$$

Also, from $(4.6.6)_1$ we find that

$$t^{-p-1}\left(-pP_0 + A_0 P_0\right) + \frac{1}{2}C_0^2 M_0 t^{-2A_0-m} + Q_4 f_4\left(t\right) = 0. \tag{4.6.15}$$

If the function $f_4\left(t\right)$ is selected as $f_4\left(t\right) = t^{-p-1} = t^{-2A_0-m}$, equation $(4.6.15)$ is reduced to

$$P_0\left(A_0 - p\right) + \frac{1}{2}C_0^2 M_0 + Q_4 = 0, \quad m - p = 1 - 2A_0. \tag{4.6.16}$$

By using $(4.6.11)_2$–$(4.6.14)_2$ and $(4.6.16)_2$, we find

$$\begin{aligned}
p &= 2 - 3A_0, \quad m = 3 - 5A_0, \quad d = 1 + A_0, \quad b = 2 - 2A_0, \\
k &= 4 - 8A_0, \quad e = 4A_0.
\end{aligned} \tag{4.6.17}$$

Therefore,

$$f_4\left(t\right) = t^{3A_0-3}. \tag{4.6.18}$$

From $(4.6.12)_1$–$(4.6.14)_1$ we have

$$\begin{aligned}
D_0 &= -\frac{4E_0 P_0}{2A_0 - 1}, \quad B_0 = \frac{6E_0 P_0^2}{\left(2A_0 - 1\right)^2}, \\
M_0 &= -\frac{4E_0 P_0^3}{\left(2A_0 - 1\right)^3}, \quad K_0 = \frac{E_0 P_0^4}{\left(2A_0 - 1\right)^4}.
\end{aligned} \tag{4.6.19}$$

Combining $(4.6.17)$ and $(4.6.19)$, equation $(4.6.16)_1$ becomes

$$2E_0 C_0^2 P_0^3 - 2\left(2A_0 - 1\right)^4 P_0 - Q_4\left(2A_0 - 1\right)^3 = 0. \tag{4.6.20}$$

Similarly, $(4.6.6)_2$ becomes

$$\left[-A_0 + A_0^2 + 6\frac{E_0 C_0^2 P_0^2}{(2A_0 - 1)^2}\right] t^{-2} + Q_1 f_1(t) = 0. \tag{4.6.21}$$

Selecting

$$f_1(t) = t^{-2}, \tag{4.6.22}$$

we have

$$A_0^2 - A_0 + 6\frac{E_0 C_0^2 P_0^2}{(2A_0 - 1)^2} + Q_1 = 0. \tag{4.6.23}$$

Similarly, by taking

$$f_2(t) = t^{-1-3A_0}, \quad f_3(t) = t^{-6A_0}, \tag{4.6.24}$$

equations $(4.6.6)_2$ and $(4.6.6)_3$ are reduced to

$$Q_2 = 6\frac{E_0 C_0^2 P_0}{2A_0 - 1}, \quad E_0 C_0^2 = -\frac{Q_3}{2}. \tag{4.6.25}$$

Combining these two equations we get

$$P_0 = \frac{1}{3}(2A_0 - 1)\frac{Q_2}{Q_3}. \tag{4.6.26}$$

Equations (4.6.20) and (4.6.23) now become

$$\frac{1}{27}\frac{Q_2^3}{Q_3^2} + \frac{2}{3}(2A_0 - 1)^2\frac{Q_2}{Q_3} - Q_4 = 0 \tag{4.6.27}$$

and

$$A_0(A_0 - 1) = -Q_1 + \frac{Q_2^2}{3Q_3}. \tag{4.6.28}$$

Finding A_0 from (4.6.28) and substituting it in (4.6.27), it is seen that the coefficients $Q_i, i = 1, ..., 4$, are not independent but must satisfy the following algebraic equation

$$\frac{25}{27}\frac{Q_2^3}{Q_3^2} - \frac{8}{3}\frac{Q_1 Q_2}{Q_3} + \frac{2}{3}\frac{Q_2}{Q_3} - Q_4 = 0. \tag{4.6.29}$$

Without loss of generality we can take $C_0 = 1$, and from $(4.6.25)_2$ it follows that

$$E_0 = -\frac{Q_3}{2}. \tag{4.6.30}$$

Therefore, finding A_0 and P_0 from (4.6.28) and (4.6.26), the group of parameters given by (4.6.19) can be expressed in terms of the coefficients Q_i. Recalling (4.6.9) and (4.6.17) we find that the functions (4.6.7) can be expressed as

$$
\begin{aligned}
A(t) &= A_0 t^{-1}, \quad P(t) = -\frac{1}{3}(2A_0 - 1)\frac{Q_2}{Q_3}t^{3A_0-2}, \\
C(t) &= t^{-A_0}, \quad K(t) = -\frac{1}{162}\frac{Q_2^4}{Q_3^3}t^{8A_0-4}, \\
M(t) &= -\frac{2}{27}\frac{Q_2^3}{Q_3^2}t^{5A_0-3}, \quad B(t) = -\frac{1}{3}\frac{Q_2^2}{Q_3}t^{2A_0-2}, \\
D(t) &= -\frac{2}{3}Q_2 t^{-A_0-1}, \quad E(t) = -\frac{Q_3}{2}t^{-4A_0}.
\end{aligned}
\tag{4.6.31}
$$

We can now formulate the following result: if the coefficients $Q_1, ..., Q_4$ in the differential equation

$$
\ddot{x} + Q_1 x t^{-2} + Q_2 x^2 t^{-3A_0-1} + Q_3 x^3 t^{-6A_0} + Q_4 t^{(3A_0-3)} = 0 \tag{4.6.32}
$$

satisfy the algebraic equation (4.6.29), the dynamical system (4.6.32) admits a conservation law in the form (4.6.3), namely,

$$
\begin{aligned}
\dot{x} &= -\frac{1}{3}(2A_0 - 1)\frac{Q_2}{Q_3}t^{(3A_0-2)} + A_0 x t^{-1} \\
&\quad + t^{-A_0}\left[I - \frac{1}{2}Q_3\left(xt^{-A_0} + \frac{1}{3}\frac{Q_2}{Q_3}t^{(2A_0-1)}\right)^4\right]^{1/2}, \tag{4.6.33}
\end{aligned}
$$

where A_0 is a solution of the equation (4.6.28) and I is an arbitrary constant.

It is of interest to note that for the classical time-dependent Duffing equation $Q_2 = Q_4 = 0$, that is,

$$
\ddot{x} + Q_1 x t^{-2} + Q_3 x^3 t^{-6A_0} = 0, \tag{4.6.34}
$$

the algebraic equation (4.6.29) is identically satisfied for the arbitrary Q_1 and Q_3. The expression (4.6.33) gives the following conservation law:

$$
\dot{x} = A_0 x t^{-1} + t^{-A_0}\left(I - \frac{1}{2}Q_3 x^4 t^{-4A_0}\right)^{1/2}, \tag{4.6.35}
$$

where A_0 is a root of

$$
A_0(A_0 - 1) + Q_1 = 0. \tag{4.6.36}
$$

As an example, consider the case treated by Ranganathan [90], for which $Q_1 = 2/9$, $Q_3 = a/9$, where a is a constant. One of the roots of (4.6.36) is $A_0 = 1/3$. Therefore, the differential equation (4.6.34) becomes

$$
\ddot{x} + \frac{2}{9}x t^{-2} + \frac{a}{9}x^3 t^{-2} = 0, \tag{4.6.37}
$$

whose conservation law, according to (4.6.35), is

$$\dot{x} = \frac{1}{3}xt^{-1} + t^{-1/3}\left(I - \frac{a}{18}x^4t^{-4/3}\right)^{1/2}, \qquad (4.6.38)$$

which is the same as the result obtained in [90], using a different method. For the second root of (4.6.36), that is, $A_0 = 2/3$, it follows that the differential equation

$$\ddot{x} + \frac{2}{9}xt^{-2} + \frac{a}{9}x^3t^{-4} = 0 \qquad (4.6.39)$$

has a conservation law

$$\dot{x} = \frac{2}{3}xt^{-1} + t^{-2/3}\left(I - \frac{a}{18}x^4t^{-8/3}\right)^{1/2}. \qquad (4.6.40)$$

Note that the analysis of the auxiliary equations (4.6.6) offers many possibilities for obtaining solutions in the class of power functions of time considered here. Therefore, the differential equations (4.6.32) and (4.6.34) and their conservation laws (4.6.33) and (4.6.35) are just some of many possibilities. We do not believe that all possible cases and subcases can be exhaustively presented in this monograph. Thus, we will discuss some cases that should be interesting to a wide circle of readers.

4.6.1   The Case of Arbitrary Q's $(P = 0)$

For the cases considered previously in this section, it was shown that the constant coefficients $Q_i, i = 1, ..., 4$, are not independent; that is, they have to satisfy the algebraic equation (4.6.29). In addition, it is evident that the conservation law is not valid for the case $Q_3 = 0$. To avoid this drawback, we seek a conservation law of the dynamical system (4.6.1) in the form (4.6.3) with $P = 0$ in the auxiliary equations (4.6.6). Repeating exactly the same procedure as in the previous case, we arrive at the following result: the differential equation

$$\ddot{x} + Q_1 xt^{-2} + Q_2 x^2t^{-5/2} + Q_3 x^3t^{-3} + Q_4 t^{-3/2} = 0 \qquad (4.6.41)$$

has the conservation law

$$\begin{aligned}\dot{x} &= \frac{1}{2}xt^{-1} + t^{-1/2}\left[I - 2Q_4 xt^{-1/2} + \left(\frac{1}{2} - Q_1\right)x^2t^{-1}\right. \\ &\qquad \left. -\frac{2}{3}Q_2 x^3t^{-3/2} + \frac{1}{2}Q_3 x^4t^{-2}\right]^{1/2}, \qquad (4.6.42)\end{aligned}$$

where $Q_1, ..., Q_4$ are arbitrary constants.

For example, considering again the case $Q_2 = Q_4 = 0$, we find that the differential equation

$$\ddot{x} + Q_1 xt^{-2} + Q_3 x^3t^{-3} = 0 \qquad (4.6.43)$$

has a conservation law in the form

$$\dot{x} = \frac{1}{2}xt^{-1} + t^{-1/2}\left[I + \left(\frac{1}{4} - Q_1\right)x^2t^{-1} - \frac{1}{2}Q_3x^4t^{-2}\right]^{1/2}, \qquad (4.6.44)$$

where Q_1 and Q_2 are arbitrary.

4.6.2   The Case $Q_3 = 0$ ($E = 0$)

We mentioned previously that the case $Q_3 = 0$ should be considered separately, which follows from (4.6.27) and (4.6.28). It is evident from (4.6.6)$_4$ that if $Q_3 = 0$, then $E(t) = 0$, and the equation (4.6.6)$_{10}$ is identically satisfied. Thus, the relevant equations for finding the functions $P(t), A(t), C(t), K(t), M(t)$, $B(t)$, and $D(t)$ are (4.6.6)$_1$–(4.6.6)$_3$ and (4.6.6)$_5$–(4.6.6)$_9$. By supposing that these functions are of the form (4.6.7) and repeating the same procedure as before, we find

$$
\begin{aligned}
P(t) &= P_0 t^{(5A_0-3)}, \quad A(t) = A_0 t^{-1}, \quad C(t) = t^{-A_0}, \\
K(t) &= \frac{Q_2 P_0^3}{12(2A_0-1)^3}t^{(12A_0-6)}, \quad M(t) = -\frac{Q_2 P_0^2}{(2A_0-1)^2}t^{(7A_0-4)}, \\
B(t) &= \frac{Q_2 P_0}{(2A_0-1)}t^{(2A_0-2)}, \quad D(t) = -\frac{2}{3}Q_2 t^{-3A_0}.
\end{aligned}
\qquad (4.6.45)
$$

The functions of time figuring in the differential equation (4.6.1) are found to be

$$f_1(t) = t^{-2}, \quad f_2(t) = t^{-5A_0}, \quad f_4(t) = t^{(5A_0-4)}, \qquad (4.6.46)$$

where A_0 and P_0 are roots of the algebraic equations

$$
\begin{aligned}
Q_2 P_0^2 - 12P_0(2A_0-1)^3 - 4Q_4(2A_0-1)^2 &= 0, \\
A_0^2 - A_0 + \frac{Q_2 P_0}{2A_0-1} + Q_1 &= 0,
\end{aligned}
\qquad (4.6.47)
$$

and the coefficients Q_1, Q_2, and Q_3 are arbitrary.

Entering with (4.6.45)–(4.6.47) into (4.6.1) and (4.6.3), we can formulate the following result.

The differential equation

$$\ddot{x} + Q_1 xt^{-2} + Q_2 x^2 t^{-5A_0} + Q_4 t^{(5A_0-4)} = 0, \qquad (4.6.48)$$

has a conservation law in the form

$$
\dot{x} = P_0 t^{(5A_0-3)} + A_0 xt^{-1}
$$
$$
+ t^{-A_0}\left\{I - \frac{2}{3}Q_2\left[xt^{-A_0} - \frac{P_0}{2(2A_0-1)}t^{2(2A_0-1)}\right]^3\right\}^{1/2},
$$
$$(4.6.49)$$

where P_0 and A_0 are roots of the algebraic equations (4.6.47), I is an arbitrary constant to be determined from the given initial conditions prescribed to (4.6.48), and Q_1, Q_2, and Q_4 are a given set of constant parameters.

As a special case, let us consider the case $Q_1 = Q_4 = 0$ and $Q_2 = 1$. From (4.6.47) we find $P_0 = 12 \left(2A_0 - 1\right)^3$ and $49A_0^2 - 49A_0 + 12 = 0$. The solution of these equations read

$$
\begin{aligned}
A_{0(1)} &= \frac{4}{7}, \quad P_{0(1)} = \frac{12}{343}, \\
A_{0(2)} &= \frac{3}{7}, \quad P_{0(2)} = -\frac{12}{343}.
\end{aligned}
\tag{4.6.50}
$$

Therefore, from (4.6.48) and (4.6.49) it follows that the differential equation

$$
\ddot{x} + x^2 t^{-20/7} = 0
\tag{4.6.51}
$$

has a conservation law

$$
\begin{aligned}
\dot{x} &= -\frac{12}{343} t^{-6/7} + \frac{3}{7} x t^{-1} \\
&\quad + t^{-3/7} \left[I - \frac{2}{3} \left(x t^{-4/7} - \frac{6}{49} t^{2/7} \right)^3 \right]^{1/2},
\end{aligned}
\tag{4.6.52}
$$

while the differential equation

$$
\ddot{x} + x^2 t^{-15/7} = 0
\tag{4.6.53}
$$

possesses a conservation law in the form

$$
\begin{aligned}
\dot{x} &= -\frac{12}{343} t^{-6/7} + \frac{3}{7} x t^{-1} \\
&\quad + t^{-3/7} \left[I - \frac{2}{3} \left(x t^{-3/7} + \frac{6}{49} t^{-2/7} \right)^3 \right]^{1/2}.
\end{aligned}
\tag{4.6.54}
$$

Note that both conservation laws have been found by Leach, Marthens, and Maharaj [64] using the method of the Lie point symmetry analysis.

4.6.3 Reduction to Quadratures by Means of the Hamilton–Jacobi Method

In this section we shall demonstrate that the conservation laws obtained in the previous section can be used as a starting point for the reduction to quadratures of the corresponding dynamical systems.

First, we note that the dynamical system (4.6.32) can be written in the canonical form

$$
\begin{aligned}
\dot{x} &= p, \\
\dot{p} &= -Q_1 x t^{-2} - Q_2 x^2 t^{-3A_0 - 1} - Q_3 x^3 t^{-6A_0} - Q_4 t^{(3A_0 - 3)}.
\end{aligned}
\tag{4.6.55}
$$

It is easily seen that this system has the Hamiltonian structure and that the Hamilton–Jacobi partial differential equation is

$$\frac{\partial S}{\partial t} + \frac{1}{2}\left(\frac{\partial S}{\partial x}\right)^2 + \frac{1}{2}Q_1 x^2 t^{-2}$$
$$+\frac{1}{3}Q_2 x^3 t^{-3A_0-1} + \frac{1}{4}Q_3 x^4 t^{-6A_0} + Q_4 x t^{3A_0-3} = 0, \qquad (4.6.56)$$

where the linear momentum p is a gradient of the field function $S(t,x)$, that is, $p = \partial S/\partial x$.

It is easy to verify that by differentiating this equation partially with respect to x and denoting by $\partial S/\partial x = \phi(t,x)$, we arrive at the basic field equation of the system (4.6.55) for $p = \phi(t,x)$:

$$\frac{\partial \phi}{\partial t} + \phi\frac{\partial \phi}{\partial x} + Q_1 x t^{-2} + Q_2 x^2 t^{-3A_0-1} + Q_3 x^3 t^{-6A_0} + Q_4 t^{(3A_0-3)} = 0. \quad (4.6.57)$$

Taking into account this connection we can employ the incomplete solution of this quasi-linear equation in the form (4.6.33), namely,

$$\dot{x} = p = \phi(t,x) = \frac{\partial S}{\partial x} = -\frac{1}{3}(2A_0 - 1)\frac{Q_2}{Q_3}t^{(3A_0-2)} + A_0 x t^{-1}$$
$$+t^{-A_0}\left[I - \frac{1}{2}Q_3\left(xt^{-A_0} + \frac{1}{3}\frac{Q_2}{Q_3}t^{(2A_0-1)}\right)^4\right]^{1/2}, \qquad (4.6.58)$$

as a starting point for finding a complete solution of the Hamilton–Jacobi equation (4.6.56). Thus, by partially integrating (4.6.58) with respect to x, we find

$$S(t,x,I) = -\frac{1}{3}(2A_0-1)\frac{Q_2}{Q_3}xt^{(3A_0-2)} + \frac{1}{2}A_0 x^2 t^{-1}$$
$$+ \int^{xt^{-A_0}+\frac{1}{3}\frac{Q_2}{Q_3}t^{(2A_0-1)}}\left(I - \frac{1}{2}Q_3 Y^4\right)^{1/2} dY$$
$$+\theta(t,I), \qquad (4.6.59)$$

where

$$Y = xt^{-A_0} + \frac{1}{3}\frac{Q_2}{Q_3}t^{(2A_0-1)}, \quad dY = t^{-A_0}dx, \qquad (4.6.60)$$

and $\theta(t,I)$ is a function of time and the constant I. Substituting (4.6.59) into the Hamilton–Jacobi equation (4.6.56), one arrives at

$$\dot{\theta}(t,I) + \frac{1}{2}It^{-2A_0} + V(t)$$
$$-xt^{3A_0-3}\left[\frac{1}{27}\frac{Q_2^3}{Q_3^2} + \frac{2}{3}(2A_0-1)\frac{Q_2}{Q_3} - Q_4\right]$$
$$+\frac{1}{2}\left[A_0(A_0-1) + Q_1 - \frac{1}{3}\frac{Q_2^2}{Q_3^3}\right] = 0, \qquad (4.6.61)$$

where

$$V\left(t\right) = \left[\frac{1}{18}\frac{Q_2}{Q_3^2}\left(2A_0 - 1\right)^2 - \frac{1}{324}\frac{Q_2^4}{Q_3^3}\right]t^{6A_0 - 4}. \tag{4.6.62}$$

The expressions in the square brackets in (4.6.61) are equal to zero due to (4.6.27) and (4.6.28). Therefore, the equation (4.6.61) is reduced to

$$\dot{\theta}\left(t, I\right) + \frac{1}{2}It^{-2A_0} + V\left(t\right) = 0. \tag{4.6.63}$$

Note that the function $V\left(t\right)$ plays the role of a characteristic gauge function, which has no influence in our considerations. Namely, since it does not contain x or I, it will not affect the application of the Jacobi theorem $p = \partial S/\partial x, \partial S/\partial I = B = const.$ Thus, in the subsequent text, the group of terms stemming from $\int^t V\left(u\right) du$ will be denoted as irrelevant terms (i.t.) and in subsequent analysis will not be written explicitly. Integrating (4.6.63), we find that the principal function (4.6.59) represents a *complete solution* of the Hamilton–Jacobi equation

$$\begin{aligned}
S\left(t, x, I\right) &= -\frac{1}{3}\left(2A_0 - 1\right)\frac{Q_2}{Q_3}xt^{(3A_0 - 2)} + \frac{1}{2}A_0 x^2 t^{-1} \\
&\quad + \int^{xt^{-A_0} + \frac{1}{3}\frac{Q_2}{Q_3}t^{(2A_0 - 1)}}\left(I - \frac{1}{2}Q_3 Y^4\right)^{1/2} dY \\
&\quad - \frac{1}{2}I\int^t u^{-2A_0} du + \text{i.t.}
\end{aligned} \tag{4.6.64}$$

Applying the Jacobi theorem, we find that the equation $\partial S/\partial I = B = const.$ leads to

$$\int^{xt^{-A_0} + \frac{1}{3}\frac{Q_2}{Q_3}t^{(2A_0 - 1)}}\frac{dY}{\left(I - \frac{1}{2}Q_3 Y^4\right)^{1/2}}dY - \int^t u^{-2A_0} du = B = const., \tag{4.6.65}$$

which together with (4.6.58) completes the solution of the canonical system (4.6.55).

Note that all differential equations and corresponding conservation laws considered in this section can be reduced to quadratures by the process described. Since the treatment of each case is the same as for the case just considered, we will not repeat all the details here, but give only the final results.

(a) For the Duffing equation (4.6.34) and the conservation law (4.6.35),

$$\ddot{x} + Q_1 x t^{-2} + Q_3 x^3 t^{-6A_0} = 0, \tag{4.6.66}$$

and

$$\dot{x} = A_0 x t^{-1} + t^{-A_0}\left(I - \frac{1}{2}Q_3 x^4 t^{-4A_0}\right)^{1/2}, \tag{4.6.67}$$

the corresponding Hamilton–Jacobi equation is (with $Q_2 = 0$)

$$\frac{\partial S}{\partial t} + \frac{1}{2}\left(\frac{\partial S}{\partial x}\right)^2 + \frac{1}{2}Q_1 x^2 t^{-2} + \frac{1}{4}Q_3 x^4 t^{-6A_0} = 0. \tag{4.6.68}$$

A complete solution of this equation is of the form $(xt^{-A_0} = Y)$

$$S(t,x,I) = \frac{1}{2}A_0 x^2 t^{-1} + \int^{xt^{-A_0}}\left[I - \frac{1}{2}Q_3 Y^4\right]^{1/2} du$$

$$-\frac{1}{2}I\int^t \xi^{-2A_0} d\xi + \text{ i.t.,} \tag{4.6.69}$$

where A_0 is a root of (4.6.36). The Jacobi theorem, $\partial S/\partial I = B = const.$, leads to the solution of (4.6.66) in the form

$$\int^{xt^{-A_0}} \frac{dY}{\left[I - \frac{1}{2}Q_3 Y^4\right]^{1/2}} - \int^t \xi^{-2A_0} d\xi = B = const. \tag{4.6.70}$$

(b) For the differential equation (4.6.37) and its conservation law (4.6.39),

$$\ddot{x} + \frac{2}{9}xt^{-2} + \frac{a}{9}x^3 t^{-4} = 0; \quad \dot{x} = \frac{2}{3}xt^{-1} + t^{-2/3}\left(I - \frac{a}{18}x^4 t^{-8/3}\right)^{1/2}, \tag{4.6.71}$$

the Hamilton–Jacobi equation is

$$\frac{\partial S}{\partial t} + \frac{1}{2}\left(\frac{\partial S}{\partial x}\right)^2 + \frac{1}{9}x^2 t^{-2} + \frac{a}{36}x^4 t^{-4} = 0. \tag{4.6.72}$$

A corresponding complete solution is found to be $(A_0 = 1/3)$

$$S(t,x,I) = \frac{1}{3}x^2 t^{-1} + \int^{xt^{-2/3}}\left(I - aY^4\right)^{1/2} dY - \frac{3}{2}It^{1/3} + \text{ i.t.,} \tag{4.6.73}$$

where $Y = xt^{-2/3}$. According to the Jacobi theorem a general solution of $(4.6.71)_1$ is

$$\int^{xt^{-2/3}}\left(I - \frac{a}{18}Y^4\right)^{-1/2} dY - 3t^{1/3} = B = const. \tag{4.6.74}$$

Similarly, the solution of (4.6.39), $\ddot{x} + \frac{2}{9}xt^{-2} + \frac{a}{9}x^3 t^{-4} = 0$, can be represented as $\int^{xt^{-2/3}}\left(I - \frac{a}{18}Y^4\right)^{-1/2} dY - 3t^{-1/3} = B = const.$, where $Y = xt^{-2/3}$.

(c) The differential equation (4.6.41),

$$\ddot{x} + Q_1 xt^{-2} + Q_2 x^2 t^{-5/2} + Q_3 x^3 t^{-3} + Q_4 t^{-3/2} = 0, \tag{4.6.75}$$

can be transposed into the Hamilton–Jacobi equation

$$\frac{\partial S}{\partial t} + \frac{1}{2}\left(\frac{\partial S}{\partial x}\right)^2 + \frac{1}{2}Q_1 x^2 t^{-2} + \frac{1}{3}Q_2 x^3 t^{-5/2}$$

$$+\frac{1}{4}Q_3 x^4 t^{-3} + Q_4 x t^{-3/2} = 0. \tag{4.6.76}$$

By integrating (4.6.42) with respect to x, we find a complete solution of this equation in the form

$$S\left(t, x, I\right) = \frac{1}{6}x^2 t^{-1} + \int^{xt^{-1/2}}\left[I - 2Q_4 Y + \left(\frac{1}{2} - Q_1\right)Y^2\right.$$

$$\left. - \frac{2}{3}Q_2 Y^3 - \frac{1}{2}Q_3 Y^4\right]^{1/2} - \frac{1}{2}I\ln t + \text{i.t.}, \tag{4.6.77}$$

where $Y = xt^{-1/2}$. Applying the Jacobi theorem $\partial S/\partial I = B = const.$, we arrive at the general solution of (4.6.75) in the form

$$\int^{xt^{-1/2}} \frac{dY}{\left[I - 2Q_4 Y + \left(\frac{1}{2} - Q_1 Y^2\right) - \frac{2}{3}Q_2 Y^3 - \frac{1}{2}Q_3 Y^4\right]^{1/2}} - \ln t$$

$$= B = const. \tag{4.6.78}$$

(d) Repeating the same process of solution by means of the Hamilton–Jacobi method, we find that the solution of the differential equation (4.6.48), $\ddot{x} + Q_1 x t^{-2} + Q_2 x^2 t^{-5A_0} + Q_4 t^{(5A_0 - 4)} = 0$, can be reduced to quadratures. Namely, integrating (4.6.49), we find that the complete solution of the corresponding Hamilton–Jacobi equation is,

$$S\left(t, x, I\right) = P_0 x t^{(5A_0 - 3)} + \frac{1}{2}A_0 x^2 t^{-1}$$

$$+ \int^{xt^{-A_0} - \frac{P_0}{2(2A_0 - 1)}t^{2(2A_0 - 1)}}\left(I - \frac{2}{3}Q_2 Y^3\right)^{1/2}dY$$

$$- \frac{1}{2}I\int^t \xi^{-2A_0}d\xi + \text{i.t.}, \tag{4.6.79}$$

where

$$Y = xt^{-A_0} - \frac{P_0}{2\left(2A_0 - 1\right)}t^{2(2A_0 - 1)}, \tag{4.6.80}$$

and A_0 and P_0 are roots of (4.6.47). Applying the Jacobi theorem $\partial S/\partial I = B = const.$, we find that the general solution of (4.6.48) can be written as

$$\int^{xt^{-A_0} - \frac{P}{2(2A_0 - 1)}t^{2(2A_0 - 1)}} \frac{dY}{\left(I - \frac{2}{3}Q_2 Y^3\right)^{1/2}} - \int^t \xi^{-2A_0}d\xi = B = const.$$

$$\tag{4.6.81}$$

For example, for the differential equation of Emden–Fowler type given by (4.6.51), (4.6.53), for which the coefficients $A_{0(1,2)}$ and $P_{0(1,2)}$ are given by (4.6.50), we have that the general solutions are, respectively,

$$\int^{xt^{-3/7}-6t^{2/7}/49} \frac{dY}{\left(I - \frac{2}{3}Y^3\right)^{1/2}} + 7t^{-1/7} = B = const., \qquad (4.6.82)$$

with $Y = xt^{-3/7} - 6t^{2/7}/49$ for the differential equation (4.6.51) and

$$\int^{xt^{-3/7}+6t^{-2/7}/49} \frac{dY}{\left(I - \frac{2}{3}Y^3\right)^{1/2}} + 7t^{-1/7} = B = const., \qquad (4.6.83)$$

where $Y = xt^{-3/7} + 6t^{-2/7}/49$ for the differential equation (4.6.53). Note that the same results were obtained by Leach, Martens, and Maharaj in [64] using a group-theoretical method.

4.6.4 The Case When a Particular Solution of the Riccati Equation Is Available

We have seen in the preceding paragraphs that one of the vital points in applying the field method is to find a solution of the corresponding Riccati equation. However, in dealing with the linear rheonomic equations with one degree of freedom, we are frequently able to find only some particular solutions of this equation. In this part we shall demonstrate that the field method presented here can be successfully applied if some particular solutions of the Riccati equation are known.

The following two examples taken from [56] will demonstrate this case.

Example 1 [56]. Let us consider the following linear rheonomic equation:

$$\ddot{x} + \frac{1}{t}\dot{x} - \frac{1}{t^2}x = a, \quad a = const. \qquad (4.6.84)$$

Taking as a field function $\dot{x} = \phi(t, x)$, the basic equation is of the form

$$\frac{\partial \phi}{\partial t} + \phi\frac{\partial \phi}{\partial x} + \frac{1}{t}\phi - \frac{1}{t^2}x - a = 0. \qquad (4.6.85)$$

Since the problem is linear, we seek a complete solution in the form

$$\dot{x} = \phi(t, x) = f_1(t)x + f_2(t). \qquad (4.6.86)$$

Entering with this into (4.6.85) and grouping the terms with x and free terms, we obtain the following system of Riccati equations:

$$\dot{f}_1 + f_1^2 + \frac{1}{t}f_1 - \frac{1}{t^2} = 0, \quad \dot{f}_2 + f_1 f_2 + \frac{1}{t}f_2 - a = 0. \qquad (4.6.87)$$

Let us seek a particular solution of $(4.6.87)_1$ in the form

$$f_1(t) = At^m, \tag{4.6.88}$$

where A and m are unknown constants. Substituting (4.6.88) into $(4.6.87)_1$, we have

$$Amt^{m-1} + A^2t^{2m} + At^{m-1} - t^{-2} = 0. \tag{4.6.89}$$

This equation is satisfied for

$$A_1 = 1, \quad A_2 = -1, \quad m_1 = -1, \quad m_2 = -1. \tag{4.6.90}$$

Therefore, the particular solutions of (4.6.87) are

$$f_1^{(1)} = t^{-1}, \quad f_1^{(2)} = -t^{-1}. \tag{4.6.91}$$

Entering with this into $(4.6.87)_2$ and integrating, we obtain

$$f_2^{(1)} = \frac{at^3}{3} + \frac{C_1}{t^2}, \quad f_2^{(2)} = at + C_2, \tag{4.6.92}$$

where C_1 and C_2 are constants of integration. Therefore, from (4.6.86) we have the following *conservation laws*:

$$\dot{x} = xt^{-1} + \frac{at}{3} + \frac{C_1}{t^2}, \quad \dot{x} = -xt^{-1} + at + C_2. \tag{4.6.93}$$

Equating the right-hand sides of these equations, we find that the general solution of (4.6.84) is

$$x = \frac{1}{3}at^2 - \frac{C_1}{2t} + \frac{C_2t}{2}. \tag{4.6.94}$$

Example 2 [56]. Let us consider the following linear rheonomic equation:

$$\ddot{x} + \frac{a}{t^2}x = 0; \quad a = const. \tag{4.6.95}$$

Taking the field function in the form $\dot{x} = \phi(t, x) = f_1(t)x + f_2(t)$ and repeating the same process as in the previous example, we arrive at the following system of Riccati equations:

$$\dot{f}_1 + f_1^2 + at^{-2} = 0, \quad \dot{f}_2 + f_1 f_2 = 0. \tag{4.6.96}$$

Supposing the same form of a particular solution as given by (4.6.88), the first Riccati equation gives

$$f_1^{(1)}(t) = \left[\frac{1}{2} + \sqrt{\frac{1}{4} - a}\right]t^{-1}, \quad f_2^{(1)}(t) = \left[\frac{1}{2} - \sqrt{\frac{1}{4} - a}\right]t^{-1}. \tag{4.6.97}$$

Entering with this into $(4.6.96)_2$, we find after a simple integration

$$f_2^{(1)} = C_1 t^{-\left(\frac{1}{2}+ib\right)}, \quad f_2^{(2)} = C_2 t^{-\left(\frac{1}{2}-ib\right)}, \qquad (4.6.98)$$

where $i = \sqrt{-1}$ and C_1 and C_2 are arbitrary constants. Also, in writing (4.6.98), we assumed that $a > 4$, so that

$$\frac{1}{4} - a = -b^2 \qquad (4.6.99)$$

with b real. Therefore, the two conservation laws

$$\dot{x} = f_1^{(1)}(t)\, x + f_2^{(1)}, \quad \dot{x} = f_1^{(2)}(t)\, x + f_2^{(2)}, \qquad (4.6.100)$$

become

$$\dot{x} = \left(\frac{1}{2} + ib\right) t^{-1} x + C_1 t^{-1} e^{-ib \ln t}, \quad \dot{x} = \left(\frac{1}{2} - ib\right) t^{-1} x + C_1 t^{-1} e^{ib \ln t}, \qquad (4.6.101)$$

where we have used the relations $t^{ib} = e^{ib \ln t}$ and $t^{-ib} = e^{-ib \ln t}$. Equating the right-hand sides of the equations in (4.6.101), we obtain the solution of (4.6.95) in the form

$$x = t^{\frac{1}{2}}\left[A \cos\left(b \ln t\right) + B \sin\left(b \ln t\right)\right], \qquad (4.6.102)$$

where

$$A = \frac{C_2 - C_1}{2ib}, \quad B = \frac{C_1 + C_2}{2b}. \qquad (4.6.103)$$

Repeating exactly the same procedure, it can be easily verified that, for the special cases when

$$a = -6, \quad a = -12, \qquad (4.6.104)$$

the solution of equations

$$\ddot{x} - 6\frac{x}{t^2} = 0, \quad \ddot{x} - 12\frac{x}{t^2} = 0 \qquad (4.6.105)$$

are, respectively,

$$x = C_1 t^3 + C_2 t^{-2}, \quad x = C_1 t^4 + C_2 t^{-3}. \qquad (4.6.106)$$

Part II

The Hamiltonian Integral
Variational Principle

Chapter 5

The Hamiltonian Variational Principle and Its Applications

5.1   Introduction

In the first part of this monograph we considered the differential variational principles, especially the Lagrange–D'Alembert principle. This principle is based upon the *local* characteristics of motion; that is, the relations between its scalar and vector characteristics are considered simultaneously in one particular instant of time. The problem of describing the global characteristics of motion has been reduced to the integration of differential equations of motion.

In the second part of this book, we consider in some detail the integral variational principle of Hamilton (Hamilton's principle of stationary action) in which the *global characteristics* of motion occupy the central place. In addition, the Hamiltonian principle contains in itself some inherent stationary and even optimal characteristics by means of which the motion and evolution of the dynamical system can be interpreted as inseparable aspects of the Hamiltonian action. Due to these facts the Hamiltonian principle can be considered the cornerstone of modern analytical mechanics. Note that the mathematical apparatus used in applications of the Hamiltonian principle is applied variational calculus. Conversly, many problems occurring in physics and engineering, formulated with the methods of the calculus of variations, can be interpreted as characteristic formulations of the Hamiltonian principle. One important characteristic of the Hamiltonian principle is that all problems treated by it are naturally defined as the *boundary value problems* since as a prerequisite, the dynamical system should be completely specified at the initial and terminal time (begining and the end of time interval) in which the motion is taking place. Let us note also that by employing the aforementioned global and optimal properties, the Hamiltonian principle can be reliably used in determining approximate solution in many linear and nonlinear boundary value problems.

5.2   The Simplest Form of the Hamiltonian Variational Principle

Let us consider a holonomic dynamical system whose position at any instant of time t can be specified by n independent generalized coordinates, $q_1(t), ..., q_n(t)$, where n is the number of degrees of freedom of the dynamical system. We first consider the so-called Lagrangian systems in which all physical manifestations and behavior are completely describable by the Lagrangian function $L(t, q_1, ..., q_n, \dot{q}_1, ..., \dot{q}_n)$.

Let the configuration of the dynamical system be given at two instants of time, t_0 and t_1, such that

$$t_0 \quad : \quad q_i(t_0) = A_i, \quad i = 1, ..., n, \quad \text{configuration } A,$$
$$t_1 \quad : \quad q_i(t_1) = B_i, \quad i = 1, ..., n, \quad \text{configuration } B, \qquad (5.2.1)$$

where A_i and B_i are a given set of constants and $[t_1, t_0]$ is a given interval of time in which the motion of the dynamical system is taking place. For example, we can imagine that these two configurations are the result of an experimental determination of the position of the dynamical system at the initial and terminal instants of time. Otherwise, if we select completely arbitrary configurations A and B, it can happen that there does not exist any actual trajectory joining these configurations.

Let us suppose that for a given dynamical system there exists a motion that
(a) joins configurations A and B,
(b) satisfies the differential equations of motion,
(c) is in full agreement with the holonomic constraints, and
(d) is unique.[12]

We refer to this motion as the *direct* or *actual trajectory* of the dynamical system. Parallel with the direct trajectory, we also introduce *virtual* or *varied trajectories* $\bar{q}_i(t)$, which are defined as a one-parameter bundle of curves defined for all $t \in [t_1, t_0]$:

$$\bar{q}_i(t) = q_i(t) + \delta q_i(t) = q_i(t) + \varepsilon h_i(t), \qquad (5.2.2)$$

where $h_i(t)$ are arbitrary continuously differentiable functions of time and $\delta q_i = \varepsilon h_i(t), i = 1, ..., n$, are *variations* or *virtual displacements* of the dynamical system (which are identical to already introduced "simultaneous" variations or virtual displacements in section 3.2). Note that parameter ε in (5.2.2) is an

---

[12] If the actual trajectory is not unique, that is, if between two fixed configurations A and B there exists a finite or infinite number of actual trajectories connecting the initial and terminal configurations for the same time interval $[t_0, t_1]$, we then say that the actual path contains *conjugate points* or *kinetic foci* (see Example 6.5.1, pp. 233–235).

infinitesimally small, constant. Note also that for $\varepsilon = 0$, all curves (5.2.2) are identical to the actual trajectory. Let us suppose that the actual and varied trajectories are passing through the initial and terminal configurations A and B, which is equivalent to the following requirements:

$$\delta q_i\,(t_0) = \delta q_i\,(t_1) = 0, \quad i = 1, ..., n. \tag{5.2.3}$$

This situation is depicted in Figure 5.2.1, where the actual trajectory APB and $A\bar{P}B$ are depicted together with the initial and terminal configurations A and B.

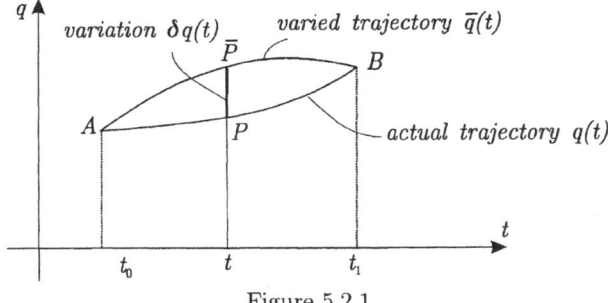

Figure 5.2.1

One of the remarkable properties of the variations introduced here and in the applied variational calculus is that the operators of variations δ and differentiation with respect to time are commutative (see also (1.3.18)):

$$\delta\dot{q}_i = \frac{d}{dt}\delta q_i. \tag{5.2.4}$$

Hamilton's action integral for the given time interval $[t_0, t_1]$ of a holonomic dynamical system is defined to be

$$I = \int_{t_0}^{t_1} L\,(t, q_1, ..., q_n, \dot{q}_1, ..., \dot{q}_n)\,dt, \tag{5.2.5}$$

where L denotes the Lagrangian for the dynamical system. Note that the expression (5.2.5) is frequently called the *functional* or *path functional*.

Hamilton's principle states: *Among all varied paths (admissible trajectories) connecting the given configurations A and B for a given time interval $[t_0, t_1]$, the actual motion makes the action integral I stationary, that is,*

$$\delta I = \delta \int_{t_0}^{t_1} L\,(t, q_1, ..., q_n, \dot{q}_1, ..., \dot{q}_n)\,dt = 0. \tag{5.2.6}$$

We will demonstrate that the Hamiltonian principle (5.2.6) leads immediately to the Euler–Lagrangian differential equations of motion. Indeed, recalling the expression for the Lagrangian variation of the action integral (3.2.17), we have

$$
\begin{aligned}
\delta \int_{t_0}^{t_1} L\left(t, \mathbf{q}, \dot{\mathbf{q}}\right) dt &= \int_{t_0}^{t_1} \left[L\left(t, \bar{\mathbf{q}}, \frac{d\bar{\mathbf{q}}}{d}\right) - L\left(t, \mathbf{q}, \dot{\mathbf{q}}\right) \right] dt \\
&= \int_{t_0}^{t_1} \left(\frac{\partial L}{\partial q_i} \delta q_i + \frac{\partial L}{\partial \dot{q}_i} \delta \dot{q}_i \right) dt = \int_{t_0}^{t_1} \delta L \, dt = 0.
\end{aligned}
\tag{5.2.7}
$$

Here, $\mathbf{q} = (q_1, ..., q_n)$. Employing (5.2.4) and integrating by parts, we find

$$
\delta I = \frac{\partial L}{\partial \dot{q}_i} \delta q_i \Big|_{t_0}^{t_1} + \int_{t_0}^{t_1} \left(\frac{\partial L}{\partial q_i} - \frac{d}{dt} \frac{\partial L}{\partial \dot{q}_i} \right) \delta q_i \, dt.
\tag{5.2.8}
$$

Taking into account the boundary conditions (5.2.3) and employing the fact that the variations δq_i are completely arbitrary inside the interval (t_0, t_1), the condition $\delta I = 0$ yields

$$
\frac{d}{dt} \frac{\partial L}{\partial \dot{q}_i} - \frac{\partial L}{\partial q_i} = 0, \quad i = 1, ..., n,
\tag{5.2.9}
$$

which are Euler–Lagrangian equations of motion already considered in the first part of this monograph (see section 1.4). However, these n differential equations of the second order should be obligatory considered together with the $2n$ boundary conditions (5.2.1). Thus, every dynamical problem formulated by means of Hamilton's principle, should be considered as a *boundary value problem* (two-point boundary value problem).

In the simplest formulation of the Hamiltonian principle, the problem essentially consists of selecting the actual trajectory $q_i(t)$, $i = 1, ..., n$, satisfying the boundary conditions (5.2.1), along which the functional (or action integral) affords an *extreme value*. Such a trajectory is termed the *extremal trajectory*. In light of these facts, we can restate Hamilton's principle in the following form: *For the dynamical systems which are completely describable by means of the Lagrangian function, the actual motion gives the extremum to the action integral* (5.2.5).

Thus far we have considered the formulation of Hamilton's principle in which the Lagrangian function L plays the leading role. It is also of interest to express the action integral in the canonical variables. By performing the Legendre transformation of the Lagrangian L (see (1.8.6)) we introduce the Hamiltonian action integral in the form

$$
I_{can} = \int_{t_0}^{t_1} \left[p_i \dot{q}_i - L\left(t, q_1, ..., q_n, \dot{q}_1, ..., \dot{q}_n\right) \right] dt.
\tag{5.2.10}
$$

This action integral should be considered under the conditions that the time interval (t_0, t_1) is completely specified and that the boundary conditions are the same as in the previous case, that is,

$$
q_i(t_0) = A_i, \quad q_i(t_1) = B_i,
\tag{5.2.11}
$$

where A_i and B_i are given constants.

It is easy to verify by the elementary calculations that the Hamiltonian principle $\delta I_{can} = 0$ leads to

$$\delta I_{can} = p_i \delta q_i \big|_{t_0}^{t_1} + \int_{t_0}^{t_1} \left[\left(\dot{q}_i - \frac{\partial H}{\partial p_i} \right) \delta p_i \right.$$

$$\left. - \left(\dot{p}_i + \frac{\partial H}{\partial q_i} \right) \delta q_i \right] dt = 0, \qquad (5.2.12)$$

which leads to the Hamiltonian canonical equation of motion

$$\dot{q}_i = \frac{\partial H}{\partial p_i}, \quad \dot{p}_i = -\frac{\partial H}{\partial q_i}, \quad i = 1, ..., n. \qquad (5.2.13)$$

Note that by determining variation of (5.2.10) we have considered the canonical variables mutually independent. Thus, these variables are independently varied in (5.2.10).

As indicated previously, the Hamiltonian variational principle provides only the necessary condition for the extremality of the Hamiltonian action integral along the actual trajectory of the dynamical system. However, the question of the character of the extremum (if it exists) can be judged on the basis of the sign of the *second variation* $\delta^2 I$. The *necessary* condition for the action integral (5.2.5) to attain a minimum is that the second variation is nonnegative; that is, $\delta^2 I \geq 0$ along the actual trajectory of the dynamical system. A sufficient condition for the functional I given by (5.2.5) to attain (a local) minimum is that the second variation is *strongly positive;* that is, $\delta^2 I \geq \gamma \left(\|\delta \mathbf{q}\| \right) \|\delta \mathbf{q}\|$, where $\gamma(\tau)$ is a nonnegative function on $(0, \infty)$ such that $\lim_{\tau \to \infty} \gamma(\tau) \to \infty$ and $\|\cdot\|$ is any norm on a linear space to which $\delta \mathbf{q}$ belongs (see [81, p. 39]).

It can be shown that for sufficiently small time intervals (t_0, t_1) the positivity of the second variation, that is, $\delta^2 I > 0$, guarantees that I is minimum on the actual trajectory of the dynamical system, when the Lagrangian L is taken to be the classical energy difference $T - \Pi$, where T is the kinetic and Π is the potential energy.

To prove this, we compute the second variation of the action integral

$$\delta^2 I = \int_{t_0}^{t_1} \delta^2 L \, dt, \qquad (5.2.14)$$

where

$$\delta^2 L = \frac{1}{2} \frac{\partial^2 L}{\partial \dot{q}_i \partial \dot{q}_j} \delta \dot{q}_i \delta \dot{q}_j + \frac{\partial^2 L}{\partial \dot{q}_i \partial q_j} \delta \dot{q}_i \delta q_j + \frac{1}{2} \frac{\partial^2 L}{\partial q_i \partial q_j} \delta q_i \delta q_j,$$

$$i, j = 1, ..., n. \qquad (5.2.15)$$

Using the following estimation for the variation of the generalized coordinates (see [68, p. 650] for $i = 1, ..., n, t_0 < t < t_1$,

$$|\delta q_i(t)| = \left| \int_{t_0}^{t_1} \delta \dot{q}_i(t) \, dt \right| \leq (t_1 - t_0) \sup_{t \in [t_0, t_1]} |\delta \dot{q}_i(t)|, \qquad (5.2.16)$$

and ignoring in (5.2.14) all terms containing factors $(t_1 - t_0)$ and $(t_1 - t_0)^2$, we obtain

$$\delta^2 I \approx \frac{1}{2} \int_{t_0}^{t_1} \frac{\partial^2 L}{\partial \dot{q}_i \partial \dot{q}_j} \delta \dot{q}_i \delta \dot{q}_j dt. \qquad (5.2.17)$$

However, for the case of the scleronomic dynamical systems, the Lagrangian is of the form (1.4.45). Thus, we have

$$\frac{\partial^2 L}{\partial \dot{q}_i \partial \dot{q}_j} = \frac{\partial^2 T}{\partial \dot{q}_i \partial \dot{q}_j} = a_{ij}. \qquad (5.2.18)$$

Therefore,

$$\delta^2 I \approx \frac{1}{2} \int_{t_0}^{t_1} a_{ij} \delta \dot{q}_i \delta \dot{q}_j dt > 0, \qquad (5.2.19)$$

and since $a_{ij} \delta \dot{q}_i \delta \dot{q}_j$ is a positive definite quadratic form with respect to variations of the generalized velocities, we conclude that $a_{ij} \delta \dot{q}_i \delta \dot{q}_j (t) > 0$ for all $t \in (t_0, t_1)$.

Let us note that this argument is strongly dependent upon the properties of the Lagrangian formed as the difference between the kinetic and potential energies.

It can be shown (see [54, pp. 78–80]) that for the case when we employ in the action integral canonical variables $q_i(t)$, $p_i(t)$, given by (5.2.10), the equivalent statement of (5.2.9) is expressed in the following min − max principle: For short time intervals, the trajectory $q_i(t), p_i(t), i = 1, ..., n$, in a conservative field of force affords a solution of the min − max problem

$$\min_{q_i} - \max_{p_i} \int_{t_0}^{t_1} [p_i \dot{q}_i - H(q_1, ..., q_n, p_1, ..., p_n)] dt.$$

Example 5.2.1. *The Brachistochrone problem.* As an illustration let us solve the famous problem of brachistochrone, where a material point slides on a smooth wire between two given points A and B in a constant gravity field. We shall restrict the curves to lie in the half-plane Axy, and the y axis is drawn vertically downward as depicted in Figure 5.2.2.

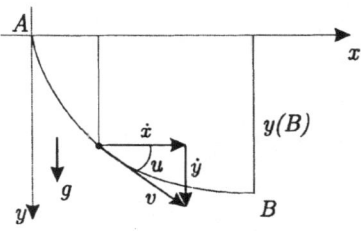

Figure 5.2.2

Let us suppose that the initial velocity of the particle is zero (for nonzero initial velocity, see [9]) and that $x_A(0) = 0, y_A(0) = 0$. The total energy $E = (mv^2/2) - mgy = 0$. Thus, we have that the velocity of the particle is

$$v = \sqrt{\dot{x}^2 + \dot{y}^2} = \sqrt{2gy}. \tag{5.2.20}$$

Now, we have $\dot{x} = \sqrt{2gy}/\sqrt{1 + (y')^2}$, where $\dot{x} = dx/dt$ and $y' = dy/dx$. Whence, $dt = \sqrt{1 + (y')^2}dx/\sqrt{2gy}$. Therefore, the problem is to find $t_{\min} = \min_y T$ of the action integral (functional)

$$T = \int_0^{x_B} L(x, y, y')\, dx = \frac{1}{\sqrt{2g}} \int_0^{x_B} \sqrt{\frac{1 + (y')^2}{y}}\, dx, \tag{5.2.21}$$

with

$$y(0) = 0, \quad y(x_B) = y_B, \tag{5.2.22}$$

and where y_B is a given constant.

Since the functional (5.2.21) does not depend upon independent variable x explicitly, the Lagrangian function L admits the Jacobi conservation law $y' \partial L/\partial y' - L = const.$, which in our case leads to

$$\frac{1}{\sqrt{y}\sqrt{1 + (y')^2}} = C = const. \tag{5.2.23}$$

This differential equation can be reduced to the form $y' = f(y)$ and then to separate variables. However, it is much simpler to introduce the angle u between the tangent at the curve $y(x)$ and the x axis as an independent variable. Thus,

$$y' = \frac{dy}{dx} = \tan u. \tag{5.2.24}$$

Employing the relation $\cos u = (1 + \tan^2 u)^{-1/2}$, equation (5.2.23) becomes

$$y = \frac{\cos^2 u}{C} = \frac{1}{2C^2}(1 + \cos 2u). \tag{5.2.25}$$

Differentiating this, we have $dy = (-2/C^2)(\sin u \cos u)\, du$. Entering with this into (5.2.24), we have $dx = (-2\cos^2 u\, du/C^2)$, whence by integration we arrive at

$$x = D - \frac{1}{2C^2}(2u + \sin 2u), \tag{5.2.26}$$

where D is an arbitrary constant of integration.

The equations (5.2.26) and (5.2.25) are the equations of the *cycloids* in the parametric form, a curve generated by a point on a circle rolling without slipping on a horizontal line.

It is easy to show that the angle $u(t)$ is a linear function of time. Indeed, from Figure 5.2.2 it follows that

$$\frac{dx}{dt} = \dot{x} = \sqrt{2gy}\cos u; \quad \frac{dy}{dt} = \sqrt{2gy}\sin u. \tag{5.2.27}$$

It was also shown that

$$\frac{dx}{du} = \left(-\frac{2}{C^2}\right)\cos^2 u; \quad \frac{dy}{du} = \left(-\frac{2}{C^2}\right)\sin u\cos u. \tag{5.2.28}$$

From the equations $(5.2.27)_1$ and (5.2.25), we have

$$\frac{dx}{du}\frac{du}{dt} = \sqrt{2g}\left(\frac{\cos^2 u}{C^2}\right)^{1/2}\cos u, \tag{5.2.29}$$

whence

$$\frac{du}{dt} = -\frac{C}{2}\sqrt{2g} = K = const., \tag{5.2.30}$$

that is,

$$u = Kt + M, \quad M = const. \tag{5.2.31}$$

Let us denote the values of the angle u at the point A and B by

$$u(A) = \alpha, \quad u(B) = \beta. \tag{5.2.32}$$

Then, by employing (8.2.28) and (8.2.27) at the boundaries A and B, we have

$$
\begin{aligned}
x_A &= 0 = D - \left(\frac{1}{2C^2}\right)(2\alpha + \sin 2\alpha), \\
y_A &= 0 = \frac{\cos^2\alpha}{C^2}, \\
x_B &= D - \left(\frac{1}{2C^2}\right)(2\beta + \sin 2\beta), \\
y_B &= D - \left(\frac{1}{2C^2}\right)(2\beta + \cos 2\beta).
\end{aligned}
\tag{5.2.33}
$$

From these four (complicated) algebraic equations we can find four constants, C, D, α, and β in terms of the given values for x_A, y_A, x_B, and y_B.

Finally, noting that

$$\sqrt{\frac{1+(y')^2}{y}} = \frac{1}{C^2}\sqrt{\frac{1+\tan^2 u}{\frac{\cos^2 u}{C^2}}}\,(-2\cos^2 u\,du) = -\frac{2}{C^2}du, \tag{5.2.34}$$

we find from (5.2.21) that the minimal time of travel of the particle is

$$\min_{y} T = t_{\min} = \sqrt{\frac{2}{gC^2}}\,(\alpha - \beta),\qquad(5.2.35)$$

and that completes the solution.

5.3 The Hamiltonian Principle for Nonconservative Force Field

In the previous section we postulated the Hamiltonian principle supposing that the behavior of the dynamical system in question can be completely described by the single function: the Lagrangian function $L(t, q_1, ..., q_n, \dot{q}_1, ..., \dot{q}_n)$. However, if the dynamical system comprises purely nonconservative forces $Q_i(t, q_1, ..., q_n, \dot{q}_1, ..., \dot{q}_n)$, $i = 1, ..., n$, the form of the Hamiltonian principle has quite a different structure, and the extremal properties of the actual trajectories are lost.

Let us begin our considerations from the Lagrange–D'Alembert differential variational principle for the holonomic nonconservative dynamical system, which is, according to (1.3.33), of the form

$$\left(\frac{d}{dt}\frac{\partial L}{\partial \dot{q}_i} - \frac{\partial L}{\partial q_i} - Q_i\right)\delta q_i = 0,\quad i = 1, ..., n,\qquad(5.3.1)$$

where, as usual, the repeated indices should be summed. Now, we have

$$\frac{d}{dt}\left(\frac{\partial L}{\partial \dot{q}_i}\delta q_i\right) - \frac{\partial L}{\partial \dot{q}_i}(\delta q_i)^{\cdot} - \frac{\partial L}{\partial q_i}\delta q_i - Q_i\delta q_i = 0.\qquad(5.3.2)$$

Employing the commutativity rule (5.2.4) and recalling, in accordance with the notation introduced in (3.2.8), that $\delta L = (\partial L/\partial q_i)\,\delta q_i + (\partial L/\partial \dot{q}_i)\,\delta \dot{q}_i$, we have, after integrating over the time interval $[t_0, t_1]$,

$$\frac{\partial L}{\partial \dot{q}_i}\delta q_i\bigg|_{t_0}^{t_1} - \int_{t_0}^{t_1}(\delta L + Q_i\delta q)\,dt = 0.\qquad(5.3.3)$$

The first term vanishes, since δq_i vanishes at t_0 and t_1, and thus

$$\int_{t_0}^{t_1}(\delta L + Q_i\delta q)\,dt = 0.\qquad(5.3.4)$$

The equation (5.3.4) is a mathematical expression of *Hamilton's principle for nonconservative dynamical systems*.

The variational equation (5.3.4) can be written in the form

$$\delta I + \int_{t_0}^{t_1} Q_i\delta q\,dt = 0,\qquad(5.3.5)$$

where $\delta I \neq 0$ is given by (5.2.6), namely,

$$\delta I = \int_{t_0}^{t_1} \delta L \, dt. \tag{5.3.6}$$

It is clear that for $Q_i = 0$, the Hamiltonian principle represents a natural problem of variational calculus, and for this case, the equations of motion of the dynamical system (the actual trajectories) are the extremals of the variational problem. These extremals make the action integral minimal (for short intervals of time). However, the variational equation (5.3.4) contains $n + 1$ functions L and $Q_i, i = 1, ..., n$, and does not possess any extremal qualities in itself. In fact, the equation (5.3.4) represents merely an integrated form of the Lagrange–D'Alembert principle. Therefore, all optimal (extremal) characteristics of the dynamical systems, for the case of nonconservative systems, are lost. We can therefore state (5.3.4), that is, Hamilton's principle for a purely nonconservative dynamical system as: *Along the actual trajectories of a purely nonconservative, holonomic dynamical system, the time integral of the function $\delta L + Q_i \delta q_i$ is equal to zero* (see [122, p. 32]).

5.4 The Functional Containing the Higher Order Derivatives

As a simple generalization of the previous considerations, let us suppose that the action integral, or, more precisely the Lagrangian function, contains the derivatives of the generalized coordinates higher than the first order, namely,

$$I = \int_{t_0}^{t_1} L\left(t, \mathbf{q}, \dot{\mathbf{q}}, \ddot{\mathbf{q}}, ..., \overset{(n)}{\mathbf{q}}\right) dt. \tag{5.4.1}$$

For the sake of simplicity, let us consider the case of a dynamical system with one degree of freedom, supposing that the Lagrangian function depends on the first and second derivatives:

$$I = \int_{t_0}^{t_1} L\left(t, q, \dot{q}, \ddot{q}\right) dt. \tag{5.4.2}$$

The variation of this action integral is

$$\delta I = \int_{t_0}^{t_1} \left(\frac{\partial L}{\partial q} \delta q + \frac{\partial L}{\partial \dot{q}} \delta \dot{q} + \frac{\partial L}{\partial \ddot{q}} \delta \ddot{q}\right) dt. \tag{5.4.3}$$

Employing the commutativity rules $\delta \dot{q} = \frac{d}{dt} \delta q$ and $\delta \ddot{q} = \frac{d^2}{dt^2} \delta q$, the second term is transformed in the usual way:

$$\frac{\partial L}{\partial \dot{q}} \frac{d}{dt} \delta q = \frac{d}{dt}\left(\frac{\partial L}{\partial \dot{q}} \delta q\right) - \frac{d}{dt} \frac{\partial L}{\partial \dot{q}} \delta q. \tag{5.4.4}$$

We next transform the third term in (5.4.3) twice, as follows:

$$
\begin{aligned}
\frac{\partial L}{\partial \ddot{q}} \frac{d}{dt} \left(\frac{d}{dt} \delta q \right) &= \frac{d}{dt} \left(\frac{\partial L}{\partial \ddot{q}} \frac{d}{dt} \delta q \right) - \frac{d}{dt} \frac{\partial L}{\partial \ddot{q}} \frac{d}{dt} \delta q \\
&= \frac{d}{dt} \left(\frac{\partial L}{\partial \ddot{q}} \delta \dot{q} \right) - \frac{d}{dt} \left(\frac{d}{dt} \frac{\partial L}{\partial \ddot{q}} \delta q \right) + \left(\frac{d^2}{dt^2} \frac{\partial L}{\partial \ddot{q}} \right) \delta q.
\end{aligned}
\tag{5.4.5}
$$

Entering with (5.4.4) and (5.4.5) into (5.4.3), and integrating, we have

$$
\begin{aligned}
\delta I &= \left[\left(\frac{\partial L}{\partial \dot{q}} - \frac{d}{dt} \frac{\partial L}{\partial \ddot{q}} \right) \delta q + \frac{\partial L}{\partial \ddot{q}} \delta \dot{q} \right] \Bigg|_{t_0}^{t_1} \\
&\quad + \int_{t_0}^{t_1} \left(\frac{\partial L}{\partial \dot{q}} - \frac{d}{dt} \frac{\partial L}{\partial \dot{q}} + \frac{d^2}{dt^2} \frac{\partial L}{\partial \ddot{q}} \right) \delta q \, dt.
\end{aligned}
\tag{5.4.6}
$$

The stationary condition $\delta I = 0$ leads to the Euler–Lagrangian equation

$$
\frac{\partial L}{\partial \dot{q}} - \frac{d}{dt} \frac{\partial L}{\partial \dot{q}} + \frac{d^2}{dt^2} \frac{\partial L}{\partial \ddot{q}} = 0,
\tag{5.4.7}
$$

and the boundary conditions at the ends of the interval, t_0 and t_1, are

$$
\left(\frac{\partial L}{\partial \dot{q}} - \frac{d}{dt} \frac{\partial L}{\partial \ddot{q}} \right) \delta q + \frac{\partial L}{\partial \ddot{q}} \delta \dot{q} = 0, \quad \text{at} \quad t = t_0 \quad \text{and} \quad t = t_1.
\tag{5.4.8}
$$

The boundary conditions will be identically satisfied if the following two boundary conditions are given in advance:

$$
q(t_0) = A, \quad \dot{q}(t_0) = B, \quad q(t_1) = C, \quad \dot{q}(t_1) = D,
\tag{5.4.9}
$$

where A, B, C, and D are given constants.

In a similar manner, we can demonstrate that for the functional (5.4.1) the Euler–Lagrangian equation reads

$$
\frac{\partial L}{\partial \dot{q}} - \frac{d}{dt} \frac{\partial L}{\partial \dot{q}} + \frac{d^2}{dt^2} \frac{\partial L}{\partial \ddot{q}} - \frac{d^3}{dt^3} \frac{\partial L}{\partial \dddot{q}} + \frac{d^4}{dt^4} \frac{\partial L}{\partial \ddddot{q}} - \cdots = 0.
\tag{5.4.10}
$$

Note, finally, that equation (5.4.7) is an ordinary differential equation of the fourth order, and the Euler–Lagrangian equation corresponding to the functional given by (5.4.1) is of $2n$th order.

5.5 The Functional Depending upon Several Independent Variables

In this section we shall demonstrate that Hamilton's principle does have important applications in setting up partial differential equations arising in various

branches of physics and engineering. We commence our considerations with a
brief discussion of the classical Hamilton's principle adjusted for use in field theory. In fact, the following discussion can be regarded as a natural generalization
of the Hamiltonian principle describing the behavior of a holonomic dynamical
system with a finite number of degrees of freedom with the given boundary
conditions, as considered in a previous section of the monograph.

Let the Hamiltonian action integral be formulated as

$$I = \int_{t_0}^{t_1} \int_V L\left(t, x_1, x_2, x_3, u, \frac{\partial u}{\partial x_1}, \frac{\partial u}{\partial x_2}, \frac{\partial u}{\partial x_2}, \frac{\partial u}{\partial t}\right) dV dt, \qquad (5.5.1)$$

where L denotes a given Lagrangian function depending on the time t, on
the spatial orthogonal coordinates x_1, x_2, x_3, and also on the field function
$u = u(t, x_1, x_2, x_3)$, characterizing the physical field in question and its partial derivatives. We suppose that the time interval (t_0, t_1) is specified and also
that the volume V over which the physical process is taking place is fixed. This
volume is bounded by a given surface S.

We introduce a small arbitrary increment δu of the field function $u(t, \mathbf{x})$,
$\mathbf{x} = (x_1, x_2, x_3)$, without any changes in the independent variables t and $\mathbf{x}$ by
writing

$$\bar{u}(t, \mathbf{x}) = u(t, \mathbf{x}) + \delta u \equiv u(t, \mathbf{x}) + \varepsilon h(t, \mathbf{x}). \qquad (5.5.2)$$

We call $\varepsilon h(t, \mathbf{x})$ *the variation of the field function.* Here $|\varepsilon| \ll 1$ is a small
constant and $h(t, \mathbf{x})$ is a continuous function with continuous derivatives in
$(t_0, t_1) \times V$. Considering the variation as a differential operator we postulate
commutativity properties of the Lagrangian variation δ with the partial derivatives

$$\frac{\partial}{\partial t}\delta(\cdot) - \delta\frac{\partial}{\partial t}(\cdot) = 0, \quad \frac{\partial}{\partial x_i}\delta(\cdot) - \delta\frac{\partial}{\partial x_i}(\cdot) = 0, \quad i = 1, 2, 3, \qquad (5.5.3)$$

and with the operator of integration

$$\delta \int_V (\cdot)\, dV - \int_V \delta(\cdot)\, dV = 0. \qquad (5.5.4)$$

With these rules in mind, we can easily calculate the variation of the action
integral (5.5.1) for unchangeable and given boundaries

$$\begin{aligned}
\delta I &= I(\bar{u}) - I(u) \\
&= \int_{t_0}^{t_1} \int_V \left[\frac{\partial L}{\partial u} - \frac{\partial}{\partial t}\left(\frac{\partial L}{\partial(\partial u/\partial t)}\right)\right. \\
&\qquad\qquad \left. - \sum_{i=1}^{3} \frac{\partial}{\partial x_i}\left(\frac{\partial L}{\partial(\partial u/\partial x_i)}\right)\right] \delta u\, dV dt \\
&\quad + \int_{t_0}^{t_1} \int_V \left[\frac{\partial}{\partial t}\left(\frac{\partial L}{\partial(\partial u/\partial t)}\delta u\right)\right. \\
&\qquad\qquad \left. + \sum_{i=1}^{3} \frac{\partial}{\partial x_i}\left(\frac{\partial L}{\partial(\partial u/\partial x_i)}\delta u\right)\right] dV dt. \qquad (5.5.5)
\end{aligned}$$

The condition for stationarity of the action integral is defined by the variational equation

$$\delta I = 0. \tag{5.5.6}$$

This condition can be analyzed in a variety of ways, depending on the boundary conditions of the physical problem in question. Here, we will focus our attention on the simplest boundary conditions. Namely, let us suppose that the function $u(t, \mathbf{x})$ is fully specified at the instants of time t_0 and t_1, which is equivalent to the equations

$$
\begin{aligned}
\delta u(t, \mathbf{x}) &= 0, \quad \text{on the boundary } S \text{ of } V \text{ for every } t, \\
\delta u(t_0, \mathbf{x}) &= \delta u(t_1, \mathbf{x}) = 0, \quad \text{everywhere in } V \text{ including boundary } S.
\end{aligned}
\tag{5.5.7}
$$

For this case, it is easy to verify, by applying the divergence theorem (see [49]), that the last two terms in (5.5.5) vanish. Hence, we have

$$\int_{t_0}^{t_1} \int_V \left[\frac{\partial L}{\partial u} - \frac{\partial}{\partial t} \left(\frac{\partial L}{\partial (\partial u / \partial t)} \right) - \sum_{i=1}^{3} \frac{\partial}{\partial x_i} \left(\frac{\partial L}{\partial (\partial u / \partial x_i)} \right) \right] dV dt \delta u = 0. \tag{5.5.8}$$

However, since δu is arbitrary, away from boundaries, this equation can be satisfied only when

$$\frac{\partial L}{\partial u} - \frac{\partial}{\partial t} \left(\frac{\partial L}{\partial (\partial u / \partial t)} \right) - \sum_{i=1}^{3} \frac{\partial}{\partial x_i} \left(\frac{\partial L}{\partial (\partial u / \partial x_i)} \right) = 0. \tag{5.5.9}$$

This is the Euler–Lagrangian equation for the field $u = u(t, x_1, x_2, x_3)$ that makes the action integral (5.5.1) stationary. It represents a partial differential equation with the partial derivatives of the second order, and the variational problem is reduced to the problem of finding a solution of this equation with given boundary and initial conditions.

For the sake of simplicity, let us suppose that the Lagrangian function is of the form

$$L = L\left(t, x, u, \frac{\partial u}{\partial t}, \frac{\partial u}{\partial x} \right). \tag{5.5.10}$$

It is easy to verify that the explicit form of the Euler–Lagrangian equation (5.5.9) is

$$
\begin{aligned}
&\frac{\partial^2 L}{\partial (\partial u / \partial t) \, \partial (\partial u / \partial t)} \frac{\partial^2 u}{\partial t^2} + 2 \frac{\partial^2 L}{\partial (\partial u / \partial t) \, \partial (\partial u / \partial x)} \frac{\partial^2 u}{\partial t \partial x} \\
&+ \frac{\partial^2 L}{\partial (\partial u / \partial x) \, \partial (\partial u / \partial x)} \frac{\partial^2 u}{\partial x^2} + \frac{\partial^2 L}{\partial u \, \partial (\partial u / \partial t)} \frac{\partial u}{\partial t} + \frac{\partial^2 L}{\partial u \, \partial (\partial u / \partial x)} \frac{\partial u}{\partial x} \\
&+ \frac{\partial^2 L}{\partial t \, \partial (\partial u / \partial t)} + \frac{\partial^2 L}{\partial x \, \partial (\partial u / \partial x)} - \frac{\partial L}{\partial u} = 0. \tag{5.5.11}
\end{aligned}
$$

In a similar way we can analyze the case in which the Lagrangian function depends upon the partial derivatives of the second and higher orders.

For example, it is easy to verify that for the Lagrangian function of the form

$$L = L\left(x, y, u, \frac{\partial u}{\partial x}, \frac{\partial u}{\partial y}, \frac{\partial^2 u}{\partial x^2}, \frac{\partial^2 u}{\partial x \partial y}, \frac{\partial^2 u}{\partial y^2}\right), \tag{5.5.12}$$

the Euler–Lagrangian equation is of the form

$$\frac{\partial L}{\partial u} - \frac{\partial}{\partial x}\frac{\partial L}{\partial (\partial u/\partial x)} - \frac{\partial}{\partial y}\frac{\partial L}{\partial (\partial u/\partial y)} + \frac{\partial^2}{\partial x^2}\frac{\partial L}{\partial (\partial^2 u/\partial x^2)}$$

$$+ \frac{\partial^2}{\partial x \partial y}\frac{\partial L}{\partial (\partial^2 u/\partial x \partial y)} + \frac{\partial^2}{\partial y^2}\frac{\partial L}{\partial (\partial^2 u/\partial y^2)} = 0, \tag{5.5.13}$$

which is a partial differential equation with partial derivatives up to the fourth order.

Example 5.4.1. *Heat conduction with finite wave speed.* Consider the generalized heat conduction equation given as

$$\tau \frac{\partial^2 T}{\partial t^2} + \frac{\partial T}{\partial t} = a\nabla^2 T, \tag{5.5.14}$$

where $T = T(t, x_1, x_2, x_3)$ is a nonstationary temperature field in a solid body, a denotes the constant thermal diffusivity, τ represents the material thermal relaxation time (that we assume to be constant), and ∇^2 is the Laplace differential operator defined as $\nabla^2(\cdot) \equiv \partial^2/\partial x_1^2(\cdot) + \partial^2/\partial x_2^2(\cdot) + \partial^2/\partial x_3^2(\cdot)$. Such an equation is hyperbolic in nature and has a finite speed C of propagation of thermal disturbance, equal to

$$C = \sqrt{\frac{a}{\tau}}. \tag{5.5.15}$$

If $\tau \to 0$, we have the classical Fourier parabolic differential equation of heat transfer

$$\frac{\partial T}{\partial t} = a\nabla^2 T, \tag{5.5.16}$$

which has the physically inconsistent property that the effects of a thermal disturbance are instantaneously manifested at a distance infinitely far from the thermal disturbance, that is, $C \to \infty$. For example, if one end of an infinitely long rod is touched by a hot body, all points of the rod are effected instantaneously, although by very small amounts, at large distances from the place of the disturbance.

It is easy to verify that the Euler–Lagrangian equation (5.5.9) for the Lagrangian function of the form (see [122] and the references cited therein)

$$L = \left[\frac{\tau}{2}\left(\frac{\partial T}{\partial t}\right)^2 - \frac{a}{2}\sum_{i=1}^{3}\left(\frac{\partial T}{\partial x_i}\right)^2\right]e^{t/\tau} \tag{5.5.17}$$

will generate the equation

$$\left(\tau\frac{\partial^2 T}{\partial t^2} + \frac{\partial T}{\partial t} - a\nabla^2 T\right)e^{t/\tau} = 0. \tag{5.5.18}$$

Since for the hyperbolic equation, $e^{t/\tau} \neq 0$, we arrive at the equation (5.5.14).

However, it was undoubtedly proved by numerous authors that the parabolic heat conduction equation (5.5.16), together with the given boundary and initial conditions, do not permit the variational description in the strict sense of the Hamiltonian variational principle.

For the heat conduction equation (5.5.14) we can construct a time invariant by using the results of section 1.8 (see (1.8.36)). The details of the procedure are given in [10]. The main result is crucially dependent on the orthogonality property of certain eigenfunctions. As an application for the construction of the time invariant, consider a rod of length l in which heat conduction is taking place according to the equation (5.5.14). Suppose that boundary and initial conditions are specified, so that the process is described by the following system of equations (the one-dimensional variant of (5.5.14)):

$$
\begin{aligned}
\tau\frac{\partial^2 T}{\partial t^2} + \frac{\partial T}{\partial t} &= a\frac{\partial^2 T}{\partial x^2}, \quad 0 < x < l, \quad t > 0, \\
\frac{\partial T(0,t)}{\partial x} &= 0, \quad T(l,t) = 0, \\
T(x,0) &= f_1(x), \quad \frac{\partial T(x,0)}{\partial t} = f_2(x),
\end{aligned}
\tag{5.5.19}
$$

where $f_1(x)$ and $f_2(x)$ are given functions. We assume the solution of (5.5.19) in the form $T = X(x)U(t)$. By substituting this into (5.5.19) we obtain

$$\frac{\tau\dfrac{d^2 U}{dt^2} + \dfrac{dU}{dt}}{U} = \frac{a\dfrac{d^2 X}{dx^2}}{X} = const. = -\omega^2. \tag{5.5.20}$$

The temporal evolution $U(t)$ is governed by the solution of the system of equations

$$\tau\frac{d^2 U}{dt^2} + \frac{dU}{dt} + \omega_n^2 U = 0, \tag{5.5.21}$$

where $\omega_n^2, n = 1, 2, ...,$ are eigenvalues determined from

$$a\frac{d^2 X}{dx^2} + \omega_n^2 X = 0, \tag{5.5.22}$$

subject to

$$\frac{dX}{dx}(0) = 0, \quad X(l) = 0. \tag{5.5.23}$$

Equation (5.5.21) is of the type considered earlier (see (1.8.23)). We found that (see (1.8.36) with $1/\tau = 2k, a/\tau = \omega^2$)

$$\frac{1}{2}e^{t/\tau}\left[\tau\left(\frac{dU}{dt}\right)^2 + \frac{a}{\tau}U^2 + \frac{1}{\tau}U\left(\frac{dU}{dt}\right)\right] = C_n, \qquad (5.5.24)$$

where $C_n, n = 1, 2, ...$, are constants.

Since (5.5.22), (5.5.23) generates the set of orthogonal functions $X_n = D_n \cos\left(\frac{2n-1}{2l}\pi x\right)$ with $D_n = const.$ $\int_0^l X_n X_m dx = 0$, if $n \neq m$ the expression (5.5.24) leads to the following time invariant (see [10]):

$$B(t) = \frac{1}{2}\int_0^l e^{t/\tau}\left[\tau\left(\frac{\partial T}{\partial t}\right)^2 + \frac{a}{\tau}\left(\frac{\partial T}{\partial x}\right)^2 + \frac{1}{\tau}T\left(\frac{\partial T}{\partial t}\right)\right]dx. \qquad (5.5.25)$$

It is easy to verify that

$$\frac{dB}{dt} = 0 \qquad (5.5.26)$$

on the solution of (5.5.19). The property (5.5.26) could serve as a basis for obtaining approximate solutions to (5.5.19). An example of such an application is presented in [10].

Example 5.4.2. *Generalized wave equation.* Consider a partial differential equation of the form

$$\tau\frac{\partial^2 u}{\partial t^2} + \frac{\partial u}{\partial t} - k\frac{\partial^2 u}{\partial x^2} + k_1\frac{\partial u}{\partial x} = 0, \quad 0 < x < l, \quad t > 0,$$

$$\frac{\partial u(0,t)}{\partial x} = 0, \quad \frac{\partial u(l,t)}{\partial x} = 0, \qquad (5.5.27)$$

and suppose that certain initial conditions are specified. Equation (5.5.27)$_1$ arises in the wave theory [128]. It can be verified by applying (5.4.9) that (5.5.27) is the Euler–Lagrangian equation of the following functional:

$$I = \int_0^t \int_0^l \frac{e^{\frac{t}{\tau} - \frac{k_1}{k}x}}{2}\left[\tau\left(\frac{\partial u}{\partial t}\right)^2 - k\left(\frac{\partial u}{\partial x}\right)^2\right]dxdt, \qquad (5.5.28)$$

so that $\delta I = 0$ reproduces (5.5.27). By applying the procedure as presented in [10] (separation of variables method) or by using the Noether-type theory developed in [11], one can show that the following functional is a time invariant for the system (5.5.27):

$$B = \int_0^l \frac{e^{\frac{t}{\tau} - \frac{k_1}{k}x}}{2}\left[\tau\left(\frac{\partial u}{\partial t}\right)^2 + k\left(\frac{\partial u}{\partial x}\right)^2 + u\frac{\partial u}{\partial t}\right]dx. \qquad (5.5.29)$$

Thus, for (5.5.29), we have

$$\frac{dB}{dt} = 0. \qquad (5.5.30)$$

Example 5.4.3. *Axially loaded elastic rod with external friction.* Consider now the system of equations describing the lateral motion of an elastic rod that is simply supported at both ends and axially loaded by a constant force. We assume also that there is external damping linearly proportional to the velocity (see [65, p. 158]). The differential equation of motion reads

$$EI\frac{\partial^4 u}{\partial x^4} + \lambda\frac{\partial^2 u}{\partial x^2} + \beta\frac{\partial u}{\partial t} + \rho\frac{\partial^2 u}{\partial t^2} = 0, \quad 0 < x < l, \quad t > 0, \tag{5.5.31}$$

with the following boundary conditions, corresponding to simply supported ends:

$$\begin{aligned} u(0,t) &= 0, \quad u(l,t) = 0, \\ \frac{\partial^2 u}{\partial x^2}(0,t) &= 0, \quad \frac{\partial^2 u}{\partial x^2}(l,t) = 0. \end{aligned} \tag{5.5.32}$$

In (5.5.31), EI is the bending rigidity of the rod, λ is the axial force, ρ is the line density, and β is the viscous friction coefficient. All quantities are assumed to be constant. The system (5.5.31), (5.5.32) could be derived from the stationary conditions of the functional

$$I = \frac{1}{2}\int_0^t \int_0^l e^{\frac{\beta}{\rho}t}\left[EI\left(\frac{\partial^2 u}{\partial x^2}\right)^2 - \lambda\left(\frac{\partial u}{\partial x}\right)^2 - \rho\left(\frac{\partial u}{\partial t}\right)^2\right] dx dt. \tag{5.5.33}$$

Again, we may construct a time invariant for the system (5.5.31), (5.5.32). We shall use the procedure presented in [10]. Thus, we assume solution of (5.5.31), (5.5.32) in the form $u = X(x)T(t)$, so that

$$\frac{EI\dfrac{d^4 X}{dx^4} + \lambda\dfrac{dX^2}{dx^2}}{X} = -\frac{\beta\dfrac{dT}{dt} + \rho\dfrac{d^2 T}{dt^2}}{T} = \omega^2, \tag{5.5.34}$$

where ω^2 is a separation constant that is determined from

$$EI\frac{d^4 X}{dx^4} + \lambda\frac{dX^2}{dx^2} - \omega^2 X, \tag{5.5.35}$$

subject to

$$\begin{aligned} X(0) &= 0; \quad X(l) = 0, \\ \frac{d^2 X(0)}{dx^2} &= 0; \quad \frac{d^2 X(l)}{dx^2} = 0. \end{aligned} \tag{5.5.36}$$

The temporal part $T(t)$ has a conservation law of the form (1.8.36) and since (5.5.35), (5.5.36) generates an orthogonal set of functions, we conclude that

$$B = \frac{1}{2}\int_0^l e^{\frac{\beta}{\rho}t}\left[\frac{EI}{\rho}\left(\frac{\partial^2 u}{\partial x^2}\right)^2 - \frac{\lambda}{\rho}\left(\frac{\partial u}{\partial x}\right)^2 + \left(\frac{\partial u}{\partial t}\right)^2 + \frac{\beta}{\rho}u\left(\frac{\partial u}{\partial t}\right)\right] dx. \tag{5.5.37}$$

It can be easily shown by direct calculation that $dB/dt = 0$.

Chapter 6

Variable End Points, Natural Boundary Conditions, Bolza Problems

6.1 Introduction

In this chapter we discuss some generalizations of the Hamiltonian variational principle concerning the various that that can arise in applications. We shall consider in particular the cases in which the initial or terminal configurations (or both) are not specified. Also, it may happen that the time interval in which the evolutionary process is taking place is not given. For these cases the Hamiltonian principle usually produces characteristic information about the boundary conditions and the transversality conditions. In order to express a physical process correctly, we sometimes must add some terms outside the sign of the action integral, and these types of optimization problems are usually referred to as the problems of Bolza.

6.2 Time Interval (t_0, t_1) Specified, $q_i(t_0)$, $q_i(t_1)$ Free

It can frequently happen that the time interval (t_0, t_1) in which the dynamical process is taking place is specified, but the initial or terminal position, (configuration A and B or both) denoted by (5.2.1) are *not given*. For these cases we shall demonstrate that the Hamiltonian variational principle will automatically produce the necessary boundary conditions, which are usually named *natural boundary conditions*.

(a) Consider first the case for which

$$q_i(t_0) = A_i, \quad A_i \text{ given constants} \quad q_i(t_1) \text{ are not specified.} \tag{6.2.1}$$

It is clear from the general expression (5.2.8) that $\delta q_i(t_0) = 0$, but $\delta q_i(t_1) \neq 0$ and is not specified. Thus, in order to satisfy the stationary condition $\delta I = 0$, besides the Euler–Lagrangian equations (5.2.9), we must also have

$$\left.\frac{\partial L}{\partial \dot{q}_i}\right|_{t=t_i} = 0, \quad i = 1,...,n. \tag{6.2.2}$$

This is a necessary set of natural boundary conditions that must be satisfied. It serves for finding the constants of integration during the process of solution of the Euler–Lagrangian equations (5.2.9). Let us stress, as we mentioned already, that *all types of boundary conditions* considered in this chapter *will never change the form of the Euler–Lagrangian differential equations.*

(b) Similarly, if

$$q_i(t_0) \text{ are not specified}, \quad q_i(t_1) = B_i, \quad B_i \text{ given constants}, \tag{6.2.3}$$

the natural boundary conditions are

$$\left.\frac{\partial L}{\partial \dot{q}_i}\right|_{t=t_0} = 0, \quad i = 1,...,n. \tag{6.2.4}$$

(c) Finally, if the initial and terminal configurations A and B are not specified, the $2n$ natural boundary conditions are

$$\left.\frac{\partial L}{\partial \dot{q}_i}\right|_{t=t_0} = 0, \quad \left.\frac{\partial L}{\partial \dot{q}_i}\right|_{t=t_1} = 0, \quad i = 1,...,n. \tag{6.2.5}$$

Combinations of these cases are depicted in Figure 6.2.1.

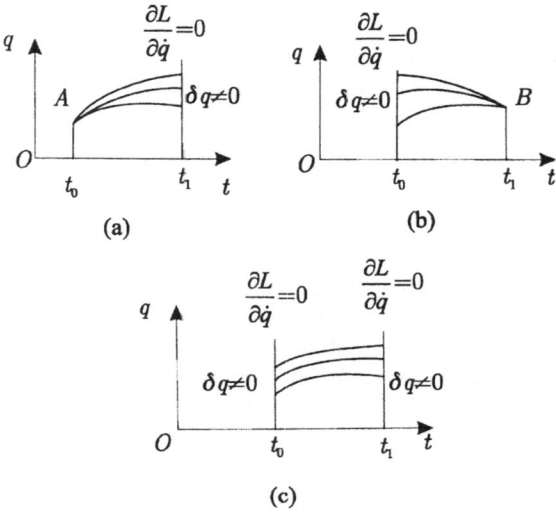

Figure 6.2.1

If the Hamiltonian principle is expressed in terms of canonical variables $q_i(t)$, $p_i(t)$, we find from $\delta I_{can} = 0$, given by (5.2.12), that the natural boundary conditions for the cases discussed above are, respectively,

$$
\begin{align}
\text{(a)} \quad & p_i(t_0) & = & \quad 0, \\
\text{(b)} \quad & p_i(t_1) & = & \quad 0, \\
\text{(c)} \quad & p_i(t_0) & = & \quad 0, \quad p_i(t_1) = 0.
\end{align}
\tag{6.2.6}
$$

Example 6.2.1. *The cylindrical brachistochrone.* Let us consider the problem of finding a curve on a vertical circular cylinder, a curve that evolves from a given point A and terminates on the vertical line that is the generator of the cylinder. Under the influence of gravity, a particle should move on this curve in the least time. We shall assume that the particle starts from rest, that friction force between the particle and the cylinder is negligible, and that the terminal vertical line is located by a given angle θ_1, as shown in Figure 6.2.2.

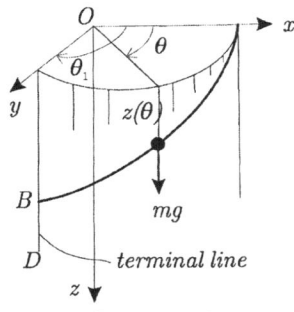

Figure 6.2.2

Since the particle has two degrees of freedom we shall select the cylindrical coordinates $q_1 = \theta, q_2 = z$. As in Example 5.2.1 the total energy of the particle is equal to zero, namely,

$$
E = \frac{m}{2}\left(r^2\dot{\theta}^2 + \dot{z}^2\right) - mgz = 0,
\tag{6.2.7}
$$

where r is the radius of the cylinder. Taking the angle θ as an independent variable, we find $dt = (\sqrt{r^2 + (z')^2})/(\sqrt{2gz})$, where $(\cdot)' = \frac{d}{d\theta}(\cdot)$. Therefore, the problem is reduced to find the minimum $t_{\min} = \min_z T$ of the following functional:

$$
T = \int_0^{\theta_1} L(\theta, z, z')\, d\theta = \frac{1}{\sqrt{2g}} \int_0^{\theta_1} \sqrt{\frac{r^2 + (z')^2}{z}}\, d\theta.
\tag{6.2.8}
$$

We shall suppose that the location of the terminal line CD is specified by the angle θ_1, while the position of the point B on this line is not given. Thus, a boundary condition is missing, although both boundaries in the integral (6.2.8) are specified. From this, it follows that the missing boundary condition is given by the case (a); that is, with equation (6.2.2)

$$\left.\frac{\partial L}{\partial z'}\right|_{\theta=\theta_1} = \frac{z'}{\sqrt{2gz\left(r^2 + (z')^2\right)}} = 0, \qquad (6.2.9)$$

namely, we find the following *natural boundary condition*:

$$z'\left(\theta_1\right) = 0. \qquad (6.2.10)$$

Since the Lagrangian L defined by (6.2.8) does not depend upon the independent variable θ, we employ the Jacobi conservation law $(\partial L/\partial z')\, z' - L = const.$ which leads to the relation

$$r^2 + \left(z'\right)^2 = \frac{K}{2}, \quad K = const. \qquad (6.2.11)$$

We can easily integrate this differential equation in parametric form. Let

$$z' = \frac{dz}{d\theta} = r \cot \frac{\lambda}{2}, \qquad (6.2.12)$$

where λ is a new parameter. From (6.2.11) we find that

$$z = \frac{K}{2r^2} \left(1 - \cos \lambda\right), \qquad (6.2.13)$$

and from this, we have $dz = \left(K/2r^2\right) \sin \lambda d\lambda$. Substituting this into (6.2.12) it follows that $rd\theta = \tan\left(\lambda/2\right) dz = \left(K/2r^2\right)\left(1 - \cos \lambda\right) d\lambda$. Integrating, we arrive at

$$r\theta = \frac{K}{2r^2} \left(\lambda - \sin \lambda\right) + K_1, \qquad (6.2.14)$$

where K_1 is a new constant.

Let us select λ in such a way that $\lambda = 0$ when $\theta = 0$. Therefore, the constant $K_1 = 0$. In order to determine K we apply the natural boundary condition (6.2.10) from which we easily find $K = 2r^3\theta_1/\pi$. Therefore, the brachistochrone on the vertical cylinder is found to be

$$z = \frac{r\theta_1}{\pi} \left(1 - \cos \lambda\right), \quad \theta = \frac{\theta_1}{\pi} \left(\lambda - \sin \lambda\right). \qquad (6.2.15)$$

Example 6.2.2. Transversal vibrations of a beam [68], [130]. Let us consider a straight uniform beam having a length l whose mass per unit length (the line density) is $m\,(kg/m)$. A cross-sectional moment of inertia is I_x, and a modulus of elasticity is E. The beam performs small transverse bending vibrations in the vertical plane xAy. Using the Hamiltonian principle we shall find the differential equation of motion and the corresponding boundary conditions. Let $y(t, x)$ be the transversal displacement of the beam. The kinetic energy of the beam is

$$T = \frac{1}{2} \int_0^l m \left(\frac{\partial y}{\partial t} \right)^2 dx, \tag{6.2.16}$$

where t is the time and x is an independent variable oriented along the axis of the rod, that is, $x \in [0, l]$ (see Figure 6.2.3).

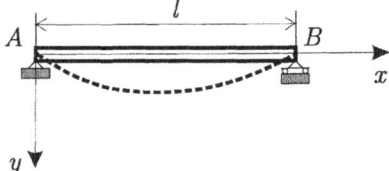

Figure 6.2.3

It is well known from the elasticity that the potential energy of bending is

$$\Pi = \frac{1}{2} \int_0^l \frac{M^2}{EI_x} dx = \frac{1}{2} \int_0^l EI_x \left(\frac{\partial^2 y}{\partial x^2} \right)^2 dx, \tag{6.2.17}$$

where M is the bending moment of the beam, and $M = EI_x \left(\partial^2 y / \partial x^2 \right)$. Note that the influence of the shear forces on the deformation of the beam is neglected. Thus (6.2.17) corresponds to the Bernoulli–Euler beam theory.

The action integral now becomes

$$I = \int_{t_0}^{t_1} \left\{ \int_0^l \left[\frac{1}{2} m \left(\frac{\partial y}{\partial t} \right)^2 - \frac{1}{2} EI_x \left(\frac{\partial^2 y}{\partial x^2} \right)^2 \right] dx \right\} dt. \tag{6.2.18}$$

Performing a variation of I, we find

$$\delta I = \int_{t_0}^{t_1} \int_0^l \left[m \left(\frac{\partial y}{\partial t} \right) \delta \left(\frac{\partial y}{\partial t} \right) - EI_x \left(\frac{\partial^2 y}{\partial x^2} \right) \delta \left(\frac{\partial^2 y}{\partial x^2} \right) \right] dx\, dt. \tag{6.2.19}$$

Using the commutativity rules $\delta (\partial y / \partial t) = (\partial / \partial t)\, \delta y$ and $\delta (\partial^2 y / \partial x^2) = (\partial^2 / \partial x^2)\, \delta y$, we have

$$\int_{t_0}^{t_1} \int_0^l m \frac{\partial y}{\partial t} \frac{\partial}{\partial t} (\delta y)\, dx\, dt = \int_0^l \left[m \frac{\partial y}{\partial t} \delta y \right]_{t_0}^{t_1} dx - \int_{t_0}^{t_1} \int_0^l m \frac{\partial^2 y}{\partial t^2} (\delta y)\, dx\, dt. \tag{6.2.20}$$

Next we integrate the second term in (6.2.19) by parts twice, as follows:

$$\int_{t_0}^{t_1} \int_0^l EI \left(\frac{\partial^2 y}{\partial x^2} \right) \frac{\partial^2}{\partial x^2} (\delta y)\, dx\, dt$$

$$= \int_{t_0}^{t_1} \left[EI \frac{\partial^2 y}{\partial x^2} \delta \left(\frac{\partial y}{\partial x} \right) - EI \frac{\partial^3 y}{\partial x^3} \delta y \right]_0^l dt + \int_{t_0}^{t_1} \int_0^l EI \frac{\partial^4 y}{\partial x^4} \delta y\, dx\, dt. \tag{6.2.21}$$

Therefore, the stationary condition $\delta I = 0$ becomes

$$\delta I = \int_0^l \left[m \frac{\partial y}{\partial t} \delta y \right]\Bigg|_{t_0}^{t_1} dx + EI_x \int_{t_0}^{t_1} \left[\frac{\partial^2 y}{\partial x^2} \delta \left(\frac{\partial y}{\partial x} \right) - \frac{\partial^3 y}{\partial x^3} \delta y \right]\Bigg|_0^l dt$$

$$- \int_{t_0}^{t_1} \int_0^l \left[m \frac{\partial^2 y}{\partial t^2} + EI_x \frac{\partial^4 y}{\partial x^4} \right] \delta y\, dx\, dt = 0. \tag{6.2.22}$$

Since by the definition of the varied path, $\delta y\,(x, t_0) = \delta y\,(x, t_1) = 0$, the first expression on the right-hand side of (6.2.22) is equal to zero. Similarly, since $\delta y\,(x, t)$ is arbitrary, we conclude that the third expression on the right-hand side vanishes if

$$\frac{\partial^2 y}{\partial t^2} + b^2 \frac{\partial^4 y}{\partial x^4} = 0, \quad b^2 = \frac{EI_x}{m}. \tag{6.2.23}$$

This equation is at the same time the Euler–Lagrangian equation for the Hamiltonian principle, whose action integral is defined by (6.2.18).

Therefore, equation (6.2.22) becomes

$$\delta I = EI_x \int_{t_0}^{t_1} \left[\frac{\partial^2 y}{\partial x^2} \delta \left(\frac{\partial y}{\partial x} \right) - \frac{\partial^3 y}{\partial x^3} \delta y \right]\Bigg|_0^l dt = 0. \tag{6.2.24}$$

Naturally, since this expression should be satisfied for every $t \in [t_0, t_1]$, we have

$$\frac{\partial^2 y}{\partial x^2} \delta \left(\frac{\partial y}{\partial x} \right) - \frac{\partial^3 y}{\partial x^3} \delta y = 0, \quad \text{for } x = 0 \text{ and } x = l. \tag{6.2.25}$$

These two equations will serve for analyzing the boundary conditions, which can be prescribed in accordance to the various end conditions of the beam.

It is useful to note that by applying the end conditions the following relations can be employed with respect to the geometric variable x.

- The deflection is equal to y and is zero at any rigid support.

- The slope is equal to $\partial y / \partial x$ and is zero at any clamped end.

- The moment is proportional to $\partial^2 y / \partial x^2$ and is zero at any simply supported end.

- The shear force is proportional to $\partial^3 y / \partial x^3$ and is zero at any free end.

Since the Euler–Lagrangian equation (6.2.23) is a fourth-order differential equation, four boundary conditions must be specified for a correctly formulated problem of transversal vibrations of the beam.

(a) *Simply supported ends.* This case is shown in Figure 6.2.3, and it is obvious that the imposed conditions are

$$y\,(0, t) = 0, \quad y\,(l, t) = 0. \tag{6.2.26}$$

Since the variations $\delta\left(\partial y/\partial x\right)$ are arbitrary, then vanishing of (6.2.25) requires two additional boundary conditions

$$\frac{\partial^2 y\left(0,t\right)}{\partial x^2} = 0, \quad \frac{\partial^2 y\left(l,t\right)}{\partial x^2} = 0. \tag{6.2.27}$$

(b) *Clamped ends.* The boundary conditions imposed on the beam are

$$y\left(0,t\right) = 0, \quad y\left(l,t\right) = 0, \quad \frac{\partial y\left(0,t\right)}{\partial x} = 0; \quad \frac{\partial y\left(l,t\right)}{\partial x} = 0. \tag{6.2.28}$$

Therefore all of the boundary conditions are specified, and no natural boundary conditions have to be added. Note that (6.2.25) is identically satisfied; namely, from (6.2.28) we have $\delta y\left(0,t\right) = \delta y\left(l,t\right) = 0$ and $\delta\left(\partial y\left(0,t\right)/\partial x\right) = \delta\left(\partial y\left(l,t\right)/\partial x\right) = 0$.

(c) *Left end built in; right end free.* Let the beam be clamped at the point $x = 0$ and free at the $x = l$. The boundary conditions are

$$y\left(0,t\right) = 0, \quad \frac{\partial y\left(0,t\right)}{\partial x} = 0, \tag{6.2.29}$$

while $y\left(l,t\right)$ and $\partial y\left(l,t\right)/\partial x$ are not specified. In order to have (6.2.25) satisfied we must have

$$\frac{\partial^2 y\left(l,t\right)}{\partial x^2} = 0, \quad \frac{\partial^3 y\left(l,t\right)}{\partial x^3} = 0. \tag{6.2.30}$$

The conditions (6.2.30) have the important mechanical interpretation. Since $M\left(l,t\right) = EI_x\frac{\partial^2 y(l,t)}{\partial x^2}$ the condition $(6.2.30)_1$ implies $M\left(l,t\right) = 0$. Also, the transversal force is given as (see [14]) $F_T\left(l,t\right) = EI_x\frac{\partial^3 y(l,t)}{\partial x^3}$. Therefore, $(6.2.30)_2$ implies $F_T\left(l,t\right) = 0$. The conditions $M\left(l,t\right) = 0$ and $F_T\left(l,t\right) = 0$ define a free end of a beam.

6.3   The Problem of Bolza

In section 1.4 we have seen that a given Lagrangian function $L\left(t,\mathbf{q},\dot{\mathbf{q}}\right)$ can always be replaced by a new Lagrangian L^* in the following way:

$$L^* = L + \frac{d}{dt}\Psi\left(t,q_1,...,q_n\right), \tag{6.3.1}$$

where to given L is added a total time derivative of an arbitrary function Ψ depending on time t and generalized coordinates q_i, without changing the form of the Euler–Lagrangian equations formed by means of Lagrangian function L, namely,

$$\frac{d}{dt}\frac{\partial L^*}{\partial \dot{q}_i} - \frac{\partial L^*}{\partial q_i} \equiv \frac{d}{dt}\frac{\partial L}{\partial \dot{q}_i} - \frac{\partial L}{\partial q_i} = 0, \quad i = 1,...,n. \tag{6.3.2}$$

Let us form the action integral $I = \int_{t_0}^{t_1} L^* dt$. Integrating (6.3.1) we have

$$I = \Psi\left(t, q_1, ..., q_n\right)\big|_{t_0}^{t_1} + \int_{t_0}^{t_1} L\left(t, q_1, ..., q_n, \dot{q}_1, ..., \dot{q}_n\right) dt. \qquad (6.3.3)$$

This form of the action integral is usually named the Hamiltonian principle in *Bolza form*. It is clear that the first term of this functional contributes to the variation of the action integral at the end points and will therefore influence the boundary conditions only.

Performing the variation of (6.3.3) and using (5.2.8) for the variation of $\int_{t_0}^{t_1} L dt$, one has

$$\delta I = \left(\frac{\partial \Psi}{\partial q_i} + \frac{\partial L}{\partial \dot{q}_i}\right) \delta q_i \bigg|_{t_0}^{t_1} + \int_{t_0}^{t_1} \left(\frac{\partial L}{\partial q_i} - \frac{d}{dt}\frac{\partial L}{\partial \dot{q}_i}\right) \delta q_i dt. \qquad (6.3.4)$$

Thus, the stationary condition $\delta I = 0$ generates the Euler–Lagrangian equations (6.3.2) and

$$\left(\frac{\partial \Psi}{\partial q_i} + \frac{\partial L}{\partial \dot{q}_i}\right) \delta q_i = 0 \quad \text{at} \quad t = t_0, \quad \text{and} \quad t = t_1. \qquad (6.3.5)$$

If the dynamical problem is formulated in such a way that the initial and terminal configurations A and B are given in the form (5.2.1), then the expression (6.3.5) vanishes since the variations $\delta q_i\left(t_0\right)$ and $\delta q_i\left(t_1\right)$ are equal to zero (see (5.2.3)).

If, however, the initial and terminal configurations are not given we have for the specified time interval (t_0, t_1) the following $2n$ natural boundary conditions:

$$\frac{\partial \Psi}{\partial q_i} + \frac{\partial L}{\partial \dot{q}_i} = 0 \quad \text{at} \quad t = t_0, \quad \text{and} \quad t = t_1, \quad i = 1, ..., n. \qquad (6.3.6)$$

It can frequently happen that the Bolza problem should be formulated in a slightly different form. For example,

$$I = \Psi_1\left(t_0, \mathbf{q}\left(t_0\right)\right) + \Psi_2\left(t_1, \mathbf{q}\left(t_1\right)\right) + \int_{t_0}^{t_1} L\left(t, \mathbf{q}, \dot{\mathbf{q}}\right) dt \qquad (6.3.7)$$

or

$$I = \Psi\left(t_0, t_1, \mathbf{q}\left(t_0\right), \mathbf{q}\left(t_1\right)\right) + \int_{t_0}^{t_1} L\left(t, \mathbf{q}, \dot{\mathbf{q}}\right) dt. \qquad (6.3.8)$$

For these more complex cases the essential features remain unchanged.

Example 6.3.1. Sturm–Liouville problem. As an illustration we shall demonstrate that the Sturm-Liouville equation

$$-\frac{d}{dt}\left[f_1\left(t\right)\frac{dx}{dt}\right] + f_0\left(t\right) x = \lambda g_0\left(t\right) x, \quad t_0 < t < t_1, \qquad (6.3.9)$$

together with the Sturm boundary conditions (see [29])

$$c_1 \dot{x}(t_0) + c_2 x(t_0) = 0, \quad c_3 \dot{x}(t_1) + c_4 x(t_1) = 0 \tag{6.3.10}$$

can be represented as the extremal problem of Hamilton's variational principle. Here, the time interval (t_0, t_1) is specified, λ is a constant, $f_1(t), f_0(t)$, and $g_0(t)$ are given functions of time, and $c_1, ..., c_4$ are given constants.

Before forming the corresponding action integral, we note that the Sturm–Liouville differential equation (6.3.9) can be derived as an Euler–Lagrangian equation of the functional whose Lagrangian function is

$$L = \frac{\dot{x}^2}{2} f_1(t) + \frac{1}{2} \left[\lambda g_0(t) - f_0(t) \right] x^2, \tag{6.3.11}$$

which is easy to verify.

Let us consider the following action integral in the Bolza form:

$$
\begin{aligned}
I &= \int_{t_0}^{t_1} \left\{ \frac{\dot{x}^2}{2} f_1(t) + \frac{1}{2} \left[\lambda g_0(t) - f_0(t) \right] x^2 \right\} dt \\
&\quad - \frac{1}{2} f_1(t_0) \left(\frac{c_2}{c_1} \right) \left[x(t_0) \right]^2 + \frac{1}{2} f_1(t_1) \left(\frac{c_4}{c_3} \right) \left[x(t_1) \right]^2 . \tag{6.3.12}
\end{aligned}
$$

Since the functions f_0, g_0, and f_1 are *given functions of time*, they are not affected by the process of variation of (6.3.12). Thus, the variation of (6.3.12) leads to

$$
\begin{aligned}
\delta I &= \int_{t_0}^{t_1} \left\{ f_1(t) \dot{x} \delta \dot{x} + \left[\lambda g_0(t) - f_0(t) \right] x \delta x \right\} dt - f_1(t_0) \left(\frac{c_2}{c_1} \right) x(t_0) \delta x(t_0) \\
&\quad + f_1(t_1) \left(\frac{c_4}{c_3} \right) x(t_1) \delta x(t_1) . \tag{6.3.13}
\end{aligned}
$$

Using the commutativity relation $\delta \dot{x} = (\delta x)^{\cdot}$ and noting that $f_1(t) \dot{x} \frac{d}{dt} \delta x = \frac{d}{dt} \left[f_1(t) \dot{x} \delta x \right] - \left[f_1(t) \ddot{x} + f_1(t) \dot{x} \right] \delta x$, we have

$$
\begin{aligned}
\delta I &= \int_{t_0}^{t_1} \left\{ -\frac{d}{dt} \left[f_1(t) \frac{dx}{dt} \right] - x \left[\lambda g_0(t) - f_0(t) \right] \right\} \delta x dt \\
&\quad + \frac{f_1(t_1)}{c_3} \left[c_3 \dot{x}(t_1) + c_4 x(t_1) \right] \delta x(t_1) \\
&\quad - \frac{f_1(t_0)}{c_1} \left[c_1 \dot{x}(t_0) + c_2 x(t_0) \right] \delta x(t_0) . \tag{6.3.14}
\end{aligned}
$$

The requirement for stationary I, that is, $\delta I = 0$, gives

(a) the differential equation (6.3.9), since the variation δx is arbitrary in the interval (t_0, t_1);

(b) the Sturm boundary conditions (6.3.10), since $\delta x(t_0)$ and $\delta x(t_1)$ are not specified at the boundaries and $f_1(t_1) / c_3$ and $f_2(t_0) / c_1$ are different from zero.

Example 6.3.2. Elastic vibration of a string. Consider as a second example the problem of vibrating string. We take the problem of a uniform and slightly elastic string, stretched at a large tension P between its two points The end points of the string, by means of two massless smooth rings A and B, are forced to move along two vertical rods, as shown in Figure 6.3.1a.

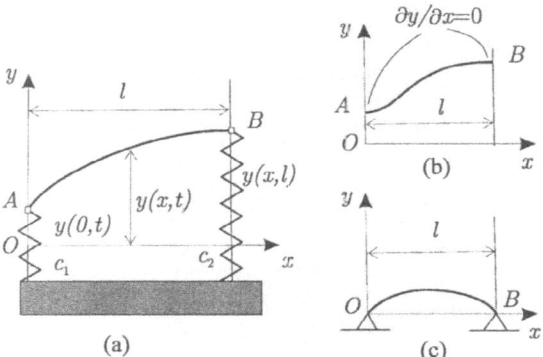

Figure 6.3.1

The rings are connected with two elastic springs having the spring constants c_1 and c_2, respectively. The length of the stretched string is l and the springs are not deformed when the string is horizontal and rings A and B are situated at the horizontal axis Ox. The string is assumed to be perfectly flexible, and it performs small transverse oscillations in the vertical plane xOy. We suppose that the tension P is permanently constant throughout the motion.

The string is a continuous system and the displacement of an arbitrary point of the string is a function of time t and the horizontal coordinate x, which are taken as independent variables.

Let the mass per unit length of the string be ρ. Then, if $y(x,t)$ is the displacement of a generic point, the kinetic energy of the string is

$$T = \frac{1}{2} \int_0^l \rho \left(\frac{\partial y}{\partial t} \right)^2 dx, \qquad (6.3.15)$$

while the potential energy, under the condition that the transversal displacement of the string is small, reads (see [84, p. 39])

$$\Pi_s = \frac{P}{2} \int_0^l \left(\frac{\partial y}{\partial x} \right)^2 dx. \qquad (6.3.16)$$

The potential energy of elastic springs at the end points $(0,0)$ and $(0,l)$ is

$$\Pi_o = \frac{1}{2} c_1 y(0,t)^2 + \frac{1}{2} c_2 y(l,t)^2. \qquad (6.3.17)$$

The Lagrangian function of the system consisting of string and two springs is therefore $L = T - \Pi_s - \Pi_o$, and the action integral becomes

$$I = \int_0^{t_1} L \, dt = \frac{1}{2} \int_{t_0}^{t_1} \int_0^l \left[\rho \left(\frac{\partial y}{\partial t} \right)^2 - P \left(\frac{\partial y}{\partial x} \right)^2 \right] dx \, dt$$

$$-\frac{1}{2} c_1 \int_{t_0}^{t_1} y(0,t)^2 \, dt - \frac{1}{2} c_2 \int_{t_0}^{t_1} y(l,t)^2 \, dt, \qquad (6.3.18)$$

where the time interval (t_0, t_1) is specified. Further, we suppose that variation of the initial and terminal configuration is equal to zero; that is, $\delta y(x, t_0) = \delta y(x, t_1) = 0$.

Calculating the first variation of I given by (6.3.18), we have

$$\delta I = \int_{t_0}^{t_1} \int_0^l \left[\rho \frac{\partial y}{\partial t} \delta \left(\frac{\partial y}{\partial t} \right) - P \frac{\partial y}{\partial x} \delta \left(\frac{\partial y}{\partial x} \right) \right] dx \, dt$$

$$-c_1 \int_{t_0}^{t_1} y(0,t) \, \delta y(0,t) \, dt - c_2 \int_{t_0}^{t_1} y(l,t) \, \delta y(l,t) \, dt \quad (6.3.19)$$

Employing the commutativity rules (5.4.3) and noting that

$$\rho \frac{\partial y}{\partial t} \frac{\partial}{\partial t} \delta y = \frac{\partial}{\partial t} \left(\rho \frac{\partial y}{\partial t} \delta y \right) - \rho \frac{\partial^2 y}{\partial t^2} \delta y,$$

$$P \frac{\partial y}{\partial x} \frac{\partial}{\partial x} \delta y = \frac{\partial}{\partial x} \left(P \frac{\partial y}{\partial x} \delta y \right) - P \frac{\partial^2 y}{\partial x^2} \delta y, \qquad (6.3.20)$$

we have by integration and grouping terms with the same variations

$$\delta I = J_1 + J_2 + J_3 + \int_{t_0}^{t_1} \int_0^l \left(-\rho \frac{\partial^2 y}{\partial t^2} + P \frac{\partial^2 y}{\partial x^2} \right) \delta y \, dx \, dt, \qquad (6.3.21)$$

where

$$J_1 = \int_0^l \rho \left(\frac{\partial y}{\partial t} \delta y \bigg|_{t_1} - \frac{\partial y}{\partial t} \delta y \bigg|_{t_0} \right) dx,$$

$$J_2 = -\int_{t_0}^{t_1} \left[c_1 y(0,t) - P \frac{\partial y(0,t)}{\partial x} \right] \delta y(0,t) \, dt,$$

$$J_3 = -\int_{t_0}^{t_1} \left[c_2 y(l,t) + P \frac{\partial y(l,t)}{\partial x} \right] \delta y(l,t) \, dt. \qquad (6.3.22)$$

Since $\delta y(x, t_0) = \delta y(x, t_1) = 0$, if follows that $J_1 = 0$.

Taking into account that the variation $\delta y(x, t) = 0$ is arbitrary inside the intervals of independent variables, that is, for $0 < x < l$ and $t_0 < t < t_1$ and at the places A and B where the rings are connected with the string, we conclude that the variations $\delta y(0, t)$ and $\delta y(l, t)$ are also arbitrary. Therefore, the condition for stationary I, that is, $\delta I = 0$ which follows from (6.3.21) generates

(i) the Euler–Lagrangian equation of the vibration of the string

$$\frac{\partial^2 y}{\partial t^2} = k^2 \frac{\partial^2 y}{\partial x^2}, \quad k^2 = \frac{P}{\rho};$$

(6.3.23)

(ii) the following two natural boundary conditions:

$$c_1 y(0,t) - P\frac{\partial y(0,t)}{\partial x} = 0, \quad c_2 y(l,t) + P\frac{\partial y(l,t)}{\partial x} = 0.$$

(6.3.24)

The partial differential equation (6.3.23) is of hyperbolic type and is known as the *wave equation*. Writing the boundary conditions (6.3.24) in the form [49, p. 158]

$$\alpha y(0,t) + \frac{\partial y(0,t)}{\partial x} = 0, \quad \beta y(l,t) + \frac{\partial y(l,t)}{\partial x} = 0,$$
$$\alpha = -\frac{c_1}{P}, \quad \beta = \frac{c_2}{P},$$

(6.3.25)

several special cases are in order.

(a) If the springs are removed ($c_1 = c_2 = 0$) and the rings are able to move freely along the vertical rods $x = 0$ and $x = l$, the boundary conditions become

$$\frac{\partial y(0,t)}{\partial x} = 0, \quad \frac{\partial y(l,t)}{\partial x} = 0;$$

(6.3.26)

that is, the tangents at the points A and B to the string are during the whole motion normal to the rods at $x = 0$ and $x = l$.

(b) If the ends of the string are held fixed at $x = 0$ and $x = l$, then putting $c_1 \to \infty$, and $c_2 \to \infty$, and dividing (6.3.24)$_{1,2}$ by α and β, we arrive at the boundary conditions

$$y(0,t) = 0, \quad y(l,t) = 0.$$

(6.3.27)

The cases (a) and (b) are shown in Figures 6.3.1b and 6.3.1c.

Let us note that the wave equation (6.3.23) besides the transverse vibration of a taut string considered in this section has the same mathematical form in the study of longitudinal vibration of a bar, the longitudinal vibration of a helical spring, and the torsional vibration of a shaft. It has also been indicated that the gravitational waves traveling along an open channel are governed by the same equation. In each of these cases, the constant k represents the velocity of wave propagation in the direction of the coordinate x.

6.4   Unspecified Initial and Terminal Time, Variable End Points

In deriving conditions for the stationary I, we have thus far assumed that the time interval (t_0, t_1) in which the physical process is taking place is completely specified. We now consider a generalization of the variational principles formulated above, in which the boundaries of the action integral,

$$I = \int_{t_0}^{t_1} L\left(t, q_1, ..., q_n, \dot{q}_1, ..., \dot{q}_n\right) dt, \tag{6.4.1}$$

are permitted to change freely.

In search of the optimal trajectory of the Hamilton principle (6.4.1), we must suppose that the time should be varied together with the variation of the generalized coordinates and generalized velocities. Consequently, we must introduce the generalized or nonsimultaneous variations, which have already been introduced in section 3.2 of this monograph.

We have seen in section 3.2 that the relation between the generalized (nonsimultaneous) variations Δ and Lagrangian (simultaneous) variations δ of the generalized coordinates are of the form (see equations (3.2.5) and Figure 3.2.1)

$$\Delta q_i = \delta q_i + \dot{q}_i \Delta t. \tag{6.4.2}$$

However, here we will suppose that Δq_i and Δt are continuous, infinitesimal arbitrary functions of time, that is,

$$\Delta q_i = \varepsilon F_i\left(t\right), \quad \Delta t = \varepsilon f\left(t\right). \tag{6.4.3}$$

In fact, the functional structure of variations $\delta \mathbf{q}$, $\Delta \mathbf{q}$, and Δt can vary depending upon the purpose for which we use them.

As pointed out in section 3.2 the relation (6.4.2) can serve as a useful pattern for finding the nonsimultaneous variations of any scalar, vector, or tensor quantity. This formal analogy will be used here. Thus, if we put instead of q_i in (6.4.2), the integral (6.4.1), we shall have

$$\Delta I = \Delta \int_{t_0}^{t_1} L\left(t, \mathbf{q}, \dot{\mathbf{q}}\right) dt = \delta \int_{t_0}^{t_1} L\left(t, \mathbf{q}, \dot{\mathbf{q}}\right) dt + L\left(t, \mathbf{q}, \dot{\mathbf{q}}\right) \Delta t\big|_{t_0}^{t_1}. \tag{6.4.4}$$

Note that this expression is identical to equation (3.2.15). The Lagrangian variation of the integral on the right-hand side of this equation means that we should vary this integral considering the t_0 and t_1 as given constants. Therefore, from (5.2.7) and (5.2.8), we have

$$\delta \int_{t_0}^{t_1} L\left(t, \mathbf{q}, \dot{\mathbf{q}}\right) dt = \frac{\partial L}{\partial \dot{q}_i} \delta q_i \bigg|_{t_0}^{t_1} + \int_{t_0}^{t_1} \left(\frac{\partial L}{\partial q_i} - \frac{d}{dt}\frac{\partial L}{\partial \dot{q}_i}\right) \delta q_i dt. \tag{6.4.5}$$

Substituting $\delta q_i = \Delta q_i - \dot{q}_i \Delta t$ into the first term on the right-hand side of this equation, and entering with (6.4.5) into (6.4.4), we have after collecting terms the following *basic formula* of the nonsimultaneous variation of the action integral (6.4.1) whose end points are not specified:

$$\Delta I = \frac{\partial L}{\partial \dot{q}_i} \Delta q_i \bigg|_{t_0}^{t_1} + \left(L - \frac{\partial L}{\partial \dot{q}_i}\dot{q}_i\right) \Delta t \bigg|_{t_0}^{t_1} + \int_{t_0}^{t_1} \left(\frac{\partial L}{\partial q_i} - \frac{d}{dt}\frac{\partial L}{\partial \dot{q}_i}\right) \delta q_i dt. \tag{6.4.6}$$

Note that the integral term represents the partial variation of I caused by the Lagrangian (simultaneous) variation δq_i inside the interval (t_0, t_1) and the rest of the terms in (6.4.6) are the partial variation caused by the deferences in the end points. Together, these partial variations make up the general (nonsimultaneous) variation of the action integral I.

In order for the motion of the dynamical system defined by the action integral (6.4.1) with unspecified initial and terminal time to represent an extremal process, the integral term in (6.4.6) must be equal to zero which leads to the Euler–Lagrangian equations

$$\frac{\partial L}{\partial q_i} - \frac{d}{dt}\frac{\partial L}{\partial \dot{q}_i} = 0, \quad i = 1, ..., n. \tag{6.4.7}$$

Therefore, the boundary conditions at the initial and terminal time are specified by the relationship

$$\Delta I = \frac{\partial L}{\partial \dot{q}_i}\Delta q_i\bigg|_{t_0}^{t_1} + \left(L - \frac{\partial L}{\partial \dot{q}_i}\dot{q}_i\right)\Delta t\bigg|_{t_0}^{t_1} = 0. \tag{6.4.8}$$

Equations (6.4.7) and (6.4.8) are the key equations, since they summarize necessary conditions that must be satisfied by an extremal curve.

It is of interest to note that the equation (6.4.8) can be conveniently written in the canonical form, that is, expressed in terms of the canonical variables $q_1, ..., q_n, p_1, ..., p_n$. Namely, since the generalized momenta p_i are defined by equation (1.8.1) as $p_i = \partial L/\partial \dot{q}_i$ and the Hamiltonian function is given in the form (1.8.4), $H = \frac{\partial L}{\partial \dot{q}_i}\dot{q}_i - L$, equation (6.4.8) becomes

$$\Delta I = p_i\Delta q_i\big|_{t_0}^{t_1} - H\Delta t\big|_{t_0}^{t_1} = 0. \tag{6.4.9}$$

The boundary conditions can be obtained by making the appropriate substitution in (6.4.8) or (6.4.9).

We will now discuss the following examples of these substitutions for determination of natural boundary conditions. For the sake of simplicity we will suppose that *the initial time t_0 and initial position $q_i(t_0) = A_i$ are specified*, where A_i is a set of n given constants.

1. *The terminal position is specified and the terminal time is unspecified.* For this case, we obviously have that $\Delta q_i(t_0) = \Delta q_i(t_1) = 0$ and $\Delta t(t_0) = 0$, but $\Delta t(t_1)$ is arbitrary. Substituting this into (6.4.8) we have $2n$ prescribed boundary conditions

$$q_i(t_0) = A_i, \quad q_i(t_1) = B_i, \quad A_i, B_i \text{ given constants}, i = 1, ..., n, \tag{6.4.10}$$

and

$$L - \frac{\partial L}{\partial \dot{q}_i}\dot{q}_i = 0, \quad \text{for} \quad t = t_1, \tag{6.4.11}$$

or

$$H = 0 \quad \text{for} \quad t = t_1. \tag{6.4.12}$$

Therefore, (6.4.10) and (6.4.11), or (6.4.12) constitute the system of $(2n + 1)$ equations to determine $2n$ constants of integration of the Euler–Lagrangian equations (6.4.7) and unknown terminal time t_1.

2. *The terminal time and terminal position are not specified and are independent.* For this case we have

$$q_i(t_0) = A_i, \quad i = 1, ..., n, \tag{6.4.13}$$

and $\Delta t(t_0) = 0$, but $\Delta t(t_1) \neq 0$ and $\Delta q_i(t_1) \neq 0$ are independent of one another and arbitrary, hence their coefficients must each be zero. From (6.4.8) we therefore have

$$\frac{\partial L}{\partial \dot{q}_i} = 0, \quad i = 1, ..., n, \quad \text{for} \quad t = t_1 \tag{6.4.14}$$

and

$$L - \frac{\partial L}{\partial \dot{q}_i} \dot{q}_i = 0 \quad \text{for} \quad t = t_1. \tag{6.4.15}$$

However, the last two equations imply that

$$L(t, q_1, ..., q_n, \dot{q}_1, ..., \dot{q}_n) = 0 \quad \text{for} \quad t = t_1. \tag{6.4.16}$$

Thus, equations (6.4.13), (6.4.14), and (6.4.16) constitute a system of $(2n + 1)$ equations to determine $2n$ constants of integration of the Euler–Lagrangian equations (6.4.7) and unknown terminal time t_1.

3. *The terminal time and terminal position are functionally related in a prescribed way.* Let us consider the case where the initial time t_0 and initial position are given:

$$q_i(t_0) = A_i, \quad i = 1, ..., n, \tag{6.4.17}$$

but the terminal time t_1 and the terminal position $q_i(t_1)$ are not independent, though they are constrained to satisfy the following relation:

$$q_i(t_1) = \theta_i(t_1), \quad i = 1, ..., n. \tag{6.4.18}$$

In other words, the problem is to find the optimal trajectory of the dynamical system $q_i(t), i = 1, ..., n$, which minimizes the action integral (6.4.1), where t_0 and $q_i(t_0)$ are known and t_1 is the first time for which the optimal trajectory intersects the *"target set"* $\theta_i = \theta_i(t_1)$.

Applying the generalized variation to (6.4.18), we have

$$\Delta q_i(t_1) = \frac{d\theta_i(t_1)}{dt_1} \Delta t_1. \tag{6.4.19}$$

Substituting (6.4.19) into (6.4.8) and taking into account that $\Delta q_i(t_0) = 0$, $\Delta t(t_0) = 0$, we obtain, after collecting terms,

$$\frac{\partial L}{\partial \dot{q}_i}(t_1, \mathbf{q}(t_1), \dot{\mathbf{q}}(t_1)) \left[\frac{d\theta_i(t_1)}{dt_1} - \dot{q}_i(t_1)\right] + L(t_1, \mathbf{q}(t_1), \dot{\mathbf{q}}(t_1)) = 0. \quad (6.4.20)$$

Therefore, equations (6.4.17), (6.4.18), and (6.4.20) constitute the system of ($2n$ +1) equations to determine $2n$ constants of integration of the Euler–Lagrangian equations (6.4.7) and unknown terminal time t_1.

Example 6.4.1. *Motion of a two-degree-of-freedom system.* Find the motion of a dynamical system whose action integral is

$$I = \int_0^{t_1} L dt = \int_0^{t_1} (\dot{q}_1 \dot{q}_2 + t q_1 + q_2) \, dt, \quad (6.4.21)$$

and determine the relationship required to evaluate the constants of integration. The boundary conditions are specified as

$$q_1(0) = 0, \quad q_1(t_1) = 1, \quad q_2(0) = -4, \quad q_2(t_1) = 12, \quad (6.4.22)$$

and t_1 is not specified.

From (6.4.7) the Euler–Lagrangian equations are found to be

$$\ddot{q}_1 - 1 = 0, \quad \ddot{q}_2 - t = 0. \quad (6.4.23)$$

The general solution of these equations are

$$q_1 = \frac{t^2}{2} + C_1 t + C_2, \quad q_2 = \frac{t^3}{6} + C_3 t + C_4. \quad (6.4.24)$$

Substituting (6.4.22) into (6.4.24) we find $C_1 = (1/t_1) - (t_1/2), C_2 = 0, C_3 = (12/t_1) - (t_1^2/6) + (4/t_1), C_4 = -4$. Thus,

$$q_1 = \frac{t^2}{2} + \left(\frac{1}{t_1} - \frac{t_1}{2}\right) t, \quad q_2 = \frac{t^3}{6} + \left(\frac{12}{t_1} - \frac{t_1^2}{6} + \frac{4}{t_1}\right) t - 4. \quad (6.4.25)$$

Since the initial and terminal configurations are specified, the natural boundary condition (6.4.11) is of the form $L - (\partial L/\partial \dot{q}_1) \dot{q}_1 - (\partial L/\partial \dot{q}_2) \dot{q}_2 = 0$ for $t = t_1$. Therefore,

$$t_1 q_1(t_1) + q_2(t_1) - \dot{q}_1(t_1) \dot{q}_2(t_1) = 0. \quad (6.4.26)$$

Substituting (6.4.25) into (6.4.26), we have

$$t_1^5 - 4t_1^3 - 24t_1^2 + 96 = 0. \quad (6.4.27)$$

The positive root of (6.4.27) is $t_1 = 2$. Therefore, the motion of the dynamical system is described by the equations

$$q_1 = \frac{t^2}{2} - \frac{t}{2}, \quad q_2 = \frac{t^3}{6} + \frac{22}{3} t - 4. \quad (6.4.28)$$

Example 6.4.2. A system with one degree of freedom. We consider the problem of determining the curve that is extremal for the functional

$$I = \int_0^{t_1} L dt = \int_0^{t_1} \frac{1}{2} \left(\dot{q}^2 - q^2 \right) dt, \tag{6.4.29}$$

if $q(0) = 0$, and the terminal position is neither specified nor free but constrained, so that it satisfies the following relation:

$$q(t_1) = \theta(t_1), \tag{6.4.30}$$

where

$$\theta(t) = (2 - t)^{1/3}. \tag{6.4.31}$$

From (6.4.7) the Euler–Lagrangian equation is $\ddot{q} + q = 0$, whose solution, which satisfied the condition $q(0) = 0$, is

$$q = C \sin t, \tag{6.4.32}$$

where C is a constant.

Since $q(t_1)$ and t_1 are not specified, but they are connected with the given relationship (6.4.30), we must apply the condition (6.4.20), namely,

$$\frac{\partial L}{\partial \dot{q}} \left[\frac{d\theta}{dt} - \dot{q} \right] + L = 0 \quad \text{for} \quad t = t_1, \tag{6.4.33}$$

whence

$$C \cos t_1 \left[-\frac{1}{3} (2 - t_1)^{-2/3} - C \cos t_1 \right] + \frac{1}{2} C^2 \cos^2 t_1 - \frac{1}{2} \sin^2 t_1 = 0. \tag{6.4.34}$$

From this relation, we find

$$C = -\frac{2}{3} (2 - t_1)^{-2/3} \cos t_1, \tag{6.4.35}$$

and the equation (6.4.32) becomes

$$q = -\frac{2}{3} (2 - t_1)^{-2/3} \cos t_1 \sin t. \tag{6.4.36}$$

To find the terminal time t_1, we must find the point of intersection of the extremal (6.4.36) and the "target set" (6.4.31); that is, the equation (6.4.30) gives after simple calculation

$$\sin 2t_1 = 3t_1 - 6. \tag{6.4.37}$$

A numerical solution of this equation gives $t_1 = 1.83$. Thus $C = 0.557$ and the equation of the extremal is $q = 0.557 \sin t$.

6.5   Jacobi's Form of the Variational Principle Describing the Paths of Conservative Dynamical Systems

In this section, we shall demonstrate that Hamilton's principle can be transformed into a form suitable for studying the paths (orbits) of a dynamical system, independent of the motion of the system. The considerations that follow rest upon the supposition that the dynamical system is holonomic and conservative, that is, that the total mechanical energy is conserved:

$$T + \Pi = E = const., \qquad (6.5.1)$$

where T denotes the kinetic energy given by (1.4.15), $T = \frac{1}{2}a_{ij}\dot{q}_i\dot{q}_j$, and $\Pi = \Pi(q_1, ..., q_n)$ is the potential energy. The main idea underlying our study is the elimination of time as the independent variable by means of generalized coordinates, say q_1. This is known as the *Jacobi method*.

Taking into account the structure of kinetic energy, the conservation law (6.5.1) can be represented in the form

$$2T(dt)^2 = a_{ij}(q_1, ..., q_n)\,dq_i dq_j = 2(E - \Pi)(dt)^2, \qquad (6.5.2)$$

whence

$$dt = \sqrt{\frac{a_{ij}dq_i dq_j}{2(E - \Pi)}}. \qquad (6.5.3)$$

Taking q_1 as the independent, we can write (6.5.3) in the form

$$\sqrt{\frac{G}{E - \Pi}}\,dq_1 = dt, \qquad (6.5.4)$$

where

$$G = \frac{1}{2}\left[a_{11} + 2a_{1s}(q_s)' + a_{sj}(q_s)'(q_j)'\right], \quad s, j = 2, 3, ..., n, \qquad (6.5.5)$$

and $\frac{d}{dq_1}(\cdot) = (\cdot)'$. By combining Hamilton's action integral,

$$I = \int_{t_0}^{t_1} (T - \Pi)\,dt, \qquad (6.5.6)$$

with (6.5.1), we can write it as

$$I = \int_{t_0}^{t_1} 2T\,dt - E(t_1 - t_0). \qquad (6.5.7)$$

Now, by means of (6.5.1) and (6.5.4), we have

$$W = \int_{q_1(0)}^{q_1(t_1)} 2T\sqrt{\frac{G}{E - \Pi}}\,dq_1 = 2\int_{q_1(0)}^{q_1(t_1)} \sqrt{G(E - \Pi)}\,dq_1, \qquad (6.5.8)$$

where $W = I + E(t_1 - t_0)$ and $q_1(0)$ and $q_1(t_1)$ are the initial and terminal values of the independent variable q_1.

The integral (6.5.8) is usually called the *Lagrangian action* and *Jacobi's form of the principle of least (stationary) action*. It states that the integral

$$W = \int_{q_1(0)}^{q_1(t_1)} \sqrt{R}\,dq_1 \tag{6.5.9}$$

is stationary (i.e., $\delta W = 0$) for the actual path, as compared with the neighboring paths joining the same two end points in the $q_1, ..., q_n$ space. The integrand in (6.5.9) is given as

$$R = 4G(E - \Pi) = 2(E - \Pi)\left[a_{11} + 2a_{1s}(q_s)' + a_{sj}(q_s)'(q_j)'\right]. \tag{6.5.10}$$

Therefore, the problem of finding the paths (orbits) of a holonomic conservative dynamical system is reduced to the problem of finding the extremals of Lagrangian integral (6.5.9). In addition, the derivation of the differential equations of extremals is not necessary, since it is possible to use the following analogy in the notation: the role of the Lagrangian L in Hamilton's principle is replaced by the function $\sqrt{R}$, while the independent variable t is replaced by q_1. In this way we arrive at the differential equations for the paths in the form of Jacobi:

$$\frac{d}{dq_1}\frac{\partial\sqrt{R}}{\partial(q_s)'} - \frac{\partial\sqrt{R}}{\partial q_s} = 0. \tag{6.5.11}$$

Finally, we can determine the time by using the energy integral (6.5.3)

$$t_1 - t_0 = \int_{q_1(0)}^{q_1(t_1)} \frac{\sqrt{R}}{2(E - \Pi)}dq_1. \tag{6.5.12}$$

The general solution to the problem contains $2n$ constants: t_0, E, and $2n - 2$ arbitrary constants, arising as the result of integration of (6.5.11).

Example 6.5.1. *Motion of a heavy particle in a vertical plane.* Consider the motion of a heavy particle of mass m in a vertical plane xOy, with x axis positioned at the ground and z axis oriented vertically upward. By applying the Jacobi form of the variational principle, we wish to find the path of motion of a particle joining two points (x_0, z_0) and (x_1, z_1).

The particle has two degrees of freedom, and we take for the generalized coordinates $q_1 = x, q_2 = z$. The total energy of the particles

$$\frac{m}{2}\left(\dot{x}^2 + \dot{z}^2\right) + mgz = E = const., \tag{6.5.13}$$

where g is the acceleration of gravity. Taking z as an independent variable, the function R defined by (6.5.10) becomes $R = 2m(E - mgz)\left[(x')^2 + 1\right]$, where $x' = dx/dz$. Therefore

$$\sqrt{R} = \sqrt{2m(E - mgz)\left[(x')^2 + 1\right]}. \tag{6.5.14}$$

Since the coordinate x is an ignorable coordinate (i.e., x does not figure in $\sqrt{R}$), we have the cyclic conservation law (see (1.4.40)) $\partial\sqrt{R}/\partial x' = C = const.$ Hence

$$\sqrt{\frac{E - mgz}{1 + (x')^2}}\, x' = C. \tag{6.5.15}$$

By squaring and separating variables we have $dx = dz/\sqrt{-\frac{mg}{C^2}z + \left(\frac{E}{C^2} - 1\right)}$. Integrating, one has $z + D = -\frac{2C^2}{mg}\sqrt{-\frac{mg}{C^2}z + \left(\frac{E}{C^2} - 1\right)}$, where D is a constant of integration. Squaring again and calculating $z = f(x)$, we have that the path of a particle is of the form

$$z = -\frac{mg}{4C^2}x^2 - \frac{mgD}{2C^2}x - \frac{mgD^2}{4C^2} + \frac{E}{mg} - \frac{C^2}{mg}. \tag{6.5.16}$$

Since $-mg/4C^2 < 0$, we see that the path is a parabola, which is open downward.

To find two constants of integration C and D, we first suppose for simplicity that $x(0) = 0$ and $z(0) = 0$. Then, from (6.5.16), we have

$$\frac{E}{mg} - \frac{C^2}{mg} - \frac{mgD^2}{4C^2} = 0, \tag{6.5.17}$$

and the path is reduced to the form

$$z = -\frac{mg}{4C^2}x^2 - \frac{mgD}{2C^2}x. \tag{6.5.18}$$

At this point it is convenient to introduce the initial elevation angle α between the horizontal axis Ox and initial velocity vector $\mathbf{v}_0$. Thus, $\tan\alpha = (dz/dx)_{x=0}$. Differentiating (6.5.18) with respect to x we get $z' = -(mg/2C^2)x - (mgD/2C^2)$. Now by evaluating this expression at $x_0 = z_0 = 0$, we obtain $\tan\alpha = -mgD/2C^2$, that is, $D = -\frac{2C^2}{mg}\tan\alpha$. Entering with this into (6.5.17) we find $C = \sqrt{E}\cos\alpha$, so that

$$D = -\frac{2E}{mg}\sin\alpha\cos\alpha. \tag{6.5.19}$$

Therefore, the path (6.5.18) becomes

$$z = x\tan\alpha - \frac{mg}{4E\cos^2\alpha}x^2. \tag{6.5.20}$$

This parabola has to pass through the point (x_1, z_1), and we can write the last equation in the form

$$\tan^2\alpha - \frac{4E}{mgx_1}\tan\alpha + \frac{4Ez_1}{mgx_1^2} + 1 = 0. \tag{6.5.21}$$

The selection of terminal point $A\left(x_1, z_1\right)$ is not arbitrary if the total energy E is given. Solving (6.5.21) with respect to $\tan\alpha$ we can find the elevation angle α for which the path is going to pass through A:

$$\tan\alpha = \frac{2E}{mgx_1} \pm \sqrt{\frac{4E^2}{m^2g^2x_1^2} - \frac{4Ez_1}{mgx_1^2} - 1}. \qquad (6.5.22)$$

This solution depends upon the discriminant $\Delta = \frac{4E^2}{m^2g^2x_1^2} - \frac{4Ez_1}{mgx_1^2} - 1$. Namely, if $\Delta > 0$ it follows that the terminal point $A\left(x_1, z_1\right)$ is the point of intersection of two parabolas 1 and 2 depicted in Figure 6.5.1. From $\Delta > 0$ it follows that $4E\left(E - mgz_1\right) > m^2g^2x_1^2$, and from this we obtain the upper bound

$$z_1 < \frac{E}{mg}. \qquad (6.5.23)$$

However, if $\Delta = 0$, we have two identical roots, and it follows that for this case a parabolic curve

$$x_1^2 = \frac{4E}{mg}\left(\frac{E}{mg} - z_1\right), \qquad (6.5.24)$$

which represents the *envelope* of all possible parabolas emerging from the point O. Thus, the point $A\left(x_1, z_1\right)$ for the equal roots lays at the envelope.

The least-action path in Figure 6.5.1 is parabola 2, since the time needed to traverse from O to A is less than that needed to travel along parabola 1. This confirms the fact, stated earlier, that the long trajectories need not minimize Hamilton's action integral. It is obvious that a particle shot upward from the origin with an initial inclination $\alpha = \pi/2$ cannot cross the envelope TB. A point on an extremal through O that touches the envelope (like the point B on parabola 1) is said to be *conjugate to* O or to be a *kinetic focus*. On the other hand, according to Jacobi's necessary condition (see, for example, [83, p. 118]), a minimizing arc OA cannot contain a point conjugate to O. Therefore, since the point B on parabola 1 is a conjugate point, the path OBA is not a minimal arc.

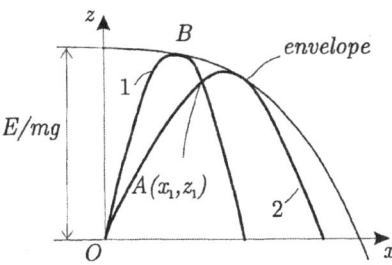

Figure 6.5.1

6.6   Piecewise Continuous Extremals. The Weierstrass–Erdmann Corner Conditions

Thus far in our development of Hamilton's variational principle we have considered only the so-called smooth extremals for which the $q_i(t)$ and $\dot{q}_i(t)$ are continuous in the whole time interval (t_0, t_1). However, frequently we may be confronted with the situation in which the extremal trajectory $q_i(t)$ is continuous while the generalized velocity vector $\dot{q}_i(t)$ can be only piecewise continuous, that is, it may consist of segments of trajectories joined at points called *corners* at which $\dot{q}_i(t)$ is discontinuous. In this section we shall summarize the conditions that should hold at these corners. These conditions are usually referred to as the Weierstrass–Erdmann corner conditions. For the sake of simplicity we will consider the case with one generalized coordinate $q(t)$ only.

Let us consider the action integral with corresponding boundary conditions

$$I = \int_{t_0}^{t_1} L(t, q, \dot{q})\, dt,$$
$$q(t_0) = A, \quad q(t_1) = B. \tag{6.6.1}$$

Let us suppose that the extremal of this problem has a discontinuous derivative at the time τ as depicted in Figure 6.6.1. Note that τ is not fixed, nor it is usually known in advance.

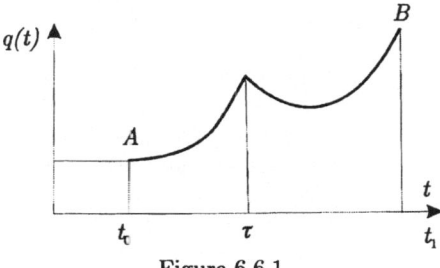

Figure 6.6.1

It is obvious that for $t \in [t_0, \tau]$ and $t \in [\tau, t_1]$ the extremal must satisfy the Euler–Lagrangian equation

$$\frac{\partial L}{\partial q} - \frac{d}{dt}\frac{\partial L}{\partial \dot{q}} = 0. \tag{6.6.2}$$

Let us now express the action integral (6.6.1) as a sum of two action integrals,

$$I = \int_{t_0}^{\tau} L(t, q, \dot{q})\, dt + \int_{\tau}^{t_1} L(t, q, \dot{q})\, dt = I_1 + I_2. \tag{6.6.3}$$

Since the time τ is not fixed in calculating the variation of (6.6.3) we must apply the nonsimultaneous variations of the action integrals I_1 and I_2. Therefore, on the basis of (6.6.4) we have $\Delta I = \Delta I_1 + \Delta I_2$, where

$$
\begin{aligned}
\Delta I_1 &= \left.\frac{\partial L}{\partial \dot{q}}\Delta q\right|_{t_0}^{\tau_-} + \left.\left(L - \frac{\partial L}{\partial \dot{q}}\dot{q}\right)\Delta t\right|_{t_0}^{\tau_-} + \int_{t_0}^{\tau_-}\left(\frac{\partial L}{\partial q} - \frac{d}{dt}\frac{\partial L}{\partial \dot{q}}\right)\delta q\, dt, \\
\Delta I_2 &= \left.\frac{\partial L}{\partial \dot{q}}\Delta q\right|_{\tau_+}^{t_1} + \left.\left(L - \frac{\partial L}{\partial \dot{q}}\dot{q}\right)\Delta t\right|_{\tau_+}^{t_1} + \int_{\tau_+}^{t_1}\left(\frac{\partial L}{\partial q} - \frac{d}{dt}\frac{\partial L}{\partial \dot{q}}\right)\delta q\, dt,
\end{aligned}
$$

$$(6.6.4)$$

where $\tau_- = \tau - 0$ and $\tau_+ = \tau + 0$. Since the Euler–Lagrangian equations are satisfied for both time intervals, the extremality condition $\Delta I = 0$ becomes

$$
\begin{aligned}
\Delta I_1 &= \frac{\partial L}{\partial \dot{q}}\Delta q(\tau) + \left(L - \frac{\partial L}{\partial \dot{q}}\dot{q}\right)\Delta t(\tau) \quad \text{for} \quad \tau = \tau_-, \\
\Delta I_2 &= -\frac{\partial L}{\partial \dot{q}}\Delta q(\tau) - \left(L - \frac{\partial L}{\partial \dot{q}}\dot{q}\right)\Delta t(\tau) \quad \text{for} \quad \tau = \tau_+. \quad (6.6.5)
\end{aligned}
$$

From (6.6.5) we conclude that the extremal solution must satisfy

$$
\frac{\partial L}{\partial \dot{q}}\Delta q(\tau_-) = \frac{\partial L}{\partial \dot{q}}\Delta q(\tau_+) \tag{6.6.6}
$$

and

$$
\left(L - \frac{\partial L}{\partial \dot{q}}\dot{q}\right)_{\tau_-} = \left(L - \frac{\partial L}{\partial \dot{q}}\dot{q}\right)_{\tau_+}. \tag{6.6.7}
$$

The last two equations constitute the Weierstrass–Erdmann conditions. As already mentioned, the Euler–Lagrangian equations must be satisfied at each part of the extremals between (t_0, τ) and (τ, t_1). Since these equations are second-order differential equations, their solution contains four constants of integration. These constants should be determined from the given boundary conditions

$$
q(t_0) = A = const., \quad q(t_1) = B = const., \tag{6.6.8}
$$

and two Weierstrass–Erdmann conditions (6.6.6), (6.6.7), which require that $\frac{\partial L}{\partial \dot{q}}$ and $L - \frac{\partial L}{\partial \dot{q}}\dot{q}$ must be continuous across the corner.

Note that if we use the canonical variables q and p, then the Weierstrass–Erdmann conditions are $p_{\tau_-} = p_{\tau_+}$ and $H_{\tau_-} = H_{\tau_+}$.

Example 6.6.1. *Newton's problem.* In this example we consider Newton's famous problem of finding the optimal shape of a solid body of revolution, which moves through a resisting medium in the direction of its axis of revolution with the least possible resistance. The problem was set up and solved by Newton in late 1685. However, this problem happens to be relevant in modern times since its results agree very well with experimental data at hypersonic speeds (see [26] and [73]).

We shall consider the case when the solid is at rest in a steady stream. The pressure drag of a body of revolution at zero angle of attack in hypersonic flow is given by the expression

$$\text{Drag force} = 2\pi Q \int_0^l C_p\,(\theta)\,y\,dy, \tag{6.6.9}$$

where Q denotes the dynamical pressure, $y = y(x)$ is the radius of the body, $dy/dx = \tan\theta$, and the *pressure coefficient* $C_p\,(\theta)$ is given by

$$C_p\,(\theta) = 2\sin^2\theta = \frac{2\,(y')^2}{1 + (y')^2}, \tag{6.6.10}$$

where θ is the angle between the tangent at the point (x, y) and the x axis (see Figure 6.6.2).

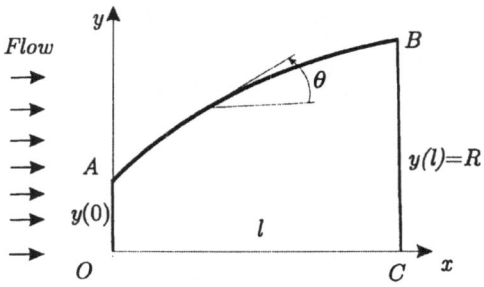

Figure 6.6.2

We shall suppose that the body has a blunt tip whose radius at $x = 0$ is denoted by $y\,(0)$. Note that this radius in *not given,* and the problem is to determine $y\,(0)$ and the form of the *Newton curve* AB for which the resistance is minimal. We suppose that the length l and the maximal radius R of the body of revolution are given.

The action integral of this problem is found to be

$$\frac{\text{Drag force}}{2\pi Q} = I = y^2\,(0) + \int_0^l L\,(y, y')\,dx, \tag{6.6.11}$$

where the Lagrangian function is

$$L\,(y, y') = \frac{2y\,(y')^3}{1 + (y')^2}. \tag{6.6.12}$$

Since the radius $y\,(0)$ is not given, but its place $x = 0$ is given, we see that the solution curve OAB has a corner at A; that is, it has a discontinuous slope.

Since the Lagrangian function does not depend upon the independent variable x, we have the Jacobi conservation law

$$\frac{\partial L}{\partial y'} y' - L = \frac{4y\,(y')^3}{\left(1 + (y')^2\right)^2} = const. \tag{6.6.13}$$

Let us now apply the Weierstrass–Erdmann corner condition at the corner A, where the slope of the extremal curve is discontinuous. Thus, by applying (6.6.6) we have

$$\left.\frac{\partial L}{\partial y'}\right|_{x=0} = \left.\frac{\partial L}{\partial y'}\right|_{x=0+}, \tag{6.6.14}$$

whence,

$$\frac{\dfrac{6y\,(0)}{(y'\,(0))^2} + 2y\,(0)}{\dfrac{1}{(y'\,(0))^4} + \dfrac{2}{(y'\,(0))^2} + 1} = \frac{\dfrac{6y\,(0+)}{(y'\,(0+))^2} + 2y\,(0+)}{\dfrac{1}{(y'\,(0+))^4} + \dfrac{2}{(y'\,(0+))^2} + 1}. \tag{6.6.15}$$

This expression will be satisfied for $y\,(0) = y\,(0+)$ and $y'\,(0) \to \infty, y'\,(0+) = 1$.

We see from this that at the corner a a minimizing curve must cut the vertical line OA at an angle of 45 degrees.

Note that we can draw the same conclusion by calculating the variation of the action integral (6.6.11). Namely, since the Euler–Lagrangian equation has the conservation law (6.6.13) and the boundaries of the action integral $(0, l)$ in (6.6.11) are fixed, the first variation of (6.6.11) reads

$$\delta I = \left[\left.\frac{\partial L}{\partial y'}\right|_{x=0} + 2y\,(0) \right] \delta y\,(0) = \frac{2y\,(0)\,[y'\,(0+) - 1]}{\left[1 + \left(y'\,(0+)^2\right)\right]^2} \delta y\,(0+) = 0. \tag{6.6.16}$$

Since $y\,(0)$ is not given, $\delta y\,(0)$ is arbitrary, and we have, again, that $y'\,(0+) = 1$.

Therefore, from (6.6.13) we conclude that the arbitrary constant is equal to $y\,(0)$. By using this in (6.6.13) we obtain the radius of the body in terms of the slope $q = dy/dx = \tan\theta$ as

$$y\,(q) = \frac{y\,(0)\,\left(1 + q^2\right)^2}{4} \frac{}{q^3}, \quad q\,(0) = 1. \tag{6.6.17}$$

To find x in terms of q, we note that $dx = \frac{1}{q} dy\,(q)$, that is,

$$dx = \frac{y\,(0)}{4}\left(-\frac{3}{q^5} - \frac{2}{q^3} + \frac{1}{q}\right) dq. \tag{6.6.18}$$

Integrating, we find

$$x\,(q) = \frac{y\,(0)}{4}\left(\frac{3}{4q^4} + \frac{1}{q^2} + \ln q\right) + D, \tag{6.6.19}$$

where $D = const$. Since for $x = 0$ we have $q = 1$, the constant D is found to be $D = -7y(0)/16$. Hence, the equation (6.6.17) and the equation

$$x(q) = \frac{y(0)}{4}\left(\frac{3}{4q^4} + \frac{1}{q^2} + \ln q - \frac{7}{4}\right) \tag{6.6.20}$$

represent the parametric equations for the optimal shape in terms of the slope q.

It is convenient to introduce a new parameter $p = 1/q = dx/dy = \cot\theta$ so that the optimal shape of the body is expressed in the following way:

$$x(p) = \frac{y(0)}{4}\left(p^2 + \frac{3}{4}p^4 - \ln p - \frac{7}{4}\right), \quad y(p) = \frac{y(0)\left(1+p^2\right)^2}{4}{p}. \tag{6.6.21}$$

Finally, the tip radius $y(0)$ and the slope p_l at $x = l$ are obtained by solving the following system of transcendental equations, for specified l and R:

$$\frac{R}{y(0)} = \frac{\left(1+p_l^2\right)^2}{p_l}, \quad \frac{l}{y(0)} = \frac{1}{4}\left(p_l^2 + \frac{3}{4}p_l^4 - \ln p_l - \frac{7}{4}\right). \tag{6.6.22}$$

For example, let us suppose that $R = l$. From the last two equations we have

$$(p_l)^2 + \frac{3}{4}(p_l)^4 - \ln p_l - \frac{7}{4} = \frac{\left(1+p_l^2\right)^2}{p_l}, \tag{6.6.23}$$

whose solution is found to be $p_l = 1.917$. Thus, the parameter $p = 1/q = dx/dy$ changes from 1 to 1.917. From the first equation of the system (6.6.22) it follows that $y(0) = 0.352R$, and with this the optimal profile is completely specified.

The minimum-drag coefficient can be easily calculated from (6.6.11) and (6.6.12):

$$\frac{\text{Drag force}}{\pi Q} = 2y^2(0) + \int_0^l \frac{4y(y')^3}{1+(y')^2}\,dx. \tag{6.6.24}$$

Remembering that $y' = q$ and substituting (6.6.17) and (6.6.18) into the last equation, we obtain

$$\frac{\text{Drag force}}{\pi Q} = 2y^2(0) + \frac{y(0)^2}{4}\int_1^{q_l}\left(-\frac{3}{q^5} - \frac{5}{q^3} - \frac{1}{q} + q\right)dq, \tag{6.6.25}$$

where q_l denotes the slope at $x = l$. From (6.6.17), evaluated at $x = l$, it follows that $y(0) = 4Rq_l^3/\left(1+q_l^2\right)^2$. Entering with this into the last equation and integrating, we obtain the minimum drag coefficient as

$$C_D = \frac{\text{Drag force}}{\pi Q R^2} = \frac{q_l^2}{\left(1+q_l^2\right)^4}\left(17q_l^4 + 3 + 10q_l^2 - 4q_l^4\ln q_l + 2q_l^6\right). \tag{6.6.26}$$

For example, if $R = l, q_l = 1/p_l = 0.522$, the minimum drag coefficient is $C_D = 0.7494$.

Chapter 7

Constrained Problems

7.1 Introduction

In the previous sections we considered Hamilton's variational principle in terms of *independent* generalized coordinates $q_i, i = 1, ..., n$, where n is the number of degrees of freedom of a dynamical system. In this chapter we will consider several important situations in which the generalized coordinates are *not independent* but are restricted by given auxiliary conditions. Namely, it is not uncommon in the analysis of applied variational problems to be faced with the task of finding an extremal dynamical trajectory within the framework of a certain number of restrictions that have physical origin

7.2 Isoperimetric Constraints

An isoperimetric[13] problem is the one in which one seeks the extremal of the given action integral

$$I = \int_{t_0}^{t_1} L\left(t, q_1, ..., q_n, \dot{q}_1, ..., \dot{q}_n\right) dt \qquad (7.2.1)$$

for the class of trajectories for which the auxiliary conditions occur as a set of definite integrals which must have specified *constant values* $C_k, k = 1, ..., r$, with $r \lesseqgtr n$, namely,

$$\int_{t_0}^{t_1} G_k\left(t, q_1, ..., q_n, \dot{q}_1, ..., \dot{q}_n\right) dt = C_k; \quad C_k \text{ are given constants.} \qquad (7.2.2)$$

As indicated before, we might have an arbitrary number of conditions (7.2.2) prescribed as auxiliary conditions.

---

[13]The term *isoperimetric* comes from one of the oldest problems of variational calculus for which one has to find a simple closed curve of given length which closes the largest area.

To find the extremal of isoperimetric variational problems, we shall employ the method of Lagrangian undetermined multipliers, or the so-called *Euler rule.* Namely, the constraints (7.2.2) are accounted for by introducing r unknown *constant Lagrangian multipliers* $\lambda_k, k = 1, ..., r$, and defining the augmented functional

$$I_{aug.} = \int_{t_0}^{t_1} L_{aug.}\,(t, \mathbf{q}, \dot{\mathbf{q}}, \boldsymbol{\lambda})\, dt \equiv \int_{t_0}^{t_1} [L\,(t, \mathbf{q}, \dot{\mathbf{q}}) + \lambda_k G_k\,(t, \mathbf{q}, \dot{\mathbf{q}})]\, dt, \quad (7.2.3)$$

where $\mathbf{q} = (q_1, ..., q_n)$, $\dot{\mathbf{q}} = (\dot{q}_1, ..., \dot{q}_n)$, and $\boldsymbol{\lambda} = (\lambda_1, ..., \lambda_r)$.

Calculating the first variation of the expression (7.2.3), we obtain

$$\begin{aligned}
\delta I_{aug.} &= \int_{t_0}^{t_1} \left[\left(\frac{\partial L}{\partial q_i} + \lambda_k \frac{\partial G_k}{\partial q_i} \right) \delta q_i + \left(\frac{\partial L}{\partial \dot{q}_i} + \lambda_k \frac{\partial G_k}{\partial \dot{q}_i} \right) \delta \dot{q}_i \right] dt, \\
i &= 1, ..., n, \quad k = 1, ..., r.
\end{aligned} \quad (7.2.4)$$

Employing the commutativity rule $\delta \dot{q}_i = (d/dt)\,\delta q_i$ and integrating the second group of terms by parts, we arrive at

$$\begin{aligned}
\delta I_{aug.} &= \left. \left(\frac{\partial L}{\partial \dot{q}_i} + \lambda_k \frac{\partial G_k}{\partial \dot{q}_i} \right) \delta q_i \right|_{t_0}^{t_1} + \int_{t_0}^{t_1} \left[\left(\frac{\partial L}{\partial q_i} + \lambda_k \frac{\partial G_k}{\partial q_i} \right)_i \right. \\
&\quad + \left. \frac{d}{dt} \left(\frac{\partial L}{\partial \dot{q}_i} + \lambda_k \frac{\partial G_k}{\partial \dot{q}_i} \right) \right] \delta q_i dt.
\end{aligned} \quad (7.2.5)$$

Let us suppose that the boundary conditions are given in the form of (5.2.1); that is, $q_i\,(t_0) = A_i = const., q_i\,(t_1) = B_i = const.$, and the time interval (t_0, t_1) is prescribed. The condition for extremal I, that is, $\delta I_{aug.} = 0$, leads to the Euler–Lagrangian equations

$$\frac{\partial L_{aug.}}{\partial q_i} - \frac{d}{dt} \frac{\partial L_{aug.}}{\partial \dot{q}_i} = 0, \quad i = 1, ..., n, \quad (7.2.6)$$

which together with the boundary conditions and isoperimetric constraints characterized the solution.

Example 7.2.1. *One-degree-of-freedom isoperimetric problem.* Find the extremal of the action integral

$$I = \int_0^{\pi/2} \left(\dot{q}^2 - q^2 \right) dt, \quad (7.2.7)$$

subject to the isoperimetric constraint

$$\int_0^{\pi/2} q \sin t\, dt = 1, \quad (7.2.8)$$

if the boundary conditions are

$$q\,(0) = 0, \quad q\,(\pi/2) = 0. \quad (7.2.9)$$

Let us introduce the constant Lagrangian multiplier λ and form the augmented Lagrangian $L_{aug.} = \left(\dot{q}^2 - q^2\right) + \lambda q \sin t$. The Euler–Lagrangian equation (7.2.6) is

$$\ddot{q} + q = \frac{\lambda}{2} \sin t. \tag{7.2.10}$$

The general solution of this nonhomogeneous differential equation is found to be $q = C_1 \cos t + C_2 \sin t - (\lambda t/4)$, where C_1 and C_2 are arbitrary constants. Applying the boundary conditions (7.2.9) we find $q = -(\lambda t/4) \cos t$. Substituting this into (7.2.8) we find

$$\lambda = \frac{-4}{\int_0^{\pi/2} t \sin t \cos t \, dt} = -\frac{32}{\pi}. \tag{7.2.11}$$

Therefore, the extremal function is $q = (8t/\pi) \cos t$.

Example 7.2.2. *Equilibrium configuration of a flexible rope.* To illustrate the proceeding discussion, let us determine the configuration (the form) in which a perfectly flexible, uniform rope that is fixed at both ends will hang in a uniform gravitational field, in equilibrium. The rope has a uniform mass per unit length (constant line density) denoted by ρ. We propose that the rope will hang in such a way that its potential energy is in minimum, subject to the constraint that the length of the rope remain constant (inextensibility condition). The potential energy of a rope, taking $y = 0$ as the zero energy level, is $\Pi = \pi g \int_{x_0}^{x_1} y \, ds$, where g is the acceleration of gravity and x_0, x_1 are coordinates of the points A and B where the rope is hanged on the horizontal axis x (see Figure 7.2.1) and is the s arc length of the rope axis.

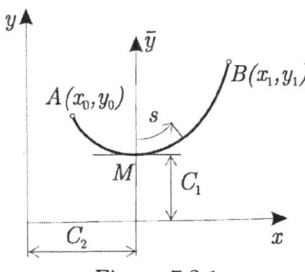

Figure 7.2.1

To simplify calculations, we introduce the action integral in the form

$$I = \frac{\Pi}{\rho g} = \int_{x_0}^{x_1} y \, ds. \tag{7.2.12}$$

The isoperimetric constraint is of the form

$$\int_{x_0}^{x_1} ds = L, \quad \text{given the length of the rope.} \tag{7.2.13}$$

Here g denotes the acceleration of gravity and $ds = \sqrt{1 + (y')^2}\,dx$ is the elementary arc length of the rope. Here $(\cdot)' = \frac{d}{dx}(\cdot)$. The augmented Lagrangian function (7.2.3) is

$$L_{aug.} = (y - \lambda)\sqrt{1 + (y')^2}, \qquad (7.2.14)$$

where λ is a constant Lagrangian multiplier.

Instead of writing the Euler–Lagrangian equation in accordance with (7.2.6), we notice that the augmented Lagrangian function (7.2.14) does not depend explicitly upon variable x. Thus, according to (1.4.44), there exists the Jacobi conservation law $L_{aug.} - (\partial L_{aug.}/\partial y')\, y' = C_1 = const.$, which yields

$$(y - \lambda)\sqrt{1 + (y')^2} - \frac{(y - \lambda)\,(y')^2}{\sqrt{1 + (y')^2}} = C_1 = const., \qquad (7.2.15)$$

whence

$$y' = \sqrt{\frac{(y - \lambda)^2 - C_1^2}{C_1^2}}. \qquad (7.2.16)$$

Separating variables and integrating, we find that

$$C_1 \ln\left[\frac{y - \lambda}{C_1} + \sqrt{\left(\frac{y - \lambda}{C_1}\right)^2 - 1}\right] = x - C_2, \qquad (7.2.17)$$

where C_2 is a new constant of integration. Therefore,

$$\frac{y - \lambda}{C_1} + \sqrt{\left(\frac{y - \lambda}{C_1}\right)^2 - 1} = e^{\left(\frac{x - C_2}{C_1}\right)}. \qquad (7.2.18)$$

The reciprocal of this expression leads to

$$\frac{y - \lambda}{C_1} - \sqrt{\left(\frac{y - \lambda}{C_1}\right)^2 - 1} = e^{\left(-\frac{x - C_2}{C_1}\right)}. \qquad (7.2.19)$$

Combining, the last two expressions, we have

$$y - \lambda = C_1 \cosh\frac{x - C_2}{C_1}. \qquad (7.2.20)$$

Three constants C_1, C_2, and λ should be determined from the following three conditions: (1) the curve must pass through the point $A\,(x_0, y_0)$; (2) the curve must pass through the point $B\,(x_1, y_1)$; and (3) the length of the curve as computed by equation (7.2.13) must have prescribed value L.

At this point the following two remarks are in order. First, by introducing the coordinate transformation

$$\bar{x} = x, \quad \bar{y} = y - \lambda, \quad \bar{y}' = y', \qquad (7.2.21)$$

the curve (7.2.20) is represented in the form

$$\bar{y} = C_1 \cosh \frac{x - C_2}{C_1}. \qquad (7.2.22)$$

The axis $\bar{y}$ is depicted in Figure 7.2.1 and it passes through the lowest point M of the curve. The constants C_1 and C_2 are also depicted. The constant C_1 is referred to as the directriss, and the shape of the rope is called a *catenary*.

Second, it is easy to calculate the arc length of the catenary starting from the lowest position, that is, from the point M. Since $ds = \sqrt{1 + (y')^2} dx$ we find from (7.2.13) that $ds = C_1 \cosh\left(\frac{x-C_2}{C_1}\right) d\left(\frac{x-C_2}{C_1}\right)$, where we have used the relation $\cosh^2 x - \sinh^2 x = 1$. Integrating from the lowest point M, we have

$$s = C_1 \sinh\left(\frac{x - C_2}{C_1}\right). \qquad (7.2.23)$$

In order to determine the constants C_1, C_2, and λ, we shall suppose that $y_1 > y_0$ and $x_1 > x_0$, as shown in Figure 7.2.1. Now we have from (7.2.23) that

$$
\begin{aligned}
L &= C_1 \left[\sinh\left(\frac{x_1 - C_2}{C_1}\right) - \sinh\left(\frac{x_0 - C_2}{C_1}\right)\right] \\
&= 2 C_1 \sinh\frac{x_1 - x_0}{2C_1} \cosh\frac{x_1 + x_0 - 2C_2}{2C_1},
\end{aligned} \qquad (7.2.24)
$$

and from (7.2.20) one has

$$
\begin{aligned}
y_1 - y_0 &= C_1 \left(\cosh\frac{x_1 - C_2}{C_1} - \cosh\frac{x_0 - C_2}{C_1}\right) \\
&= 2 C_1 \sinh\frac{x_1 - x_0}{2C_1} \sinh\frac{x_1 + x_0 - 2C_2}{C_1},
\end{aligned} \qquad (7.2.25)
$$

where we used the well-known relations $\sinh X - \sinh Y = 2\sinh\frac{1}{2}(X+Y) \times \cosh\frac{1}{2}(X-Y)$ and $\cosh X - \cosh Y = 2\sinh\frac{1}{2}(X+Y)\sinh\frac{1}{2}(X-Y)$. Combining (7.2.24) and (7.2.25), we have

$$\tanh\frac{x_1 + x_0 - 2C_2}{C_1} = \frac{y_1 - y_0}{L}. \qquad (7.2.26)$$

Since the right-hand side of this transcendental equation is known, we can find a unique solution of it in the form

$$\frac{x_1 + x_0 - 2C_2}{C_1} = \theta = const. \qquad (7.2.27)$$

From (7.2.24) and (7.2.25) we have

$$\sqrt{L^2 - (y_1 - y_0)^2} = 2C_1 \sinh \frac{x_1 - x_0}{2C_1}. \tag{7.2.28}$$

Dividing both sides of this equation by

$$\phi = \frac{x_1 - x_0}{2C_1}, \tag{7.2.29}$$

we have the second transcendental equation

$$\frac{\sin \phi}{\phi} = \frac{\sqrt{L^2 - (y_1 - y_0)^2}}{x_1 - x_0} = k \quad (k > 1), \tag{7.2.30}$$

where k is a given number.[14] Therefore, by determining ϕ from the transcendental equation

$$\sinh \phi = k\phi, \tag{7.2.31}$$

we can easily find the constants C_1 and C_2 from two algebraic equations (7.2.27) and (7.2.29):

$$C_1 = \frac{1}{2\phi} (x_1 - x_0), \quad C_2 = \frac{1}{2} \left[x_0 + x_1 - \frac{\theta}{\phi} (x_1 - x_0) \right]. \tag{7.2.32}$$

Finally, the Lagrangian multiplier λ can be easily determined from the equation (7.2.20).

7.3   Algebraic (Holonomic) Constraints

Let us determine a set of necessary conditions for the actual motion of a dynamical system whose trajectory is $\mathbf{q} = (q_1(t), ..., q_n(t))$ to be an extremal for the functional of the form

$$I = \int_{t_0}^{t_1} L(t, q_1, ..., q_n, \dot{q}_1, ..., \dot{q}_n) \, dt, \tag{7.3.1}$$

in the presence of the algebraic constraints

$$f_s(t, q_1, ..., q_n) = 0, \quad s = 1, ..., k, \quad k < n. \tag{7.3.2}$$

We suppose that the time interval (t_0, t_1) is given. It is clear that the appearance of these k constraint relations means that only $(n - k)$ of n components of $\mathbf{q}$ are independent.

---

[14] For proof that $k > 1$, the reader can consult [6]. Determination of C_1, C_2, and λ in [83], which we follow, takes this fact for granted.

Generally, it would be difficult or frequently impossible to calculate k dependent coordinates, say $q_1, ..., q_k$, from (7.3.2) in terms of $(n - k)$ independent coordinates $q_{k+1}, ..., q_n$ and to use these equations to eliminate k dependent coordinates from I.

As a better approach we can use the Lagrangian multiplier method. First, we form the augmented functional by adjoining the constraint relations to I, and thus we have

$$I_{aug.} = \int_{t_0}^{t_1} [L(t, \mathbf{q}, \dot{\mathbf{q}}) + \lambda_s(t) f_s(t, \mathbf{q})] \, dt, \quad s = 1, ..., k. \tag{7.3.3}$$

Note that for the case when auxiliary equations (7.3.2) are satisfied, $I_{aug.} = I$ for all unknown *Lagrangian multipliers* $\lambda_s(t)$, $s = 1, ..., k$.

Calculating the first variation in the usual manner (7.3.3), that is, introducing variations in the functions $\mathbf{q}, \dot{\mathbf{q}}$, and λ_s, we find

$$\delta I_{aug.} = \int_{t_0}^{t_1} \left[\left(\frac{\partial L}{\partial q_i} + \lambda_s \frac{\partial f_s}{\partial q_i} \right) \delta q_i + \frac{\partial L}{\partial \dot{q}_i} \delta \dot{q}_i + f_s \delta \lambda_s \right] dt. \tag{7.3.4}$$

Employing the commutativity rule $\delta \dot{q}_i = (d/dt) \, \delta q_i$, integrating by parts, and retaining only the terms inside the integral (supposing that the boundary conditions $q_i(t_0) = A_i, q_i(t_1) = B_i$ hold), we obtain

$$\begin{aligned} \delta I_{aug.} &= \int_{t_0}^{t_1} \left[\left(\frac{\partial L}{\partial q_i} - \frac{d}{dt} \frac{\partial L}{\partial \dot{q}_i} + \lambda_s \frac{\partial f_s}{\partial q_i} \right) \delta q_i + f_s \delta \lambda_s \right] dt, \\ i &= 1, ..., n. \end{aligned} \tag{7.3.5}$$

On an extremal, the variation $\delta I_{aug.} = 0$. Also, the auxiliary equations (7.3.2) must be satisfied by an extremal, that is, $f_s(t, \mathbf{q}) = 0, s = 1, ..., k$. Since the holonomic constraints are satisfied, we can select k Lagrangian multipliers $\boldsymbol{\lambda}$ arbitrarily. Let us select $\boldsymbol{\lambda}$'s in such a way that the coefficients of k components of δq_i, say $i = 1, ..., k$, are zero over the whole time interval (t_0, t_1). The remaining $n - k$, components of $\delta \mathbf{q}$, that is, $\delta q_i, i = k - 1, ..., n$, are then independent. Thus, the coefficients of these components of $\delta \mathbf{q}$ must be equal to zero. The final result is that in addition to k auxiliary equations (7.3.2), the equations

$$\frac{d}{dt} \frac{\partial L}{\partial \dot{q}_i} - \frac{\partial L}{\partial q_i} = \lambda_s \frac{\partial f_s}{\partial q_i}, \quad i = 1, ..., n; \quad s = 1, ..., k, \tag{7.3.6}$$

must be satisfied.

Therefore, n Euler–Lagrangian equations (7.3.6) and k algebraic constrained equations (7.3.2) constitute a system of $n + k$ equations for determining $n + k$ unknown generalized coordinates $q_i = q_i(t)$ and k Lagrangian multipliers $\lambda_s(t)$.

Note that the Euler–Lagrangian system can be written in a more concise way if we introduce the augmented Lagrangian function

$$L_{aug.}(t, \mathbf{q}, \dot{\mathbf{q}}) = L(t, \mathbf{q}, \dot{\mathbf{q}}) + \lambda_s(t) f_s(t, \mathbf{q}). \tag{7.3.7}$$

Then, the system (7.3.6) can be written as

$$\frac{d}{dt} \frac{\partial L_{aug.}}{\partial \dot{q}_i} - \frac{\partial L_{aug.}}{\partial q_i} = 0, \quad i = 1, ..., n. \tag{7.3.8}$$

7.4 Differential Equations Constraints

In this section we examine the following constrained problem: Find the extremals of the action integral

$$I = \int_{t_0}^{t_1} L\left(t, q_1, ..., q_n, \dot{q}_1, ..., \dot{q}_n\right) dt, \tag{7.4.1}$$

subject to constraints in the form of $k < n$ differential equations[15]

$$h_s\left(t, q_1, ..., q_n, \dot{q}_1, ..., \dot{q}_n\right) = 0, \quad s = 1, ..., k. \tag{7.4.2}$$

It can be shown that the solution procedure of this constrained problem proceeds along the same lines as the procedure for the problems considered in the previous section. Namely, we can use the method of Lagrangian multipliers.

Let us form the augmented action integral

$$I_{aug.} = \int_{t_0}^{t_1} \left[L\left(t, \mathbf{q}, \dot{\mathbf{q}}\right) + \lambda_s\left(t\right) h_s\left(t, \mathbf{q}, \dot{\mathbf{q}}\right)\right] dt, \tag{7.4.3}$$

where $\lambda_s\left(t\right)$ are unknown Lagrangian multipliers and $\mathbf{q} = \left(q_1, ..., q_n\right)$. Performing the variation of $\mathbf{q}$, $\dot{\mathbf{q}}$, and $\boldsymbol{\lambda}$, we find that the variation of $I_{aug.}$ is given as

$$\delta I_{aug.} = \int_{t_0}^{t_1} \left[\frac{\partial L}{\partial q_i}\delta q_i + \frac{\partial L}{\partial \dot{q}_i}\delta \dot{q}_i \right.$$
$$\left. + \lambda_s\left(\frac{\partial h_s}{\partial q_i}\delta q_i + \frac{\partial h_s}{\partial \dot{q}_i}\delta \dot{q}_i\right) + h_s\delta\lambda_s\right] dt. \tag{7.4.4}$$

Employing the commutativity rule $\delta \dot{q}_i = \frac{d}{dt}\left(\delta q_i\right)$ and integrating by parts, we find

$$\delta I_{aug.} = \left(\frac{\partial L}{\partial \dot{q}_i} + \lambda_s\frac{\partial h_s}{\partial \dot{q}_i}\right)\delta q_i\Big|_{t_0}^{t_1} + \int_{t_0}^{t_1}\left[\left(\frac{\partial L}{\partial q_i} - \frac{d}{dt}\frac{\partial L}{\partial \dot{q}_i}\right.\right.$$
$$\left.\left. + \lambda_s\frac{\partial h_s}{\partial q_i} - \dot{\lambda}_s\frac{\partial h_s}{\partial \dot{q}_i} - \lambda_s\frac{d}{dt}\frac{\partial h_s}{\partial \dot{q}_i}\right)\delta q_i + h_s\delta\lambda_s\right] dt. \tag{7.4.5}$$

On an extremal, the variation should be equal to zero, that is, $\delta I_{aug.} = 0$, and the differential equation of constraints (7.4.2) must also be satisfied. Therefore, the coefficients of δq_i and $\delta\lambda_s$ in (7.4.5) must be equal to zero.

Repeating the same reasoning as in the previous section (see discussion after (7.3.5)), the final result is that the expression under the integral sign in (7.4.5) must be equal to zero, which can be written in the form

$$\frac{\partial}{\partial q_i}\left(L + \lambda_s h_s\right) - \frac{d}{dt}\frac{\partial}{\partial \dot{q}_i}\left(L + \lambda_s h_s\right) = 0, \quad i = 1, ..., n. \tag{7.4.6}$$

---

[15]Note that differential equations (7.4.2) are not supplemented by Hertz–Hölder conditions (1.5.12) or (1.5.13).

The rest of the equation (7.4.5) becomes

$$\delta I_{aug.} = \left(\frac{\partial L}{\partial \dot{q}_i} + \lambda_s \frac{\partial h_s}{\partial \dot{q}_i}\right) \delta q_i \Bigg|_{t_0}^{t_1} = 0, \quad i = 1, ..., n. \tag{7.4.7}$$

In conclusion we can state the following.

(a) The n Euler–Lagrangian differential equations of extremals (7.4.6), which are equations of the second order, together with k differential equations of constraint (7.4.2), form a set of $k + n$ differential equations for finding the n generalized coordinates (extremals) $q_1, ..., q_n$ and k Lagrangian multipliers $\lambda_s(t)$, $s = 1, ..., k$.

(b) The expression (7.4.7) obviously serves as the source for finding the boundary conditions of the variational problem (7.4.1), (7.4.2).

(i) If the initial and terminal configurations A and B are given in the form of (5.2.1), that is, if the boundary conditions are prescribed in advance, then we have that $\delta q_i(t_0) = \delta q_i(t_1) = 0$ and $\delta I_{aug.} = 0$. Thus, the constants of integration will be determined from the $2n$ given conditions (5.2.1).

(ii) However, if the both configurations A and B, are not specified, or A or B are not specified, then from (7.4.7) we obtain the following natural boundary conditions:

$$\left(\frac{\partial L}{\partial \dot{q}_i} + \lambda_s \frac{\partial h_s}{\partial \dot{q}_i}\right)\Bigg|_{t_0} = 0 \quad \text{if the initial configuration } A$$
$$\text{is not specified,}$$

$$\left(\frac{\partial L}{\partial \dot{q}_i} + \lambda_s \frac{\partial h_s}{\partial \dot{q}_i}\right) \delta q_i \Bigg|^{t_1} = 0 \quad \text{if the final configuration } B$$
$$\text{is not specified,}$$

$$\left(\frac{\partial L}{\partial \dot{q}_i} + \lambda_s \frac{\partial h_s}{\partial \dot{q}_i}\right)\Bigg|_{t_0} = 0 \quad \text{if both configurations } A \text{ and } B$$

$$\left(\frac{\partial L}{\partial \dot{q}_i} + \lambda_s \frac{\partial h_s}{\partial \dot{q}_i}\right) \delta q_i \Bigg|^{t_1} = 0 \quad \text{are not specified.} \tag{7.4.8}$$

Example 7.4.1. *Particle moving freely in a plane.* The particle of unit mass is moving in the plane Oxy freely. By means of an electronic control system the velocity of the particle in the x direction is constrained to be proportional to the position of the particle in the y direction, that is,

$$h(t, x, y, \dot{x}, \dot{y}) = \dot{x} + y = 0, \tag{7.4.9}$$

where we assumed that the proportionality factor is equal to -1. Assuming that the position of the particle is specified at the end points and the time interval $(0, T)$ is given, find the extremal of the motion, taking as the basic functional (criteria of optimality)

$$I = \int_0^T \frac{1}{2}\left(\dot{x}^2 + \dot{y}^2\right) dt. \tag{7.4.10}$$

Since the particle is moving freely, its Lagrangian consists of kinetic energy only. This was used in writing (7.4.10).

Introducing the augmented action integral, we have

$$I_{aug.} = \int_0^T \left[\frac{1}{2} \left(\dot{x}^2 + \dot{y}^2 \right) + \lambda \left(\dot{x} + y \right) \right] dt. \tag{7.4.11}$$

The Euler–Lagrangian equations (7.3.6) in our case are

$$\ddot{x} + \dot{\lambda} = 0, \quad \ddot{y} - \lambda = 0. \tag{7.4.12}$$

These two equations together with the equation (7.4.9) form a system of three differential equations for finding the extremal $x(t), y(t)$ and Lagrangian multiplier $\lambda(t)$. Since the initial and terminal configurations are specified, say,

$$x(0) = A_1, \quad y(0) = A_2, \quad x(T) = B_1, \quad y(T) = B_2, \tag{7.4.13}$$

it is evident that the expression (7.3.7) is satisfied, that is, that there are no natural boundary conditions.

The integration of (7.4.12) and (7.4.9) is simple and yields the solution

$$\begin{aligned} x &= -C_1 t - C_2 \sin t - C_3 \cos t - C_4, \quad y = C_1 - C_2 \cos t - C_3 \sin t, \\ \lambda &= C_2 \cos t + C_3 \sin t, \end{aligned} \tag{7.4.14}$$

where $C_1, ..., C_4$ are constants of integration that can be easily determined from the given boundary conditions (7.4.13).

7.5   The Simplest Form of Hamilton's Variational Principle as a Problem of Optimal Control Theory

The problem of finding the extremals of an action integral in the presence of a certain number of differential equations as constraints, considered in the previous section, can be easily translated into the language of *optimal control theory* since this contemporary part of applied mathematics, physics, and engineering is in its principal part based upon the applied variational calculus and Hamilton's variational principle.

In this section we shall demonstrate that the simplest form of the Hamiltonian variational principle considered in section 5.2 can be interpreted in terms of optimal control theory.

Let us consider the action integral

$$I = \int_{t_0}^{t_1} L\left(t, q_1, ..., q_n, \dot{q}_1, ..., \dot{q}_n \right) dt, \tag{7.5.1}$$

where the time interval (t_0, t_1) is specified. Let us also assume that the initial and terminal configurations A and B are specified, that is, $q_i(t_0) = A_i =$

$const., q_i(t_1) = B_i = const., i = 1, ..., n$. We shall now reformulate Hamilton's principle $\delta I = 0$ in the following way.

Consider the functional (criterion of optimality, objective functional, performance measure, etc.)[16]

$$\bar{I} = \int_{t_0}^{t_1} L(t, q_1, ..., q_n, u_1, ..., u_n) \, dt, \qquad (7.5.2)$$

subject to the differential constraints

$$\dot{q}_1 = u_1, \quad \dot{q}_2 = u_2, ..., \dot{q}_n = u_n. \qquad (7.5.3)$$

Here, the generalized coordinates $q_i, i = 1, ..., n$ in accordance to the usual terminology of the optimal control theory are named the *state variables*, and the u_i are called the *control variables*.

According to equation (7.4.3) let us introduce the augmented integral

$$\bar{I}_{aug.} = \int_{t_0}^{t_1} [L(t, q_1, ..., q_n, u_1, ..., u_n) + \lambda_i (u_i - \dot{q}_i)] \, dt. \qquad (7.5.4)$$

Calculating the first variation of (7.5.4), using the commutativity rule, and performing the partial integration, we find

$$\begin{aligned}
\delta \bar{I}_{aug.} &= -\lambda_i(t) \, \delta q_i |_{t_0}^{t_1} + \int_{t_0}^{t_1} \left[\left(\frac{\partial L}{\partial u_i} + \lambda_i \right) \delta u_i \right. \\
&\left. + \left(\frac{\partial L}{\partial q_i} + \dot{\lambda}_i \right) \delta q_i + \delta \lambda_i (u_i - \dot{q}_i) \right] dt.
\end{aligned} \qquad (7.5.5)$$

Since $\delta q_i(t_0) = \delta q_i(t_1) = 0$, the extremality condition $\delta \bar{I}_{aug.} = 0$ generated the following system of equations:

$$\lambda_i = -\frac{\partial L}{\partial u_i}, \quad \dot{\lambda}_i = -\frac{\partial L}{\partial q_i}, \quad \dot{q}_i = u_i. \qquad (7.5.6)$$

Differentiating equation (7.5.6)$_1$ with respect to time t and combining it with (7.5.6)$_2$ and (7.5.6)$_3$, the condition $\delta \bar{I}_{aug.} = 0$ leads to the Euler–Lagrangian equations

$$\frac{\partial L}{\partial q_i} - \frac{d}{dt} \frac{\partial L}{\partial \dot{q}_i} = 0, \quad i = 1, ..., n. \qquad (7.5.7)$$

We show next that the augmented action integral can be transformed in such a way that it generates the Hamiltonian canonical equations.

Let us introduce the Hamiltonian function as

$$\bar{H}(t, \mathbf{q}, \boldsymbol{\lambda}, \mathbf{u}) = L(t, q_1, ..., q_n, u_1, ..., u_n) + \lambda_i u_i, \qquad (7.5.8)$$

---

[16]Since the terminology in the theory of optimal control is not quite unified, we have parenthetically noted several frequently used names for the integral (7.5.2) usually employed in the literature.

thus the action integral (7.5.4) becomes

$$\bar{I}_{aug.} = \int_{t_0}^{t_1} \left[\bar{H}\left(t, \mathbf{q}, \boldsymbol{\lambda}, \mathbf{u}\right) - \lambda_i \dot{q}_i \right] dt. \tag{7.5.9}$$

Performing now the variation of this action integral, using the commutativity rule and integration by parts, we find

$$\delta \bar{I}_{aug.} = -\lambda_i\left(t\right) \delta q_i \big|_{t_0}^{t_1} + \int_{t_0}^{t_1} \left[\left(\frac{\partial \bar{H}}{\partial q_i} + \dot{\lambda}_i \right) \delta q_i \right. $$
$$\left. + \left(\frac{\partial \bar{H}}{\partial \lambda_i} + \dot{q}_i \right) \delta \lambda_i + \frac{\partial \bar{H}}{\partial u_i} \delta u_i \right] dt. \tag{7.5.10}$$

The condition $\delta \bar{I}_{aug.} = 0$ generates the following system of equations:

$$\dot{\lambda}_i = -\frac{\partial \bar{H}}{\partial q_i}, \quad \dot{q}_i = \frac{\partial \bar{H}}{\partial \lambda_i}, \quad i = 1, ..., n, \tag{7.5.11}$$

and

$$\frac{\partial \bar{H}}{\partial u_i} = 0. \tag{7.5.12}$$

It is clear that with the identification $\lambda_i \equiv p_i$, where p_i are generalized momenta, $2n$ differential equations (7.5.11) are identical with the Hamiltonian canonical equations (1.8.14). Note that in analytical mechanics Hamilton's function, according to equation (1.8.4), is defined as $H = -L + p_i \dot{q}_i$, and here, in the theory of optimal control, the Hamiltonian is traditionally defined by equation (7.5.8).

Thus, the optimal control approach yields the necessary conditions for the simplest of Hamilton's principle problems

It is easy to demonstrate that the Hamiltonian function H is a conservation law on the optimal trajectory, that is, $\partial \bar{H}/\partial u_i = 0$, if the Lagrangian function L is not an explicit function of time t

$$H = const. \text{ on the optimal trajectory.} \tag{7.5.13}$$

The proof of this statement is the same as in section 1.8.

7.6   Continuous Optimal Control Problems

In this section we will apply the Hamiltonian principle with differential equations as constraints, considered previously, to optimal control problems. We shall confine ourselves to the cases in which the components of the *control vector* $\mathbf{u} = (u_1, ..., u_n)$ are not restricted, which means that the *variation* $\delta \mathbf{u}$ is *completely arbitrary* in the space of admissible controls and in the whole time interval (t_0, t_1) in which the physical process is taking place.

We note that the problem posed in this section can be considered as a special case of the *Pontryagin maximum principle* [49], [87], [26].

Consider the dynamical system described by the following system of differential equations:

$$\dot{q}_i = f_i\left(t, \mathbf{q}, \mathbf{u}\right), \quad t_0 \leq t \leq t_1, \quad i = 1, ..., n, \tag{7.6.1}$$

where $\mathbf{q} = \left(q_1, ..., q_n\right), \mathbf{u} = \left(u_1, ..., u_m\right)$. We shall assume that the functions f_i have continuous partial derivatives with respect to $\mathbf{q}$ and $\mathbf{u}$.

Consider also the following functional (criterion of optimality) in the Bolza form:

$$I = \Psi\left[t_1, \mathbf{q}\left(t_1\right)\right] + \int_{t_0}^{t_1} L\left(t, \mathbf{q}, \mathbf{u}\right) dt, \tag{7.6.2}$$

where Ψ and L possess continuous partial derivatives with respect to $\mathbf{q}$ and $\mathbf{u}$.

In what follows, we shall suppose that the time interval (t_0, t_1) is specified, and also, the initial position of the system is given, that is,

$$q_i\left(t_0\right) = A_i = const., \tag{7.6.3}$$

so that $\delta q_i\left(t_0\right) = 0$. However, the *terminal position of the system is not specified*, thus

$$\delta q_i\left(t_1\right) \neq 0; \tag{7.6.4}$$

that is, $\delta q_i\left(t_1\right)$ is completely arbitrary. The problem is to find an optimal control vector $\mathbf{u} = \left(u_1\left(t\right), ..., u_m\left(t\right)\right)$ which induces the system of equations (7.6.1) to follow an optimal trajectory $\mathbf{q} = \left(q_1\left(t\right), ..., q_n\left(t\right)\right)$ that minimizes (or maximizes) the functional (7.6.2).

Let us adjoin the differential equations (7.6.1) to the functional I by introducing the unknown Lagrangian multipliers, which we are going to denote by $p_1\left(t\right), ..., p_n\left(t\right)$:

$$I_{aug.} = \Psi\left[t_1, \mathbf{q}\left(t_1\right)\right] + \int_{t_0}^{t_1} \left\{L\left(t, \mathbf{q}, \mathbf{u}\right) + p_i\left[f_i\left(t, \mathbf{q}, \mathbf{u}\right) - \dot{q}_i\right]\right\} dt. \tag{7.6.5}$$

As suggested by equation (7.5.8), for the sake of simplicity, we can introduce the Hamiltonian function as follows:

$$H\left(t, \mathbf{q}, \mathbf{p}, \mathbf{u}\right) = L\left(t, \mathbf{q}, \mathbf{u}\right) + p_i f_i\left(t, \mathbf{q}, \mathbf{u}\right), \tag{7.6.6}$$

and the functional (7.6.5) becomes

$$I_{aug.} = \Psi\left[t_1, \mathbf{q}\left(t_1\right)\right] + \int_{t_0}^{t_1} \left[H\left(t, \mathbf{q}, \mathbf{p}, \mathbf{u}\right) - p_i\dot{q}_i\right] dt. \tag{7.6.7}$$

Calculating the first variation of (7.6.7), recalling that $-p_i\delta\dot{q}_i = -\left(d/dt\right)\left(p_i\delta q_i\right) + \dot{p}_i\delta q_i$, and performing partial integration, we obtain

$$\begin{aligned}
\delta I_{aug.} &= \left(\frac{\partial\Psi}{\partial q_i} - p_i\right)\delta q_i\bigg|_{t=t_1} + p_i\delta q_i|_{t=t_0} \\
&+ \int_{t_0}^{t_1}\left[\left(\frac{\partial H}{\partial q_i} + \dot{p}_i\right)\delta q_i + \left(\frac{\partial H}{\partial p_i} - \dot{q}_i\right)\delta p_i + \frac{\partial H}{\partial u_s}\delta u_s\right] dt.
\end{aligned} \tag{7.6.8}$$

A necessary condition for the extremality is that the first variation of $I_{aug.}$ vanishes for arbitrary $\delta q_i, \delta p_i$, and δu_i. Hence the coefficients of these variations inside the integral sign must vanish. Thus we obtain the following Euler–Lagrangian equation in canonical form:

$$\dot{q}_i = \frac{\partial H}{\partial p_i} \quad \text{or} \quad \dot{q}_i = f_i(t, \mathbf{q}, \mathbf{u}),$$

$$\dot{p}_i = -\frac{\partial H}{\partial q_i} \quad \text{or} \quad \dot{p}_i = -\frac{\partial L}{\partial q_i} - p_j \frac{\partial f_j}{\partial q_i}, \quad i, j = 1, ..., n, \qquad (7.6.9)$$

and

$$\frac{\partial H}{\partial u_s} = 0 \quad \text{or} \quad \frac{\partial L}{\partial u_s} + p_i \frac{\partial f_i}{\partial u_s} = 0, \quad i = 1, ..., n, \quad s = 1, .., m. \qquad (7.6.10)$$

Equation (7.6.8) now becomes

$$\delta I_{aug.} = \left(\frac{\partial \Psi}{\partial q_i} - p_i \right) \delta q_i \Big|_{t=t_1} + p_i \delta q_i |_{t=t_0}, \qquad (7.6.11)$$

whence, taking into account (7.6.3) and (7.6.4), we find the following n natural boundary conditions:

$$p_i = \frac{\partial \Psi}{\partial q_i} \quad \text{for} \quad t = t_1; \quad i = 1, ..., n. \qquad (7.6.12)$$

Therefore: in order that functional I given in equation (7.6.2) may be a maximum (or minimum) for a dynamical process described by equations (7.6.1), with the initial conditions at $t = t_0$ given by (7.6.3) it is necessary that there exist a nonzero continuous vector function $\mathbf{p} = (p_1, ..., p_n)$ satisfying equations $(7.6.9)_2$ and (7.6.12) and that the control vector function $\mathbf{u} = (u_1(t), ..., u_m(t))$ is so selected that the Hamiltonian function $H(t, \mathbf{q}, \mathbf{p}, \mathbf{u})$ is a maximum (or a minimum) for every t, $t_0 \leq t \leq t_1$.

This statement actually summarizes, the famous *Pontryagin maximum principle*.

Note that the optimal control vector $u_s(t)$, $s = 1, ..., m$ is selected from the set of algebraic equations (7.6.10), that is, $\partial H / \partial u_s = 0, s = 1, ..., m$. However, these equations are valid only under supposition that the components of $\mathbf{u}$ are not subject to any restrictions. If, however, the control vector is subjected to some kind of restriction, so that it belongs to a certain set of functions $\mathbf{u} \in U$, then the condition $\partial H / \partial u_s = 0$ is not valid. The condition (7.6.10) is, in this case, replaced by a more general statement:

$$\min_{\mathbf{u} \in U} H(t, \mathbf{q}, \mathbf{p}, \mathbf{u}). \qquad (7.6.13)$$

To determine optimal control $\mathbf{u} = \mathbf{u}^* \in U$ from (7.6.13), one often has to have information about $\mathbf{q}$ and $\mathbf{p}$, and this makes the process of optimization more complex. The values of $\mathbf{u} = \mathbf{u}^*$ minimizing $H(t, \mathbf{q}, \mathbf{p}, \mathbf{u})$ are located on the boundary of the set U. The interior and boundary solutions are illustrated in Figure 7.6.1.

We note that the rigorous proof of Pontryagins maximum principle is given in [87] and will not be presented here. This principle "represents, in a sense, the culmination of the efforts of mathematicians, for considerably more than a century, to rectify the Lagrangian multiplier rule" [131, p. 230].

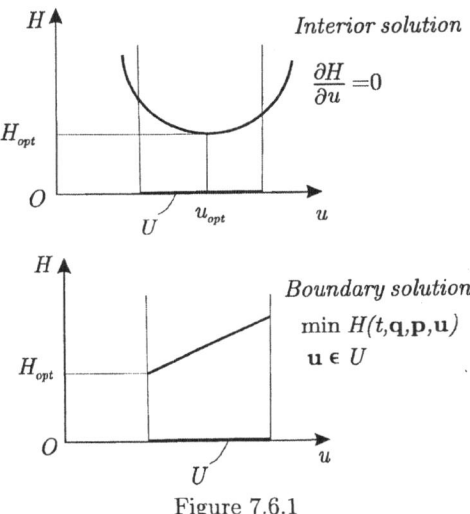

Figure 7.6.1

As in the previous section, it is easy to verify that if H (or L and f_i) is not an explicit function of time t, we have that H is a conservation law of the boundary value problem describing the minimization procedure. Namely, since $H(t, \mathbf{q}, \mathbf{p}, \mathbf{u})$, the total derivative of this function with respect to time reads

$$\frac{dH}{dt} = \frac{\partial H}{\partial q_i}\dot{q}_i + \frac{\partial H}{\partial p_i}\dot{p}_i + \frac{\partial H}{\partial u_s}\dot{u}_s + \frac{\partial H}{\partial t}. \tag{7.6.14}$$

Using the equations of motion (7.6.8) and (7.6.9) and collecting terms, we obtain

$$\frac{dH}{dt} = \left(\frac{\partial H}{\partial q_i} + \dot{p}_i\right) f_i(t, \mathbf{q}, \mathbf{u}) + \frac{\partial H}{\partial u_s}\dot{u}_s + \frac{\partial H}{\partial t}. \tag{7.6.15}$$

Along the optimal trajectory, the first term vanishes because of the differential equations (7.6.9). The second term vanishes because of (7.6.10). Thus, along the optimal trajectory,

$$\frac{dH}{dt} = \frac{\partial H}{\partial t}. \tag{7.6.16}$$

Therefore, if the problem is scleronomic (autonomous) in both L and f_i then the Hamiltonian is constant over time

$$H = const. \tag{7.6.17}$$

Example 7.6.1. *Optimal control of a two-degrees-of-freedom system.* As an illustration, let us consider the dynamical system whose behavior is described by the system of equations

$$\dot{q}_1 = u_1, \quad \dot{q}_2 = q_1 + u_2, \quad 0 \le t \le 1, \tag{7.6.18}$$

with the initial conditions

$$q_1(0) = 0, \quad q_2(0) = 0. \tag{7.6.19}$$

The problem is to determine the state variables (generalized coordinates) $q_1(t)$ and $q_2(t)$ and the optimal control variables $u_1(t)$ and $u_2(t)$ that minimize the functional

$$I = aq_1(1) + bq_2(1) + \int_0^1 \left(u_1^2 + u_2^2 + q_1\right) dt, \tag{7.6.20}$$

where a and b are given positive constants.

The Hamiltonian of this problem, according to (7.6.6), is $H = u_1^2 + u_2^2 + q_1 + p_1 u_1 + p_2(q_1 + u_2)$. The second pair of the canonical equations (7.6.9) reads

$$\dot{p}_1 = -\frac{\partial H}{\partial q_1} = -1 - p_2, \quad \dot{p}_2 = -\frac{\partial H}{\partial q_2} = 0, \tag{7.6.21}$$

while the optimal selection of the components of the control vector follows from the equations (7.6.10), namely,

$$\frac{\partial H}{\partial u_1} = 2u_1 + p_1 = 0, \quad \frac{\partial H}{\partial u_2} = 2u_2 + p_2 = 0. \tag{7.6.22}$$

In our case the Bolza term $\Psi(q_1(t_1), q_2(t_1)) = aq_1(1) + bq_2(1)$, so that the natural boundary conditions are, according to (7.6.12),

$$p_1(1) = \left.\frac{\partial \Psi}{\partial q_1}\right|_{t=1} = a, \quad p_2(1) = \left.\frac{\partial \Psi}{\partial q_2}\right|_{t=1} = b. \tag{7.6.23}$$

Integrating (7.6.21) and applying (7.6.23), we find that

$$p_1(t) = -(1+b)t + a + b + 1, \quad p_2(t) = b. \tag{7.6.24}$$

The components of the optimal control vector are therefore

$$u_1(t) = -\frac{p_1}{2} = \frac{1}{2}\left[(1+b)t + a + b + 1\right], \quad u_2(t) = -\frac{b}{2}. \tag{7.6.25}$$

Entering with this into (7.6.18), integrating, and determining the integration constants from the initial conditions (7.6.19), we find that

$$q_1(t) = \frac{1}{4}\left[(1-b)t^2 - 2(1+a+b)t\right],$$
$$q_1(t) = \frac{1}{12}\left[(1+b)^3 - 3(1+a+b)t^2 - 6bt\right]. \tag{7.6.26}$$

7.7 Optimal Control Problems with Unspecified Terminal Time

In the continuous problems of optimal control theory for which the final time t_1 is not specified, several situations may occur. For various cases see [61, pp. 192–198]. Here we are going to limit ourselves to the case whose optimality criterion is given in the Bolza form

$$I = \Psi\left[t_1, \mathbf{q}\left(t_1\right)\right] + \int_{t_0}^{t_1} L\left(t, \mathbf{q}, \mathbf{u}\right) dt, \tag{7.7.1}$$

where $\mathbf{q} = (q_1, ..., q_n)$, $\mathbf{u} = (u_1, ..., u_m)$. The behavior of the dynamical system is described by the system of differential equations

$$\dot{q}_i = f_i\left(t, \mathbf{q}, \mathbf{u}\right), \quad i = 1, ..., n. \tag{7.7.2}$$

We shall suppose that the initial time t_0 is specified and that, also, the initial position of the system is prescribed, so that

$$q_i\left(t_0\right) = A_i = const. \tag{7.7.3}$$

The terminal time t_1 is not given and the terminal configuration $q_i\left(t_1\right)$ can be given or not given.

To include the differential equations constraints (7.7.2), we form the augmented functional

$$I_{aug.} = \Psi\left[t_1, \mathbf{q}\left(t_1\right)\right] + \int_{t_0}^{t_1} \left\{L\left(t, \mathbf{q}, \mathbf{u}\right) + p_i\left[f_i\left(t, \mathbf{q}, \mathbf{u}\right) - \dot{q}_i\right]\right\} dt. \tag{7.7.4}$$

As in the previous section we introduce the Hamiltonian

$$H\left(t, \mathbf{q}, \mathbf{u}, \mathbf{p}\right) = L\left(t, \mathbf{q}, \mathbf{u}\right) + p_i f_i\left(t, \mathbf{q}, \mathbf{u}\right) \tag{7.7.5}$$

and write (7.7.4) in the form

$$I_{aug.} = \Psi\left[t_1, \mathbf{q}\left(t_1\right)\right] + \int_{t_0}^{t_1} \left\{H\left(t, \mathbf{q}, \mathbf{u}, \mathbf{p}\right) - p_i\dot{q}_i\right\} dt. \tag{7.7.6}$$

Since the terminal time t_1 is not specified, we shall use here the generalized (nonsimultaneous) variations. Recalling the equation (6.4.4), which states that the generalized variation of a functional

$$I = \int_{t_0}^{t_1} L\left(...\right) dt \tag{7.7.7}$$

is given in the form

$$\Delta I = \delta \int_{t_0}^{t_1} L\left(...\right) dt + L\Delta t\big|_{t_0}^{t_1}, \tag{7.7.8}$$

the variation of (7.7.6) becomes

$$\Delta I_{aug.} \;=\; \left(\frac{\partial \Psi}{\partial q_i} \Delta q_i + \frac{\partial \Psi}{\partial t} \Delta t \right)\Big|_{t_1} \tag{7.7.9}$$

$$+ \delta \int_{t_0}^{t_1} \{ H\left(t, \mathbf{q}, \mathbf{u}, \mathbf{p}\right) - p_i \dot{q}_i \} \, dt + \left(H\left(t, \mathbf{q}, \mathbf{u}, \mathbf{p}\right) - p_i \dot{q}_i \right) \Delta t \big|_{t_1} ,$$

where we have taken into account that the initial time t_0 and initial position are specified, so that $\Delta t_0 = \delta t_0 = 0$ and $\Delta q_i\left(t_0\right) = \delta q_i\left(t_0\right) = 0$.

Performing the variation of the integral term in (7.7.9) and noting that $-p_i \delta \dot{q}_i = -\left(d/dt\right)\left(p_i \delta q_i\right) + \dot{p}_i \delta q_i$, after partial integration, we arrive at

$$\delta \int_{t_0}^{t_1} [H - p_i \dot{q}_i] \, dt \;=\; -p_i \delta q_i |_{t_1}$$

$$+ \int_{t_0}^{t_1} \left[\left(\frac{\partial H}{\partial q_i} + \dot{p}_i \right) \delta q_i + \left(\frac{\partial H}{\partial p_i} - \dot{q}_i \right) \delta p_i + \frac{\partial H}{\partial u_s} \delta u_s \right] dt,$$

$$i \;=\; 1, ..., n, \quad s = 1, ..., m. \tag{7.7.10}$$

Employing the relation $\delta q_i = \Delta q_i - \dot{q}_i \Delta t$ (see (3.2.5)) and entering with (7.7.10) into (7.7.9), we obtain the following expression:

$$\Delta I_{aug.} \;=\; \left\{ \left(\frac{\partial \Psi}{\partial q_i} - p_i \right) \Delta q_i + \left(\frac{\partial \Psi}{\partial t} + H \right) \Delta t \right\}\Big|_{t_1}$$

$$+ \int_{t_0}^{t_1} \left[\left(\frac{\partial H}{\partial q_i} + \dot{p}_i \right) \delta q_i + \left(\frac{\partial H}{\partial p_i} - \dot{q}_i \right) \delta p_i + \frac{\partial H}{\partial u_s} \delta u_s \right] dt. \tag{7.7.11}$$

We use the same arguments as in the previous section and conclude that the optimal behavior of a dynamical system is described by the following Euler–Lagrangian equations, in the canonical form

$$\dot{q}_i \;=\; \frac{\partial H}{\partial p_i} = f_i\left(t, \mathbf{q}, \mathbf{u}\right),$$

$$\dot{p}_i \;=\; -\frac{\partial H}{\partial q_i} = -\frac{\partial L}{\partial q_i} - p_j \frac{\partial f_j}{\partial q_i}, \quad i = 1, ..., n,$$

$$\frac{\partial H}{\partial u_s} \;=\; \frac{\partial L}{\partial u_s} + p_i \frac{\partial f_i}{\partial u_s}, \quad s = 1, ..., m. \tag{7.7.12}$$

Hence, (7.7.11) becomes

$$\Delta I_{aug.} = \left(\frac{\partial \Psi}{\partial q_i} - p_i \right) \Delta q_i\left(t_1\right) + \left(\frac{\partial \Psi}{\partial t} + H \right) \Delta t\left(t_1\right). \tag{7.7.13}$$

The condition for optimality $\Delta I_{aug.} = 0$ will provide the necessary natural boundary conditions. Let us consider the following two cases which frequently arise in practical situations.

(a) *The final position is given; the final time t_1 is not specified.* Let $q_i\,(t_0) = A_i = \text{const.}, q_i\,(t_1) = B_i = \text{const.}$, and the terminal time not be specified. Therefore, $\Delta q_i\,(t_1) = 0$ and Δt_1 is arbitrary, so that we have as the natural boundary condition the following scalar equation:

$$\frac{\partial \Psi}{\partial t} + H = 0 \quad \text{for} \quad t = t_1. \tag{7.7.14}$$

If the Bolza term is equal to zero, that is, $\Psi = 0$, we obtain

$$H = 0 \quad \text{for} \quad t = t_1. \tag{7.7.15}$$

This case, for $n = 1$, is depicted in Figure 7.7.1.

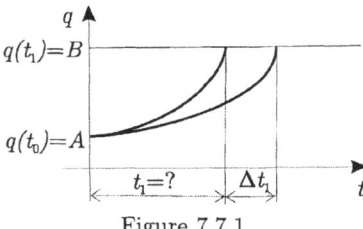

Figure 7.7.1

(b) *The final configuration $q_i\,(t_1)$ and the final time t_1 are not given and are independent.* Since $\Delta q_i\,(t_1)$ and Δt_1 are arbitrary, from (7.7.13) the requirement for extremality $\Delta I_{aug.} = 0$ generates the following natural boundary conditions:

$$\begin{aligned} p_i\,(t_1) &= \left.\frac{\partial \Psi}{\partial q_i}\right|_{t=t_1}, \quad i = 1, ..., n, \\ \frac{\partial \Psi}{\partial t} + H &= 0 \quad \text{for} \quad t = t_1. \end{aligned} \tag{7.7.16}$$

If the Bolza term is equal to zero, that is, $\Psi = 0$, we have

$$\begin{aligned} p_i\,(t_1) &= 0, \quad i = 1, ..., n, \\ H &= 0 \quad \text{for} \quad t = t_1. \end{aligned} \tag{7.7.17}$$

Note that there exist a variety of special cases for which finding the natural boundary conditions is a matter of making the appropriate substitutions into equation (7.7.13). Note also that the natural boundary condition (7.7.15) and (7.7.16)$_2$ for the case when the Hamiltonian is not an explicit function of time is equal to zero or is constant, respectively, for the whole time interval $(t_0, t_1,)$.

Example 7.7.1. The brachistochrone problem as the problem of optimal control theory. Let us consider again the problem of brachistochrone discussed in *Example 5.2.1.*

First we note that in accordance with (5.2.27), the differential equations of the motion of a heavy particle are

$$\dot{x} = \sqrt{2gy}\cos u, \quad \dot{y} = \sqrt{2gy}\sin u, \qquad (7.7.18)$$

where u (the angle between the tangent at the curve and the x axis) plays the role of the control variable. The minimum-time curve (brachistochrone) can be determined by minimizing the functional

$$I = \int_0^T 1\,dt. \qquad (7.7.19)$$

We shall suppose that the initial conditions are

$$x(0) = 0, \quad y(0) = 0, \qquad (7.7.20)$$

and

$$x(T) = x(B) = l, \qquad (7.7.21)$$

where l is a *given* constant and $y(T) = y(B)$ is *not specified,* hence the position of the point B is not known.

Since $L = 1$, the Hamiltonian of the problem is, in accordance with (7.7.5),

$$H = 1 + p_x\sqrt{2gy}\cos u + p_y\sqrt{2gy}\sin u. \qquad (7.7.22)$$

The second pair of canonical equations $(7.7.12)_2$ is

$$\dot{p}_x = -\frac{\partial H}{\partial x} = 0, \quad \dot{p}_y = -\frac{\partial H}{\partial y} = -\frac{1}{\sqrt{2gy}}\left(p_x g\cos u + p_y g\sin u\right), \qquad (7.7.23)$$

while the condition $(7.7.12)_3$ gives

$$-p_x\sqrt{2gy}\sin u + p_y\sqrt{2gy}\cos u = 0. \qquad (7.7.24)$$

The expression (7.7.13) for $\Psi = 0$, with a slight change in notation, becomes

$$\Delta I_{aug.} - p_x(T)\,\Delta x(T) - p_y(T)\,\Delta y(T) + H\Delta T = 0. \qquad (7.7.25)$$

Since $\Delta x(T) = 0$ and $y(T)$ and ΔT are not specified, it follows that we have the following two natural boundary conditions:

$$p_y(T) = 0, \quad H = 0, \quad 0 \le t \le T. \qquad (7.7.26)$$

Integrating $(7.7.23)_1$ we find that $p_x = C = const.$, and from (7.7.24) it follows that

$$p_y = C\tan u. \qquad (7.7.27)$$

Entering with this into (7.7.5), since $H = 0$, we obtain $1 + \sqrt{2gy}/\cos u = 0$, that is,

$$y = \frac{1}{2C^2 g} \cos^2 u. \tag{7.7.28}$$

Differentiating (7.7.27) with respect to time and entering with the result into (7.7.23)$_2$, one has

$$\dot{p}_y = \frac{C\dot{u}}{\cos^2 u} = -\frac{1}{\sqrt{2gy}} \left(Cg \cos u + C \tan u \sin u \right). \tag{7.7.29}$$

Combining this and (7.7.27), it follows that

$$\dot{u} = \frac{du}{dt} = Cg = const., \tag{7.7.30}$$

or

$$u = Cgt + D, \tag{7.7.31}$$

where D is a constant.

From (7.7.18)$_1$ we have

$$\frac{dx}{dt} = \frac{dx}{du}\frac{du}{dt} = \sqrt{2gy} \cos u = -\frac{\cos^2 u}{C} = -\frac{(1 + \cos 2u)}{2C}. \tag{7.7.32}$$

Separating variables and integrating, we find

$$x = K - \frac{u}{2C^2 g} - \frac{\sin 2u}{4C^2 g}, \tag{7.7.33}$$

where K is a constant.

Since $y(0) = 0$, it follows from (7.7.28) that $u(0) = \pi/2$. From (7.7.26)$_1$ and (7.7.27) it follows that $u(T) = 0$, and therefore $D = 0$ and $C = -\pi/2gT$. Finally, from $x(T) = l$, it follows that $K = l$. We obtain from $x(0) = 0$ the minimum time of travel of the particle:

$$0 = l - \frac{\pi}{4g\left(\frac{\pi^2}{4g^2 T^2}\right)}, \tag{7.7.34}$$

whence

$$T = l\sqrt{\frac{\pi}{g}}. \tag{7.7.35}$$

Example 7.7.2. Linear system with one-degree-of-freedom system. Consider a linear dynamical system described by

$$\dot{x} = \lambda x + \mu u, \quad x(0) = 0. \tag{7.7.36}$$

It is desired to minimize

$$I = AT + Bx(T) + \int_0^T \frac{1}{2}u^2 dt, \qquad (7.7.37)$$

where T and $x(T)$ are not specified, and λ, μ, A, and B are given constant parameters.

The Hamiltonian function in this case is

$$H = \frac{1}{2}u^2 + p(\lambda x + \mu u). \qquad (7.7.38)$$

Now we have

$$\dot{p} = -\frac{\partial H}{\partial x} = -\lambda p, \qquad (7.7.39)$$

and from the equation $\partial H / \partial u = 0$, it follows that

$$u = -\mu p. \qquad (7.7.40)$$

Integrating (7.7.39), one has

$$p = C_1 e^{-\lambda t}, \qquad (7.7.41)$$

where C_1 is a constant. Entering with this into (7.7.36), after integration we obtain $x(t) = \left(\mu^2 C_1 / 2\lambda\right) e^{-\lambda t} + C_2 e^{\lambda t}$, where C_2 is another constant. Matching this with the boundary condition $x(0) = 0$, we obtain

$$x(t) = -\frac{\mu^2 C_1}{2\lambda} \left(e^{\lambda t} - e^{-\lambda t}\right) = -\frac{\mu^2}{\lambda} C_1 \sinh \lambda t. \qquad (7.7.42)$$

Since the terminal time T and the terminal position $x(T)$ are not specified, according to (7.7.16) we obtain the following two natural boundary conditions $p(T) = (\partial \Psi / \partial x)|_{t=T}$ and $(\partial \Psi / \partial t)|_{t=T} + H|_{t=T} = 0$, where $\Psi = AT + Bx(T)$. Thus, we find $C_1 = Be^{\lambda T}$, from which it follows that

$$x(t) = -\frac{\mu^2}{\lambda} Be^{\lambda T} \sin \lambda t \qquad (7.7.43)$$

and

$$p(t) = Be^{\lambda T} e^{-\lambda t}. \qquad (7.7.44)$$

Combining (7.7.40) and (7.7.38) we have $H = -(1/2)\mu^2 p^2 + \lambda px$. From the second natural boundary condition given above, we have

$$A - \frac{\mu^2 p^2(T)}{2} + \lambda p(T) x(T) = 0, \qquad (7.7.45)$$

whence

$$e^{\lambda T} \sinh \lambda T = \frac{A}{\mu^2 B^2} - \frac{1}{2}, \qquad (7.7.46)$$

or, noting that $\sinh \lambda T = (1/2)\left(e^{\lambda T} - e^{-\lambda T}\right)$, the terminal time T is

$$T = \frac{1}{2\lambda} \ln \left(\frac{2A}{\mu^2 B^2}\right). \qquad (7.7.47)$$

Chapter 8

Variational Principles for Elastic Rods and Columns

8.1 Introduction

In this section we shall use the results presented so far to formulate several variational principles for the equations describing deformations and the optimal shape of elastic columns. We shall use the classical (Bernoulli–Euler) rod theory as well as generalized rod theories. The variational principles that we will formulate will be used to

(a) estimate the critical (buckling) load of a column,

(b) determine postcritical shape of the column using Ritz method,

(c) determine the optimal shape of the column, that is, the shape of the column having smallest volume and being stable against buckling. This constitutes the so-called Lagrange problem formulated in 1773 (see [31], [100]).

In formulating the variational principles for rods we shall *first* derive the corresponding differential equations and then find the variational principle for those equations. Variational principles are important because they may help develop deeper understanding of the problem under consideration. As Anthony [4] stated: "In theoretical physics a theory is often considered to be complete only if its variational principle in the sense of Hamilton is known."

Earlier, we noted that finding a Lagrangian for a given set of equations constitutes the so-called *inverse Lagrangian problem* (see discussion after (1.4.25)). Its solution (the Lagrangian that is found) is not unique. Sometimes the equations of the problem must be written in a special form in order to find a Lagrangian function for which the Euler–Lagrangian equations are equivalent to the given system of equations. One such situation was treated in section 1.8, where it was shown that the differential equation

$$\ddot{q} + 2k\dot{q} + \omega^2 q = 0 \tag{8.1.1}$$

has a Lagrangian

$$L = \frac{1}{2} \left(\dot{q}^2 - \omega^2 q^2 \right) e^{2kt}. \tag{8.1.2}$$

Note that the Euler–Lagrangian equation for (8.1.2) is $e^{2kt} \left(\ddot{q} + 2k\dot{q} + \omega^2 q \right) = 0$. Thus (8.1.1), as such, does not have a Lagrangian. However, when multiplied with a (nonzero) function e^{2kt}, it has a Lagrangian given by (8.1.2). Another example of this type was given in [27]. Namely, the equation

$$\ddot{x} - \alpha \dot{x} = 0, \quad \alpha = const. \tag{8.1.3}$$

as it stands does not have a Lagrangian, but the equation

$$\frac{1}{x} \left(\ddot{x} - \alpha \dot{x} \right) = 0, \tag{8.1.4}$$

equivalent to (8.1.3) if $x \neq 0$, has a Lagrangian function

$$L = \dot{x} \ln |\dot{x}| + \alpha x. \tag{8.1.5}$$

Thus, the Lagrangians that we formulate in this chapter are by no means the only possible variational formulation of the corresponding differential equations.

8.2 The Column with Concentrated Force at the End

The equation describing the shape of an elastic column loaded with the concentrated force at its end (see Figure 8.2.1) is

$$\frac{1}{\rho} = -\frac{Fy}{EI}, \tag{8.2.1}$$

where ρ is the radius of curvature of the central line of the column, F is the compressive force, EI is the bending rigidity, and y is the displacement of an arbitrary point on the column axis. In what follows we shall formulate three different variational principles for the equation (8.2.1) for three possible choices of dependent and independent variables. All variational principles will be formulated for the column shown in Figure 8.2.1

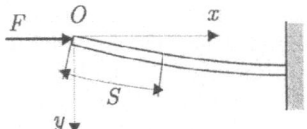

Figure 8.2.1

Case 1. Suppose that the coordinate x is taken as an independent variable. The lateral displacement of an arbitrary point on the column axis we denote by $y(x)$. Then, the curvature of the column axis $1/\rho$ can be expressed as

$$\frac{1}{\rho} = \frac{\frac{d^2 y}{dx^2}}{\sqrt{\left[1 + \left(\frac{dy}{dx}\right)^2\right]^3}}, \tag{8.2.2}$$

so that equation (8.2.1) becomes

$$\frac{\frac{d^2 y}{dx^2}}{\sqrt{\left[1 + \left(\frac{dy}{dx}\right)^2\right]^3}} + k^2 y = 0, \tag{8.2.3}$$

where $k^2 = F/EI$. The boundary conditions corresponding to the column shown in Figure 8.2.1 read

$$y(0) = 0, \quad \frac{dy}{dx}(x = x_B) = 0. \tag{8.2.4}$$

Here we consider x_B as given. With x_B given, the length of the column is $L = \int_0^{x_B}(1+(\frac{dy}{dx})^2)^{1/2}dx$. Equation (8.2.3) with the boundary conditions (8.2.4) is equivalent to the stationarity condition $\delta I_1 = 0$ for the functional

$$I_1 = \int_0^{x_B} \left[\left(1 + \left(\frac{dy}{dx}\right)^2\right)^{1/2} - \frac{k^2}{2}y^2\right] dx. \tag{8.2.5}$$

Since in the case of varying cross section, I is a function of S (not x), the functional (8.2.5) is suitable for the columns with constant cross section.

Case 2. We take S, the arc length of the column axis, measured from the point O as an independent variable and the angle θ between the tangent to the column axis and the x axis of the coordinate system xOy as the dependent variable. The curvature is then given as $1/\rho = \frac{d\theta}{dS}$ and (see [14])

$$\frac{dy}{dS} = \sin\theta, \quad \frac{dx}{dS} = \cos\theta. \tag{8.2.6}$$

By differentiating (8.2.1) with respect to S and by using (8.2.6)$_1$, we obtain

$$\frac{d}{dS}\left(EI\frac{d\theta}{dS}\right) + F\sin\theta = 0. \tag{8.2.7}$$

Boundary conditions corresponding to the column shown in Figure 8.2.1 are

$$\frac{d\theta}{dS}(S = 0), \quad \theta(S = L) = 0. \tag{8.2.8}$$

The trivial configuration in which the rod axis remains straight is described by

$$\theta_0 = 0. \tag{8.2.9}$$

The boundary value problem (8.2.7), (8.2.8) is equivalent to the stationarity conditions $\delta I_2 = 0$, where

$$I_2 = \int_0^L \left[\frac{1}{2} EI(S) \left(\frac{d\theta}{dS} \right)^2 + F \cos\theta \right] dS. \tag{8.2.10}$$

Case 3. We take S as an independent variable. By using (8.2.6) we can write the curvature as

$$\frac{1}{\rho} = \frac{\frac{d^2 y}{dS^2}}{\sqrt{1 - \left(\frac{dy}{dS} \right)^2}}. \tag{8.2.11}$$

Thus, (8.2.1) becomes

$$\frac{\frac{d^2 y}{dS^2}}{\sqrt{1 - \left(\frac{dy}{dS} \right)^2}} + k^2 y = 0, \tag{8.2.12}$$

with the boundary conditions

$$y(0) = 0, \quad \frac{dy}{dS}(L) = 0, \tag{8.2.13}$$

and L given. The coordinate x_B is determined as $x_B = \int_0^L \cos\theta(S)\, dS$. The boundary value problem (8.2.12), (8.2.13) is equivalent to the stationarity conditions $\delta I_3 = 0$, where

$$I_3 = \int_0^L \left[\left(1 - \left(\frac{dy}{dS} \right)^2 \right)^{1/2} + \frac{dy}{dS} \arcsin\left(\frac{dy}{dS} \right) - \frac{k^2(S)}{2} y^2 \right] dS. \tag{8.2.14}$$

Since the Lagrangians of all variational principles stated in Cases 1–3 do not depend explicitly on an independent coordinate if $EI = const.$, they could be used to obtain first integrals (Jacobi conservation law) of corresponding differential equations by using (1.4.44). Thus we have

$$\frac{-1}{\left(1 + \left(\frac{dy}{dx} \right)^2 \right)^{1/2}} + \frac{k^2}{2} y^2 = const. \tag{8.2.15}$$

as a conservation law for (8.2.3). For the case (8.2.7) we have

$$\frac{1}{2} \left(\frac{d\theta}{dS} \right)^2 - \frac{F}{EI} \cos\theta = const. \tag{8.2.16}$$

as a conservation law, and finally

$$-\left(1 - \left(\frac{dy}{dS}\right)^2\right)^{1/2} + \frac{k^2}{2}y^2 = const. \tag{8.2.17}$$

as a conservation law for (8.2.12). The conservation law (8.2.16) is the starting point for obtaining the exact solution of (8.2.7) by the separation of variables method.

In what follows we show an application of the principle $\delta I_3(y) = 0$. Namely, from (8.2.14) we have

$$\delta I_3 = \int_0^1 \left[\delta y' \arcsin y' - \bar{k}^2 y \delta y\right] dt = 0, \tag{8.2.18}$$

where we introduced dimensionless arc length $t = S/L$ and parameter $\bar{k}^2 = \frac{FL^2}{EI}$. Also, $\frac{d}{dt}(\cdot) = (\cdot)'$. Note that from (8.2.13) and (8.2.8) we have

$$\delta y(0) = 0, \quad \delta y'(1) = 0. \tag{8.2.19}$$

Since (8.2.18) is valid for *any* δy satisfying (8.2.19) it holds for $\delta y = y$, where y is the solution of (8.2.12). Thus, we have

$$\int_0^1 y' \arcsin y' dt = \int_0^1 \bar{k}^2 y^2 dt. \tag{8.2.20}$$

Relation (20) holds on an exact[17] solution of (8.2.12). We shall analyze the first deformation mode only. Thus we assume that

$$y(t) \geq 0; \quad y'(t) \geq 0. \tag{8.2.21}$$

Next we estimate the term $\arcsin y'$. From [12] and [74] we have[18]

$$\arcsin y' \geq y' + \frac{1}{6}(y')^3. \tag{8.2.22}$$

With (8.2.22) equation (8.2.20) becomes

$$\int_0^1 \left[(y')^2 + \frac{1}{6}(y')^4\right] dt \leq \int_0^1 \bar{k}^2 y^2 dt. \tag{8.2.23}$$

We distinguish now the following three cases.

    Case i. Suppose that $\bar{k}^2(t)$ is a decreasing function.

---

[17]Basically we interpret the first variation (8.2.18) as a weak form of the Euler–Lagrangian equations (see [81, p. 33]).

[18]We consider the function $Z(t) = \frac{\arcsin y' - y'}{(y')^3}$. It is easy to see that Z is a decreasing function ($y' \geq 0, y'' \leq 0$) so that $\min_{t \in [0,1]} Z(t) = Z(0) = 1/6$. From this the estimate (8.2.22) follows.

In this case we can use the Tchebychef inequality ($\bar{k}^2(t)$ is a decreasing and y^2 is an increasing function), so that [74]

$$\int_0^1 \left[(y')^2 + \frac{1}{6} (y')^4 \right] dt \leq \left(\int_0^1 \bar{k}^2 dt \right) \left(\int_0^1 y^2 dt \right). \tag{8.2.24}$$

Also, the Cauchy–Schwarz inequality, applied to $(y')^2$, gives

$$\|y'\|^2 = \int_0^1 (y')^2 \, dt \leq \left[\int_0^1 (y')^4 \, dt \right]^{1/2}, \tag{8.2.25}$$

where $\|z\| = \left(\int_0^1 z^2(t) \right)^{1/2}$ denotes the L_2 norm of z. Therefore, (8.2.24) becomes

$$\|y'\|^2 + \frac{1}{6} \|y'\|^4 \leq \|\bar{k}\|^2 \|y\|^2. \tag{8.2.26}$$

Finally, we need the inequality

$$\|y'\| \geq \frac{\pi}{2} \|y\|, \tag{8.2.27}$$

which is (see [74]) a consequence of the boundary conditions (8.2.13). With (8.2.27) the inequality (8.2.26) leads to

$$\|y'\|^2 \leq 6 \left[\left(\frac{2}{\pi} \right)^2 \|\bar{k}\|^2 - 1 \right]. \tag{8.2.28}$$

The boundary condition $y(0) = 0$ implies that $y(t) = \int_0^t y'(\xi) \, d\xi$, so that by applying the Cauchy–Schwarz inequality we have

$$y(t) = \int_0^t y'(\xi) \, d\xi \leq \left(\int_0^t d\xi \right)^{1/2} \left(\int_0^t (y'(\xi))^2 \right)^{1/2} \leq \|y'\|. \tag{8.2.29}$$

From (8.2.28), (8.2.29) we get the following estimate of maximal deflection:

$$f = \sup_{t \in [0,1]} y(t) \leq \left\{ 6 \left[\left(\frac{2}{\pi} \right)^2 \|\bar{k}\|^2 - 1 \right] \right\}^{1/2}. \tag{8.2.30}$$

We can also estimate the maximal value of the angle θ. From (8.2.1) we have

$$\frac{d\theta}{dS} = -\bar{k}^2 y, \tag{8.2.31}$$

so that

$$\theta(t) = \int_t^1 \bar{k}^2 y(\xi) \, d\xi \leq \left(\int_t^1 \bar{k}^2(\xi) \, d\xi \right) \left(\int_t^1 y(\xi) \, d\xi \right) \leq \|\bar{k}\|^2 \|y\|, \tag{8.2.32}$$

where Tchebychef and Cauchy–Schwartz inequalities have been used. From
(8.2.32), (8.2.27), and (8.2.28) we obtain

$$\sup_{t \in [0,1]} \theta(t) = \theta(0) \leq \frac{\|\bar{k}\|^2}{\left(\frac{\pi}{2}\right)} \left\{ 6 \left[\left(\frac{2}{\pi}\right)^2 \|\bar{k}\|^2 - 1 \right] \right\}^{1/2}. \tag{8.2.33}$$

Case ii. Suppose that $\bar{k}^2 = const.$ In this case, instead of (8.2.24), we have

$$\int_0^1 \left[(y')^2 + \frac{1}{6}(y')^4 \right] dt \leq \bar{k}^2 \left(\int_0^1 y^2 dt \right). \tag{8.2.34}$$

Following the same procedure as in Case i we obtain

$$f = \sup_{t \in [0,1]} y(t) \leq \left\{ 6 \left[\left(\frac{2}{\pi}\right)^2 \bar{k}^2 - 1 \right] \right\}^{1/2}. \tag{8.2.35}$$

Since $\left(\frac{2}{\pi}\right)^2 \bar{k}^2 = \frac{FL^2}{EI} \frac{4}{\pi^2} = \frac{F}{F_{cr}}$, with $F_{cr} = \frac{\pi^2 EI}{4L^2}$ being the Euler buckling force
for the column, equation (8.2.35) may be written as

$$f = \sup_{t \in [0,1]} y(t) \leq \left\{ 6 \left[\frac{F}{F_{cr}} - 1 \right] \right\}^{1/2}. \tag{8.2.36}$$

If $\bar{k}^2 = const.$ the estimate (8.2.33) becomes

$$\begin{aligned}
\sup_{t \in [0,1]} y(t) &= \theta(0) \leq \frac{\bar{k}^2}{\left(\frac{2}{\pi}\right)} \left\{ 6 \left[\left(\frac{2}{\pi}\right)^2 \bar{k}^2 - 1 \right] \right\}^{1/2} \\
&= \frac{\pi}{2} \frac{F}{F_{cr}} \left\{ 6 \left[\left(\frac{2}{\pi}\right)^2 \bar{k}^2 - 1 \right] \right\}^{1/2}.
\end{aligned} \tag{8.2.37}$$

Case iii. Suppose that $\bar{k}^2(t)$ is an increasing function. In this case, (8.2.23),
after the use of Cauchy–Schwartz inequality, becomes

$$\int_0^1 \left[(y')^2 + \frac{1}{6}(y')^4 \right] dt \leq \left(\int_0^1 \bar{k}^4 dt \right)^{1/2} \left(\int_0^1 y^4 dt \right)^{1/2}. \tag{8.2.38}$$

To transform (8.2.38) we need an inequality for concave functions. In [12] such
an inequality was derived. In [22] an improvement of this inequality is presented
that we shall use here. Thus, from [22] we have

$$\int_0^1 y^2(t)\, dt \geq \frac{5^{1/2}}{3} \left(\int_0^1 y^4 dt \right)^{1/2}. \tag{8.2.39}$$

or

$$\|y'\|^2 + \frac{1}{6}\|y'\|^4 \leq \|\bar{k}^2\|\frac{3}{5^{1/2}}\|y\|^2 .\tag{8.2.40}$$

Now, (8.2.40), (8.2.27), and (8.2.29) lead to

$$f = \sup_{t \in [0,1]} y(t) \leq \left\{ 6 \left[\frac{3}{5^{1/2}} \left(\frac{2}{\pi}\right)^2 \|\bar{k}^2\| - 1 \right] \right\}^{1/2}\tag{8.2.41}$$

and

$$\sup_{t \in [0,1]} \theta(t) = \theta(0) \leq \frac{\|\bar{k}\|^2}{\left(\frac{\pi}{2}\right)} \left\{ 6 \left[\frac{3}{5^{1/2}} \left(\frac{2}{\pi}\right)^2 \|\bar{k}^2\| - 1 \right] \right\}^{1/2} .\tag{8.2.42}$$

Now we apply the results obtained in this section to the important problem of estimating the critical force for a column with variable cross section. We use the inequality (8.2.30), from which we conclude that for the case when $\bar{k}^2(S)$ is a decreasing function the buckling will take place (i.e., $f \geq 0$) if

$$\left(\frac{2}{\pi}\right)^2 \|\bar{k}\|^2 - 1 \geq 0\tag{8.2.43}$$

or

$$\|\bar{k}\|^2 = \int_0^1 \frac{FL^2}{EI(t)} dt \geq \left(\frac{\pi}{2}\right)^2 .\tag{8.2.44}$$

Suppose that (see [102])

$$I(t) = I_0 e^{nt},\tag{8.2.45}$$

where n is a given constant. By substituting (8.2.45) into (8.2.44) we obtain that for buckling to take place, we must have

$$F \geq \frac{EI_0\pi^2}{4L^2}\left(1 - e^{-n}\right)\frac{1}{n}.\tag{8.2.46}$$

Note that in the limit when $n \to 0$ expression (8.2.46) reduces to the classical Euler buckling force. In [102] several approximate expressions for the critical force are presented that agree with (8.2.46). The advantage of the expression (8.2.46) is that it gives an *upper bound* for the buckling force.

8.3   Rod with Compressible Axis and the Influence of Shear Stresses on the Deformation

In this part we consider the so-called Haringx's model of an elastic rod. The constitutive equation for such a rod takes into account both compressibility of the rod axis and the fact that the cross sections in the deformed state are not normal to the rod axis due to the deformations caused by shear stresses. It could be shown that the relevant equation describing the rod (see [14, p. 129] and [40]) for the case of a rod with constant cross section reads

$$\phi'' + \frac{F}{EI}\sin\phi - \frac{F^2}{EI}\left(\frac{1}{EA} - \frac{k}{GA}\right)\sin\phi\cos\phi = 0, \qquad (8.3.1)$$

where F is the axial force in the rod, ϕ is the angle of rotation of the cross section, EI is the bending rigidity of the rod, EA is the extensional rigidity of the rod, GA is the shear rigidity of the rod, k is the Timoshenko shear correction factor $(\cdot)' = d(\cdot)/dS$, where $S \in [0, L]$ is the arc length of the rod axis in the undeformed state, and L is the length of the rod axis in the undeformed state. For the rod welded at the end $S = L$ and free at the end $S = 0$ (see Figure 8.2.1) the boundary conditions read

$$\phi'(0) = 0, \quad \phi(L) = 0. \qquad (8.3.2)$$

Introducing the dimensionless quantities

$$\lambda = \frac{FL^2}{EI}, \quad \alpha = \frac{F}{EA}, \quad \beta = \frac{Fk}{GA}, \quad t = \frac{S}{L}, \qquad (8.3.3)$$

the system (8.3.2), (8.3.3) becomes

$$\ddot{\phi} + \lambda[1 - \lambda(\alpha - \beta)\cos\phi]\sin\phi = 0, \qquad (8.3.4)$$

subject to

$$\dot{\phi}(0) = 0, \quad \phi(1) = 0, \qquad (8.3.5)$$

where $(\cdot) = d(\cdot)/dt$. Note that for the case of inextensible, unshearable rod $\alpha = \beta = 0$ with constant cross section, that is, $EI = const.$, equation (8.3.5) reduces to (8.2.7). It is easy to see that (8.3.4), (8.3.5) is derivable from the condition $\delta I = 0$, where

$$I = \int_0^1 \left\{\frac{\dot{\phi}^2}{2} - \lambda\left[1 - \left(\cos\phi - \frac{\lambda}{4}(\alpha - \beta)\cos 2\phi\right)\right]\right\} dt. \qquad (8.3.6)$$

Since the Lagrangian does not depend on t we have a first integral of (8.3.4) in the form

$$\frac{\dot{\phi}^2}{2} + \lambda\left[1 - \left(\cos\phi - \frac{\lambda}{4}(\alpha - \beta)\cos 2\phi\right)\right] = const. \qquad (8.3.7)$$

The trivial configuration of the rod is described as

$$\phi_0 = 0. \qquad (8.3.8)$$

According to the energy stability criteria (see [86]), the trivial configuration is stable if the functional I given by (8.3.6) is in weak local minimum[19] at ϕ_0. We shall determine the value of the force, that is, λ, for which $\phi_0 = 0$ is *not* a stable configuration.

First we calculate the second variation of (8.3.6) at $\phi_0 = 0$ as

$$\delta^2 I(\phi_0, \delta\phi) = \int_0^1 \left\{ \left(\delta\dot{\phi} \right)^2 - \lambda \left[1 - \lambda (\alpha - \beta) \right] (\delta\phi)^2 \right\} dt. \qquad (8.3.9)$$

The configuration $\phi_0 = 0$ is *stable* if $I(\phi_0)$ is in minimum. To examine the sign of the second variation (8.3.9) we proceed as follows. Note that (8.3.5) implies $\delta\dot{\phi}(0) = \delta\phi(1) = 0$. This enables us to get the following estimate ([74] or (8.2.27)):

$$\int_0^1 \left(\delta\dot{\phi} \right)^2 dt \geq \frac{\pi^2}{4} \int_0^1 (\delta\phi)^2 dt. \qquad (8.3.10)$$

By using (8.3.10) in (8.3.9) we obtain

$$\delta^2 I(\phi_0, \delta\phi) \geq \int_0^1 \left\{ \frac{\pi^2}{4} - \lambda \left[1 - \lambda (\alpha - \beta) \right] \right\} (\delta\phi)^2 dt. \qquad (8.3.11)$$

From (8.3.11) we conclude that the trivial configuration is stable, that is, (8.3.6) has a local minimum at $\phi_0 = 0$, if

$$\frac{\pi^2}{4} - \lambda \left[1 - \lambda (\alpha - \beta) \right] > 0. \qquad (8.3.12)$$

According to the stability definition, the condition $\delta^2 I(\phi_0, \delta\phi) = 0$ determines the stability boundary [86]. Thus, we conclude from (8.3.12) that the critical value of the dimensionless force $\lambda = \lambda_{cr}$ is determined from

$$\frac{\pi^2}{4} - \lambda_{cr} \left[1 - \lambda_{cr} (\alpha - \beta) \right] = 0. \qquad (8.3.13)$$

The result (8.3.13) is in agreement with the stability boundary obtained by the Euler method [14]. We note that the sign of $\delta^2 I$ given by (8.3.9) may be analyzed by the method used in [62].

8.4 Rotating Rod

8.4.1 Bernoulli–Euler Theory

Consider an elastic rod BC of length L, fixed at end B and free at the other end. Suppose that the rod has circular cross section, that its axis is straight, and that

---

[19]The functional I has a weak local minimum at ϕ_0 if the condition $\delta^2 I(\phi_0, \delta\phi) \geq c \|\delta\phi\|_1$ is satisfied for $c > 0$, and for $\delta\phi$ belonging to a small (in norm $\|\cdot\|_1$) neighborhood of ϕ_0, where $\|\delta\phi\|_1 = \left[\sup_{t \in [0,1]} |\delta\phi(t)| + \sup_{t \in [0,1]} \left| \delta\dot{\phi}(t) \right| \right]^{1/2}$ (see section 5.2).

it rotates with the constant angular velocity ω about its axis. Let $x-B-y$ be the rectangular Cartesian coordinate system with the axis x oriented along the rod axis in the undeformed state. Let Π be a plane defined by the system $x-B-y$ that rotates with the angular velocity ω about x axis. At a certain velocity the rod loses stability so that it could be bent under the action of centrifugal forces. If the rod is bent it will assume a relative (with respect to the rotating plane Π) equilibrium configuration (see Figure 8.4.1). Note, however, that during the *motion* between two relative equilibrium configurations (one corresponding to the initial state in which the rod axis is straight and one in which the rod axis is bent) the axis of the rod is, in general, *not* a plane curve. The problem of determining the critical rotation speed and the postcritical behavior of the rod described has been the subject of many investigations. For a review of generalized rod theories used in stability analysis of rotating rods, see [5] and [14].

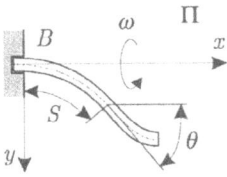

Figure 8.4.1

Suppose that the angular velocity with which the rod rotates, $\omega = \omega_0 = const.$, and length of the rod L are given. Let S be the arc length of the rod axis, so that $S \in [0, L]$. The equilibrium, geometrical, and constitutive equations for the rotating rod are

$$
\begin{aligned}
H' &= 0, \quad V' = -\rho_0 \omega^2 \bar{y}, \\
M' &= -V \cos\theta + H \sin\theta, \\
\bar{x}' &= \cos\theta, \quad \bar{y}' = \sin\theta, \quad \theta' = \frac{M}{EI},
\end{aligned} \tag{8.4.1}
$$

where H and V are components of the contact force in an arbitrary cross section, M is the bending moment, θ is the angle between the axis of rotation and tangent to the rod axis, and $\bar{x}$ and $\bar{y}$ are coordinates of an arbitrary point with respect to the rotating Cartesian frame $x-B-y$. Also in (8.4.1) we use ρ_0 to denote the line density of the rod (mass per unit length of the rod axis), E is the modulus of elasticity, I is the second moment of the cross-sectional area of the rod, and $(\cdot)' = d(\cdot)/dS$. Note that $\rho_0 = \rho A$, where ρ is the density of the rod (mass per unit volume). We assume that the cross section of the rod is circular, so that

$$
I = \alpha A^2, \tag{8.4.2}
$$

where A is the cross-sectional area and $\alpha = (1/4\pi)$. With this notation, the volume of the rod is

$$W = \int_0^L A(S)dS. \tag{8.4.3}$$

To the system (8.4.1) we adjoin the following boundary conditions corresponding to the rod shown in Figure 8.4.1:

$$H(L) = 0, \quad V(L) = 0, \quad \bar{x}(0) = 0, \quad \bar{y}(0) = 0, \quad M(L) = 0. \tag{8.4.4}$$

Suppose first that the rod cross section is constant, $I = const$. By using the dimensionless variables and parameters

$$t = \frac{S}{L}, \quad y = \frac{\bar{y}}{L}, \quad m = \frac{ML}{EI},$$

$$v = \frac{VL^2}{EI}, \quad x = \frac{\bar{x}}{L}, \quad \lambda^2 = \frac{\rho\omega^2 L^4}{\alpha EA} \tag{8.4.5}$$

and a new dependent variable

$$u = -\frac{v}{\lambda}, \tag{8.4.6}$$

we obtain from (8.4.1)

$$\dot{u} = \lambda y, \quad \dot{m} = \lambda u \cos\theta, \quad \dot{y} = \sin\theta, \quad \dot{x} = \cos\theta, \quad \dot{\theta} = m, \tag{8.4.7}$$

where $(\cdot) = \frac{d}{dt}(\cdot)$. The boundary conditions corresponding to the rod shown in Figure 8.4.1 are

$$u(1) = 0, \quad m(1) = 0, \quad y(0) = 0, \quad x(0) = 0, \quad \theta(0) = 0. \tag{8.4.8}$$

The system (8.4.7), (8.4.8) possesses a trivial solution in which the axis of the rod remains straight for any value of the dimensionless rotation speed λ. This solution is

$$u_0 = 0, \quad m_0 = 0, \quad y_0 = 0, \quad \theta_0 = 0. \tag{8.4.9}$$

The system (8.4.7) can be transformed to a system of two second-order equations by differentiating $(8.4.7)_1$ and $(8.4.7)_5$, so that

$$\ddot{u} = \lambda\sin\theta, \quad \ddot{\theta} = \lambda u \cos\theta, \tag{8.4.10}$$

with the boundary conditions

$$\dot{u}(0) = 0, \quad u(1) = 0, \quad \theta(0) = 0, \quad \dot{\theta}(1) = 0. \tag{8.4.11}$$

It is easy to see that (8.4.10) are the Euler–Lagrangian equations of the functional

$$I_0 = \int_0^1 \left[\frac{\dot{u}^2}{2} + \frac{\dot{\theta}^2}{2} + \lambda u \sin\theta\right] dt. \tag{8.4.12}$$

In [28] an analysis is presented which shows that (8.4.12) attains a local minimum on the solution of (8.4.10), (8.4.11). We note that the Lagrangian in (8.4.12) does not contain time explicitly, so that a Jacobi-type first integral exists:

$$\frac{\dot{u}^2}{2} + \frac{\dot{\theta}^2}{2} - \lambda u \sin\theta = const. \tag{8.4.13}$$

Consider the second variation of I_0 at the trivial solution $w_0 = (u_0, \theta_0) = (0,0)$ and for $\overline{\lambda} = \lambda + \Delta\lambda$, with $0 < \Delta\lambda \ll 1$. From (8.4.12) we obtain

$$\delta^2 I_0(u_0, \theta_0, \lambda + \Delta\lambda, \delta u, \delta\theta) = \int_0^1 \left[(\delta\dot{u})^2 + \left(\delta\dot{\theta}\right)^2 + 2(\lambda + \Delta\lambda)\,(\delta u)\,(\delta\theta) \right] dt. \tag{8.4.14}$$

We shall calculate $\delta^2 I_0$ for specially chosen $(\delta u, \delta\theta)$. Namely, consider a linearization of the system (8.4.10):

$$\ddot{u} = \lambda\theta, \quad \ddot{\theta} = \lambda u, \tag{8.4.15}$$

subject to (8.4.11). Eigenvalues of (8.4.15), (8.4.11) are solutions of the following equations:

$$1 + \cos\lambda^{1/2}\cosh\lambda^{1/2} = 0. \tag{8.4.16}$$

The smallest solution to (8.4.16) is $\lambda_{\min} = 3.516$. It is easy to see that all solutions of (8.4.16) are positive and that there are countably many of them. Let $\lambda_1 = \lambda_{\min}$ be the smallest solution of (8.4.16) and u_1, θ_1 corresponding eigenfunctions, that is,

$$\ddot{u}_1 = \lambda_1\theta_1, \quad \ddot{\theta}_1 = \lambda_1 u_1. \tag{8.4.17}$$

Note that by multiplying $(8.4.17)_1$ by $\dot{u}_1$ and $(8.4.17)_2$ by $\dot{\theta}_1$, integrating and using boundary conditions, we obtain

$$\int_0^1 \{\dot{u}_1^2 + \dot{\theta}_1^2 + \lambda_1 2 u_1\theta_1\}dt = 0. \tag{8.4.18}$$

Taking $\delta u = \varepsilon u_1, \delta\theta = \varepsilon\theta_1$, with $\varepsilon \ll 1$ and $\lambda = \lambda_1$ in (8.4.14), we obtain

$$\delta^2 I_0(u_0, \theta_0, \lambda + \Delta\lambda, \delta u, \delta\theta)$$
$$= \varepsilon^2 \int_0^1 \left[(\dot{u}_1)^2 + \left(\dot{\theta}_1\right)^2 + 2(\lambda_1 + \Delta\lambda)u_1\theta_1 \right] dt$$
$$= 2\varepsilon^2\Delta\lambda \int_0^1 u_1\theta_1 dt. \tag{8.4.19}$$

From (8.4.15) it is easy to see that $u_1(t)$ and $\theta_1(t)$ are of different sign:

$$u_1(t) \leq 0, \quad \theta_1(t) \geq 0, \tag{8.4.20}$$

so that[20] $\int_0^1 u_1 \theta_1 dt < 0$. Thus, we conclude that the functional (8.4.12) is *not* in a minimum at $(u_0 = 0, \theta_0 = 0)$ if $\lambda > \lambda_1$, where λ_1 is the smallest solution of (8.4.16). This implies that the configuration $(u_0 = 0, \theta_0 = 0)$ is not stable if $\lambda > \lambda_1$.

We show now how the functional (8.4.12) could be used to obtain an approximate solution to (8.4.10), (8.4.11) via the Ritz method. Namely, we assume an approximate solution to (8.4.10), (8.4.11) in the form

$$\Theta = At(2-t), \quad U = B(1-t^2), \tag{8.4.21}$$

where A and B are constants to be determined. The functions Θ and U satisfy the boundary conditions (8.4.11). Let us choose $\lambda^2 = 14$. By substituting (8.4.21) into (8.4.12) we obtain

$$I_0(A, B) = \int_0^1 \left\{ 2B^2 t^2 + \frac{1}{2}[A(2-t) - At]^2 \right.$$
$$\left. + B(14)^{1/2}(1-t^2)\sin[At(2-t)]\,dt \right\}. \tag{8.4.22}$$

Minimization of (8.4.22) with respect to A and B, that is, $\partial I_0(A, B)/\partial A = 0$, $\partial I_0(A, B)/\partial B = 0$, leads to

$$A = 0.4824, \quad B = 0.5. \tag{8.4.23}$$

The simple approximate solution (8.4.23) agrees well with the numerical solution of the nonlinear boundary value problem (8.4.10), (8.4.11) presented, as a special case, in [13].

As a generalization of the problem (8.4.10), (8.4.11), consider the rod with extensible axis. In this case, instead of (8.4.1) we have (see [13])

$$H' = 0, \quad V' = -\rho_0 \omega^2 \bar{y},$$
$$M' = -V(1 + \beta V \sin \theta)\cos \theta + H(1 + \beta V \sin \theta)\sin \theta,$$
$$\bar{x}' = \cos \theta, \quad \bar{y}' = (1 + \beta V \sin \theta)\sin \theta, \quad \theta' = \frac{M}{EI}, \tag{8.4.24}$$

where $\beta = EI/(EAL^2)$, with EA being the extensional rigidity of the rod. Proceeding as in the previous case we obtain

$$\ddot{u} = \lambda[\sin \theta - \lambda u \beta \sin^2 \theta], \quad \ddot{\theta} = \lambda u[1 - \lambda u \beta \sin \theta]\cos \theta, \tag{8.4.25}$$

and

$$I_1 = \int_0^1 \left\{ \frac{\dot{u}^2}{2} + \frac{\dot{\theta}^2}{2} + \lambda\left[u \sin \theta - \frac{\lambda u^2}{2}\beta \sin^2 \theta \right] \right\} dt. \tag{8.4.26}$$

It can be easily verified that the conditions for $\delta^2 I_1 > 0$ are the same as those formulated for $\delta^2 I_0 > 0$. Thus, *the extensibility of the rod axis does not influence the critical value of* λ.

---

[20]That $\int_0^1 u_1 \theta_1 dt < 0$ could be seen from (8.4.18) directly.

8.4.2 Rotating Rod with Shear and Compressibility: A Director Theory

Now consider again an elastic rod, as shown in Figure 8.4.1, whose axis in the undeformed state is a nonintersecting curve C lying in a plane Π. We assume that the points on the curve C, in the coordinate system $x - B - y$, could be represented as

$$X = X(S), \qquad Y = Y(S), \tag{8.4.27}$$

where S is the arc length of C and $X(S), Y(S)$ are smooth functions. We further assume that $S \in [0, L]$, where L is the length of the rod in the undeformed state. Since C is nonintersecting it has a well-defined tangent at each point. Let $\mathbf{e}_1$ and $\mathbf{e}_2$ be the unit vectors along the x and y axes, respectively. Then, if $\mathbf{T}(S)$ is the unit tangent vector at the point (X, Y), the relations

$$\cos \Theta(S) = \mathbf{T} \cdot \mathbf{e}_1, \qquad \sin \Theta(S) = \mathbf{T} \cdot \mathbf{e}_2 \tag{8.4.28}$$

determine a unique $\Theta(S)$. We assume that the rod is fixed in such a way that $\Theta(0) = 0$. With $\Theta(S)$ so determined, the curvature of C is $K_0 = d\Theta/dS$. The natural configuration of the rod is defined by two vector functions $\mathbf{R}(S) = X\mathbf{e}_1 + Y\mathbf{e}_2$ and $\mathbf{B}(S)$ characterizing the points on the rod axis and orientation of the cross section, respectively. $\mathbf{B}(S)$ is assumed to be of unit length and lying in Π. The natural configuration is stress free, so that the vector $\mathbf{B}(S)$ (also called the director) is orthogonal to $\mathbf{T}(S)$. In this case

$$\mathbf{B}(S) = -\sin\Theta(S)\mathbf{e}_1 + \cos\Theta(S)\mathbf{e}_2. \tag{8.4.29}$$

The equilibrium and geometrical equations in the system $x - B - y$ are

$$\begin{aligned}
H^{'} &= 0, \quad V^{'} = -\rho\omega^2 y, \\
M^{'} &= V(1+\varepsilon)\cos\vartheta - H(1+\varepsilon)\sin\vartheta, \\
\bar{x}^{'} &= (1+\varepsilon)\cos\vartheta, \quad \bar{y}^{'} = (1+\varepsilon)\sin\vartheta.
\end{aligned} \tag{8.4.30}$$

where we used quantities already introduced, and ε is the axial strain; that is, if $\mathbf{r} = \bar{x}\mathbf{e}_1 + \bar{y}\mathbf{e}_2$ is the position vector of an arbitrary point on the rod axis in the deformed state, then $1 + \varepsilon = |d\mathbf{r}/dS|$. To (8.4.30) we adjoin the boundary conditions (8.4.4). The deformed configuration is specified when two vector functions, that is, the position vector of an arbitrary point on the rod axis $\mathbf{r}(S)$ and the director $\mathbf{b}(S)$ (a unit vector orthogonal to the cross section in the deformed state), are specified. Let φ be the rotation angle of the cross section. Then, $\mathbf{b}(S)$ becomes

$$\mathbf{b}(S) = -\sin(\Theta + \varphi)\mathbf{e}_1 + \cos(\Theta + \varphi)\mathbf{e}_2. \tag{8.4.31}$$

Let $\mathbf{e}_3 = \mathbf{e}_1 \times \mathbf{e}_2$. We define $\mathbf{a} = \mathbf{b} \times \mathbf{e}_3$, so that

$$\mathbf{a}(S) = \cos(\Theta + \varphi)\mathbf{e}_1 + \sin(\Theta + \varphi)\mathbf{e}_2. \tag{8.4.32}$$

The *strains* are taken to be the derivative of the rotation angle φ' and two functions $\xi(S)$ and $\eta(S)$ defined by

$$\mathbf{t} = \mathbf{r}' = [1 + \xi(S)]\mathbf{a} + \eta(S)\mathbf{b}. \tag{8.4.33}$$

Introducing the shear angle γ as the angle between $\mathbf{a}(S)$ and $\mathbf{t}$, we obtain the following relation:

$$\gamma = (\vartheta - \Theta) - \varphi. \tag{8.4.34}$$

With (8.4.34) the strains $\xi(S)$ and $\eta(S)$ become [67]

$$\xi = (1 + \varepsilon) \cos\gamma, \qquad \eta = (1 + \varepsilon) \sin\gamma. \tag{8.4.35}$$

Let the internal (also called contact) force $\mathbf{F}$ and the resultant couple $\mathbf{M}$ in an arbitrary cross section of the rod be represented as $\mathbf{F} = Q\mathbf{b} + N\mathbf{a}$ and $\mathbf{M} = M\mathbf{e}_3$. Then we assume that (see [67])

$$N = EA[(1 + \varepsilon)\cos\gamma - 1], \quad Q = GA(1 + \varepsilon)\sin\gamma, \quad M = -EI\varphi', \tag{8.4.36}$$

where EA, GA, and EI are extensional, shear, and bending rigidities, respectively. They are all taken to be positive constants. From the definitions of Q and N, we have

$$Q = -H\sin(\Theta + \varphi) + V\cos(\Theta + \varphi), \quad N = H\cos(\Theta + \varphi) + V\sin(\Theta + \varphi). \tag{8.4.37}$$

Thus H and V could be expressed in terms of ξ and η. In our case (8.4.37) simplifies, since from (8.4.30)$_1$, (8.4.4)$_1$ we get $H(S) = 0$. Then, introducing a new variable $\psi = \Theta + \varphi$ in (8.4.30), we obtain

$$V'' = -\rho\omega^2 \left[\sin\psi + V \left(\frac{\sin^2\psi}{EA} + \frac{\cos^2\psi}{GA} \right) \right],$$

$$\psi'' = -\frac{V}{EI} \left[1 + V \left(\frac{1}{EA} - \frac{1}{GA} \right) \sin\psi \right] \cos\psi + K_0'. \tag{8.4.38}$$

Finally, we introduce the nondimensional quantities (8.4.5) and additionally

$$\beta = \frac{EI}{EAL^2}, \quad \mu = \frac{EI}{GAL^2}, \quad f_1 = \frac{qL^4}{EI}, \quad f_2 = K_0'L^2. \tag{8.4.39}$$

Also, let $u = -W/\lambda$ be the new dependent variable, so that the system (8.4.38) transforms to [15]

$$\ddot{u} = \lambda[\sin\psi - \lambda u(\beta \sin^2\psi + \mu \cos^2\psi)] - \frac{f_1}{\lambda},$$

$$\ddot{\psi} = \lambda u[1 - \lambda u(\beta - \mu)\sin\psi]\cos\psi + f_2. \tag{8.4.40}$$

The boundary conditions corresponding to (8.4.40) are

$$\dot{u}(0) = 0, \quad u(1) = 0, \quad \psi(0) = 0, \quad \dot{\psi}(1) = 0, \qquad (8.4.41)$$

where $(\cdot) = \frac{d}{dt}(\cdot)$. In writing $(8.4.41)_3$ we assumed that the rod is "welded" at $t = 0$. We call the rod described by (8.4.40), (8.4.41) the *imperfect rod* and we call the initially straight rod that is not loaded by external load, that is, $f_1 = 0$, the *perfect rod*. In the case of the perfect rod system, (8.4.40) becomes

$$\ddot{u} = \lambda[\sin\psi - \lambda u(\beta\sin^2\psi + \mu\cos^2\psi)],$$
$$\ddot{\psi} = \lambda u[1 - \lambda u(\beta - \mu)\sin\psi]\cos\psi. \qquad (8.4.42)$$

The functions $u_0 = \psi_0 = 0$ are solutions to (8.4.42), (8.4.41) for all values of λ. We call this solution the *trivial* solution for the perfect rod.

Consider now the functionals I_1 and I_2 defined by

$$I_1 = \int_0^1 \left\{ \frac{\dot{u}^2}{2} + \frac{\dot{\psi}^2}{2} + \lambda\left[u\sin\psi - \frac{\lambda u^2}{2}(\beta\sin^2\psi + \mu\cos^2\psi)\right] - \frac{f_1 u}{\lambda} + f_2\psi \right\} dt \tag{8.4.43}$$

and

$$I_2 = \int_0^1 \left\{ \frac{\dot{u}^2}{2} + \frac{\dot{\psi}^2}{2} + \lambda\left[u\sin\psi - \frac{\lambda u^2}{2}(\beta\sin^2\psi + \mu\cos^2\psi)\right] \right\} dt. \tag{8.4.44}$$

The functionals I_1 and I_2 represent the total potential energy of outer forces (loads) and inner forces for rotating rod and perfect rotating rod, respectively.

Note that from (8.4.44) we conclude that (8.4.42) possesses the following first integral of Jacobi type (see section 1.4):

$$K = \frac{\dot{u}^2}{2} + \frac{\dot{\psi}^2}{2} - \lambda\left[u\sin\psi - \frac{\lambda u^2}{2}(\beta\sin^2\psi + \mu\cos^2\psi)\right] = const. \tag{8.4.45}$$

From the boundary conditions it follows that $K = \dot{u}^2(1)/2$. Using this value and (see definition of u given after (8.4.39) and $(8.4.30)_2$)

$$u = -\lambda\int_t^1 y(p)dp, \tag{8.4.46}$$

we get

$$\frac{\dot{u}^2}{2} + \frac{\dot{\psi}^2}{2} - \lambda\left[u\sin\psi - \frac{\lambda u^2}{2}(\beta\sin^2\psi + \mu\cos^2\psi)\right] = \frac{\lambda^2}{2}y^2(1). \tag{8.4.47}$$

For the perfect rod without the influence of shear and extensibility ($\delta = \mu = 0$), the system (8.4.40) reduces to

$$\ddot{u} = \lambda \sin \psi, \quad \ddot{\psi} = \lambda u \cos \psi, \tag{8.4.48}$$

a problem that was treated earlier in this chapter (see (8.4.10)). In this case the first integral (8.4.47) becomes

$$\frac{\dot{u}^2}{2} + \frac{\dot{\psi}^2}{2} - \lambda u \sin \psi = \frac{\lambda^2}{2} y^2(1). \tag{8.4.49}$$

We treat the problem of determining the stability boundary for the perfect rod described by (8.4.42), (8.4.41) by the use of I_2 given by (8.4.44). The second variation of I_2 calculated at $u_0 = 0, \psi_0 = 0, \lambda + \Delta\lambda$ reads

$$\delta^2 I_2(u_0, \psi_0, \lambda + \Delta\lambda, \delta u, \delta \psi)$$
$$= \int_0^1 \left\{ (\delta \dot{u})^2 + \left(\delta \dot{\psi} \right)^2 + 2(\lambda + \Delta\lambda) \right.$$
$$\left. \times [\delta u \delta \psi - \mu(\lambda + \Delta\lambda)(\delta u)^2] \right\} dt. \tag{8.4.50}$$

Again, we consider the linearized problem (8.4.42), that is,

$$\ddot{u} = \lambda(\psi - \lambda\mu u), \qquad \ddot{\psi} = \lambda u, \tag{8.4.51}$$

subject to (8.4.41). Eigenvalues of (8.4.51), (8.4.41) are solutions of the following equation:

$$\lambda^2 - \frac{\lambda^3 \mu}{2} \sin \left[\frac{\lambda^2 \mu + \sqrt{\lambda^4 \mu^2 + 4\lambda^2}}{2} \right]^{1/2}$$
$$\times \sinh \left[\frac{-\lambda^2 \mu + \sqrt{\lambda^4 \mu^2 + 4\lambda^2}}{2} \right]^{1/2}$$
$$+ (\lambda^2 + \frac{\lambda^4 \mu}{2}) \cos \left[\frac{\lambda^2 \mu + \sqrt{\lambda^4 \mu^2 + 4\lambda^2}}{2} \right]^{1/2}$$
$$\times \cosh \left[\frac{-\lambda^2 \mu + \sqrt{\lambda^4 \mu^2 + 4\lambda^2}}{2} \right]^{1/2} = 0. \tag{8.4.52}$$

Equation (8.4.52) has a countable number of positive solutions. To see this, note that (8.4.52) could be written as $F_1(\lambda) = F_2(\lambda)$, where

$$F_1(\lambda) = \lambda^2,$$

$$F_2(\lambda) = \frac{\lambda^3 \delta}{2} \sin \left[\frac{\lambda^2 \delta + \sqrt{\lambda^4 \delta + 4\lambda^2}}{2} \right]^{1/2} \sinh \left[\frac{-\lambda^2 \delta + \sqrt{\lambda^4 \delta + 4\lambda^2}}{2} \right]^{1/2}$$

$$-(\lambda^2 + \frac{\lambda^4 \delta}{2}) \cos \left[\frac{\lambda^2 \delta + \sqrt{\lambda^4 \delta + 4\lambda^2}}{2} \right]^{1/2} \cosh \left[\frac{-\lambda^2 \delta + \sqrt{\lambda^4 \delta + 4\lambda^2}}{2} \right]^{1/2}.$$

$$(8.4.53)$$

Since $F_1(\lambda)$ is increasing and $F_2(\lambda)$ is oscillating with increasing amplitude, that is, larger than the value of $F_1(\lambda)$, it follows that (8.4.52) has a countable increasing sequence of eigenvalues. The eigenfunctions u_n and ψ_n corresponding to λ_n are

$$u_n = C_n \left\{ \cos s_{1n} t + \frac{s_{1n}^3 - \lambda_n^2 \delta}{s_{2n}^3 + \lambda_n^2 \delta} \cosh s_{2n} t + \left[\frac{s_{2n}}{s_{1n} \sinh s_{2n} - s_{2n} \sin s_{1n}} \right. \right.$$

$$\times \frac{(s_{2n}^3 + \lambda_n^2 \delta) \cos s_{1n} + (s_{1n}^3 - \lambda_n^2 \delta) \cosh s_2}{s_{2n}^3 + \lambda_n^2 \delta} \right]$$

$$\times \left. \left. [\sin s_{1n} t - \frac{s_{1n}}{s_{2n}} \sinh s_{2n} t] \right\} \right.,$$

$$\psi_n = \frac{\ddot{u}_n}{\lambda_n} + \lambda_n \delta u_n. \qquad (8.4.54)$$

In (8.4.54) we used the following notation:

$$s_{1n} = \left[\frac{\lambda_n^2 \delta + \sqrt{\lambda_n^4 \delta + 4\lambda_n^2}}{2} \right]^{1/2}, \qquad s_{2n} = \left[\frac{-\lambda_n^2 \delta + \sqrt{\lambda_n^4 \delta + 4\lambda_n^2}}{2} \right]^{1/2}.$$

$$(8.4.55)$$

Let $\lambda = \lambda_1$, where λ_1 is the smallest solution of (8.4.52). Note that (8.4.51) leads to

$$\int_0^1 \{\dot{u}_1^2 + \dot{\psi}_1^2 + \lambda_1 [2u_1 \psi_1 - \mu \lambda_1^2 u_1^2]\} dt = 0. \qquad (8.4.56)$$

Taking $\delta u = \varepsilon u_1, \delta \psi = \varepsilon \psi_1$ with $\varepsilon \ll 1$, in (8.4.50) we obtain

$$\delta^2 I_2(u_0, \psi_0, \lambda = \lambda_1 + \Delta\lambda, \delta u, \delta\psi)$$

$$= \varepsilon^2 \Delta\lambda \int_0^1 \left\{ [2u_1 \psi_1 - 2\mu \lambda_1 u_1^2 - \mu \lambda_1 u_1^2 \right.$$

$$\left. -\mu (\lambda_1 + \Delta\lambda)^2 u_1^2] \right\} dt. \qquad (8.4.57)$$

Since in the first mode, that is, for $\lambda = \lambda_1$, we have $u_1(t) \leq 0, \psi_1(t) \geq 0$, we conclude that $\delta^2 I_2 < 0$ if $\Delta\lambda > 0$. Therefore, when $\lambda > \lambda_1$, where λ_1 is the smallest root of (8.4.53) the functional (8.4.44) is not in a minimum at $u_0 = \psi_0 = 0$, and we have instability.

8.5   Rod Loaded by a Force and a Torque

The problem of determining the stability boundary of a twisted and axially compressed rod is indeed an old one. In this section we shall formulate a variational principle for such a rod and derive an equation that will be used for optimization of the rod shape.

Suppose that the rod is loaded by a compressive force of intensity P and a torsional couple of intensity M_t, as shown in Figure 8.5.1. In our analysis we shall follow [19].

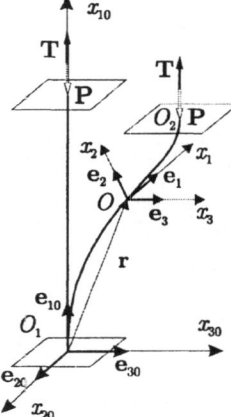

Figure 8.5.1

The compressive force $\mathbf{P} = -P\mathbf{e}_{10}$ is of constant intensity $P = const.$ and oriented along the x_{10} axis of a fixed rectangular Cartesian coordinate system $x_{10}, x_{20},$ and x_{30} with unit vectors $\mathbf{e}_{10}, \mathbf{e}_{20},$ and $\mathbf{e}_{30}$, respectively. The couple (torque) is given as $\mathbf{T} = M_t\mathbf{e}_{10}$ with $M_t = const.$ The end O_1 of the rod is fixed to an unmovable rigid plate, laying in the $x_{20} - x_{30}$ coordinate plane, so that the cross section of the rod that is in contact with the rigid plate does not have any rotation (welded end). At the end O_2, the rod is welded to a movable rigid plate that can move freely but must remain parallel to the coordinate plane $x_{20} - x_{30}$. Let S be the arc length of the rod axis in the undeformed state, so that $S \in [0, L]$, where L is the length of the rod. We specify the configuration of the rod by one vector function $\mathbf{r}(S)$, specifying the position of a point on the rod axis and by orientation of the Cartesian coordinate system with axes x_1, x_2, x_3 oriented along the normal to the cross section (tangent to the rod axis) and along the principal directions of the rod cross section at an arbitrary point O of the rod axis, respectively. Thus we have

$$\mathbf{r}(S) = x_{10}\mathbf{e}_{10} + x_{20}\mathbf{e}_{20} + x_{30}\mathbf{e}_{30}. \tag{8.5.1}$$

We denote by $\mathbf{e}_1, \mathbf{e}_2, \mathbf{e}_3$ the unit vectors along the x_1, x_2, x_3, respectively. The orientation of the system $\mathbf{e}_1, \mathbf{e}_2, \mathbf{e}_3$ with respect to the unit vectors parallel with

$\mathbf{e}_{10}, \mathbf{e}_{20}, \mathbf{e}_{30}$ and passing through point O is given by three Euler-type angles. We take ship angles (see [68]) that bring $\mathbf{e}_{10}, \mathbf{e}_{20}, \mathbf{e}_{30}$ to $\mathbf{e}_1, \mathbf{e}_2, \mathbf{e}_3$ by the sequence of the following three rotations. The first is rotation of the amount θ_1 about the $\bar{x}_{10}$ axis. The next rotation is about the ξ axis for an amount θ_3 (see Figure 8.5.2).

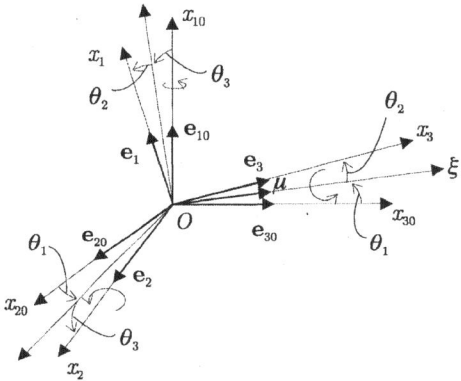

Figure 8.5.2

The last rotation is of the amount θ_2 about the axis $\bar{x}_2$. All rotations are performed counterclockwise. The vector $\boldsymbol{\omega}$ ("the angular velocity vector") is defined as

$$\boldsymbol{\omega} = \theta_1' \mathbf{e}_{10} + \theta_3' \boldsymbol{\mu} + \theta_2' \mathbf{e}_2, \tag{8.5.2}$$

where, as usual, $(\cdot)' = d/dS\,(\cdot)$ and $\boldsymbol{\mu}$ is the unit vector along the ξ axis. From (8.5.2) we obtain the components of $\boldsymbol{\omega}$ in the local coordinate system $\mathbf{e}_1, \mathbf{e}_2, \mathbf{e}_3$. Thus, $\boldsymbol{\omega} = \omega_1 \mathbf{e}_1 + \omega_2 \mathbf{e}_2 + \omega_3 \mathbf{e}_3$ with

$$\begin{aligned}
\omega_1 &= \theta_1' \cos\theta_2 \cos\theta_3 - \theta_3' \sin\theta_2, \\
\omega_2 &= \theta_2' - \theta_1' \sin\theta_3, \\
\omega_3 &= \theta_1' \cos\theta_3 \sin\theta_2 + \theta_3' \cos\theta_2.
\end{aligned} \tag{8.5.3}$$

Note that

$$\mathbf{r}' = \mathbf{e}_1. \tag{8.5.4}$$

The three quantities $(\omega_1, \omega_2, \omega_3)$ are *strains* in the classical Bernoulli–Euler rod theory that we use (see [5]). The equilibrium equations for the rod can be written as

$$\mathbf{F}' = 0, \quad \mathbf{M}' = -\mathbf{r}' \times \mathbf{F}, \tag{8.5.5}$$

where $\mathbf{F} = F_1\mathbf{e}_1 + F_2\mathbf{e}_2 + F_3\mathbf{e}_3$ is the contact force, $\mathbf{M} = M_1\mathbf{e}_1 + M_2\mathbf{e}_2 + M_3\mathbf{e}_3$ is the contact couple, and we assume that there are no distributed forces. The constitutive equation for the contact couple is taken in the form

$$\mathbf{M} = A_{11}\omega_1\mathbf{e}_1 + A_{22}\omega_2\mathbf{e}_2 + A_{33}\omega_3\mathbf{e}_3, \tag{8.5.6}$$

where A_{11}, A_{22}, A_{33} are constants. A_{11} is called torsional, while A_{22} and A_{33} are bending rigidities. In standard engineering notation we have $A_{11} = GI_0, A_{11} = EI_{22}, A_{33} = EI_{33}$, where G is the shear modulus, E is the elasticity modulus, I_0 the polar moment of inertia of the cross section, and I_{22} and I_{33} are axial moments of inertia of the cross section for the axes x_2 and x_3, respectively. By using (8.5.6) in (8.5.3), (8.5.5) and by solving (8.5.3) for $\theta_1', \theta_2', \theta_3'$ we obtain

$$F_1' + F_3\frac{M_2}{A_{22}} - F_2\frac{M_3}{A_{33}} = 0,$$

$$F_2' + F_1\frac{M_3}{A_{33}} - F_3\frac{M_1}{A_{11}} = 0,$$

$$F_3' + F_2\frac{M_1}{A_{11}} - F_1\frac{M_2}{A_{22}} = 0,$$

$$M_1' + \left(\frac{1}{A_{22}} - \frac{1}{A_{33}}\right) M_2 M_3 = 0,$$

$$M_2' + \left(\frac{1}{A_{33}} - \frac{1}{A_{11}}\right) M_1 M_3 = F_3,$$

$$M_3' + \left(\frac{1}{A_{11}} - \frac{1}{A_{22}}\right) M_1 M_2 = -F_2,$$

$$\theta_1' = \frac{c_2}{c_3}\frac{M_1}{A_{11}} + \frac{s_2}{c_3}\frac{M_3}{A_{33}},$$

$$\theta_2' = \frac{c_2 s_3}{c_3}\frac{M_1}{A_{11}} + \frac{M_2}{A_{22}} + \frac{s_2 s_3}{c_3}\frac{M_3}{A_{33}},$$

$$\theta_3' = -s_2\frac{M_1}{A_{11}} + c_2\frac{M_3}{A_{33}}, \quad x_{10}' = c_2 c_3,$$

$$x_{20}' = c_1 c_2 s_3 + s_1 s_2, \quad x_{30}' = c_2 s_1 s_3 - c_1 s_2, \tag{8.5.7}$$

where $c_1 = \cos\theta_1, s_1 = \sin\theta_1, ..., s_3 = \sin\theta_3$, and the last three equations follow from (8.5.4). Note also that in writing (8.5.7) we assumed that $c_3 \neq 0$. The boundary conditions corresponding to the rod shown in Figure 8.5.1 are

$$\begin{array}{lll}
F_1(L) = -P, & F_2(L) = 0, & F_3(L) = 0, \\
M_1(L) = M_t, & \theta_2(L) = 0, & \theta_3(L) = 0, \\
\theta_1(0) = 0, & \theta_2(0) = 0, & \theta_3(0) = 0, \\
x_{10}(0) = 0, & x_{20}(0) = 0, & x_{30}(0) = 0.
\end{array} \tag{8.5.8}$$

In the analysis that follows, we assume that the rod has axial symmetry, so that $A_{22} = A_{33}$. The system (8.5.7), (8.5.8) has two first integrals:

$$F_1 M_1 + F_2 M_2 + F_3 M_3 = -P M_t, \quad F_1^2 + F_2^2 + F_3^2 = P^2. \tag{8.5.9}$$

The first integral $(8.5.9)_1$ is the scalar product of vectors $\mathbf{F}$ and $\mathbf{M}$, while the second expresses the fact that the intensity of contact force in an arbitrary cross section is constant. By using the axial symmetry in $(8.5.7)$, we obtain

$$F_1' + \frac{1}{A_{22}}(F_3 M_2 - F_2 M_3) = 0,$$

$$F_2' + F_1 \frac{M_3}{A_{22}} - F_3 \frac{M_1}{A_{11}} = 0,$$

$$F_3' - F_1 \frac{M_2}{A_{22}} + F_2 \frac{M_1}{A_{11}} = 0,$$

$$M_1' = 0,$$

$$M_2' + \left(\frac{1}{A_{22}} - \frac{1}{A_{11}} \right) M_1 M_3 = F_3,$$

$$M_3' - \left(\frac{1}{A_{22}} - \frac{1}{A_{11}} \right) M_1 M_2 = -F_2,$$

$$\theta_1' = \frac{c_2}{c_3} \frac{M_1}{A_{11}} + \frac{s_2}{c_3} \frac{M_3}{A_{22}},$$

$$\theta_2' = \frac{c_2 s_3}{c_3} \frac{M_1}{A_{11}} + \frac{M_2}{A_{22}} + \frac{s_2 s_3}{c_3} \frac{M_3}{A_{22}};$$

$$\theta_3' = -s_2 \frac{M_1}{A_{11}} + c_2 \frac{M_3}{A_{22}}, \quad x_{10}' = c_2 c_3,$$

$$x_{20}' = c_1 c_2 s_3 + s_1 s_2, \quad x_{30}' = c_2 s_1 s_3 - c_1 s_2,$$

$$\tag{8.5.10}$$

subjected to $(8.5.8)$. Note that $(8.5.10)_4$ together with $(8.5.8)_4$ leads to

$$M_1 = M_t. \tag{8.5.11}$$

Thus, the component of a moment at every cross section in the direction of the tangent to the rod axis is equal to the applied torque. For further analysis we shall need the system $(8.5.10)$, in a slightly different form.

After the use of $(8.5.11)$, $(8.5.10)$ becomes

$$F_1' + \frac{1}{A_{22}}(F_3 M_2 - F_2 M_3) = 0,$$

$$F_2' + F_1 \frac{M_3}{A_{22}} - F_3 \frac{M_t}{A_{11}} = 0,$$

$$F_3' - F_1 \frac{M_2}{A_{22}} + F_2 \frac{M_t}{A_{11}} = 0,$$

$$M_1' = 0,$$

$$M_2' + \left(\frac{1}{A_{22}} - \frac{1}{A_{11}}\right) M_t M_3 = F_3,$$

$$M_3' - \left(\frac{1}{A_{22}} - \frac{1}{A_{11}}\right) M_t M_2 = -F_2,$$

$$\theta_1' = \frac{c_2}{c_3} \frac{M_t}{A_{11}} + \frac{s_2}{c_3} \frac{M_3}{A_{22}},$$

$$\theta_2' = \frac{c_2 s_3}{c_3} \frac{M_t}{A_{11}} + \frac{M_2}{A_{22}} + \frac{s_2 s_3}{c_3} \frac{M_3}{A_{22}},$$

$$\theta_3' = -s_2 \frac{M_t}{A_{11}} + c_2 \frac{M_3}{A_{22}}, \quad x_{10}' = c_2 c_3,$$

$$x_{20}' = c_1 c_2 s_3 + s_1 s_2, \quad x_{30}' = c_2 s_1 s_3 - c_1 s_2.$$

$$(8.5.12)$$

Also, by use of (8.5.11) the first integrals (8.5.9) become

$$F_1 M_t + F_2 M_2 + F_3 M_3 = -P M_t, \quad F_1^2 + F_2^2 + F_3^2 = P^2. \tag{8.5.13}$$

We introduce next the variables

$$X_2 = M_2^2 + M_3^2, \quad X_3 = F_2 M_2 + F_3 M_3, \quad X_4 = F_2 M_3 - F_3 M_2. \tag{8.5.14}$$

By differentiating $(8.5.13)_1$ and using (8.5.12) we get

$$F_1' = \frac{X_4}{A_{22}}. \tag{8.5.15}$$

Further, by differentiating (8.5.14) and using (8.5.12) we obtain

$$X_2' = -2X_4, \quad X_3' = -\frac{M_t}{A_{22}} X_4,$$

$$X_4' = -\frac{F_1}{A_{22}} X_2 + \frac{M_t}{A_{22}} X_3 + \left((F_1)^2 - P^2\right). \tag{8.5.16}$$

It can be seen that the system (8.5.15), (8.5.16) has the following first integrals:

$$X_3^2 + X_4^2 = \left(P^2 - (F_1)^2\right) X_2, \quad X_3 = -M_t (F_1 + P); \tag{8.5.17}$$

From system (8.5.15)–(8.5.17) we derive the following second-order equation:

$$\left[\left(A_{22} F_1'\right)'\right] \left(P^2 - (F_1)^2\right) + A_{22}(F_1')^2 F_1$$

$$+ \frac{P M_t^2}{A_{22}} (P + F_1)^2 + \left(P^2 - (F_1)^2\right)^2 = 0. \tag{8.5.18}$$

The boundary conditions corresponding to (8.5.18) follow from (8.5.8) and the condition of global equilibrium of forces in the x_{10} direction. Thus, we have

$$F_1(0) = -P, \quad F_1(L) = -P. \tag{8.5.19}$$

Let θ be the Euler angle of nutation, that is, the angle between $\mathbf{e}_1$ and $\mathbf{e}_{10}$. Then,

$$F_1 = -P\cos\theta. \tag{8.5.20}$$

With (8.5.20), equation (8.5.18) becomes

$$P^3 \sin^3\theta \left[\left(A_{22}\theta' \right)' + \frac{M_t^2}{A_{22}} \frac{\sin\theta}{(1+\cos\theta)^2} + P\sin\theta \right] = 0. \tag{8.5.21}$$

If $|\theta| < \pi$, from (8.5.21) we obtain

$$\left(A_{22}\theta' \right)' + \frac{M_t^2}{A_{22}} \frac{\sin\theta}{(1+\cos\theta)^2} + P\sin\theta = 0. \tag{8.5.22}$$

Note that in writing (8.5.22) we assumed $P \neq 0$. The boundary conditions corresponding to (8.5.22) are obtained from (8.5.19) and (8.5.20) so that

$$\theta(0) = 0, \quad \theta(L) = 0. \tag{8.5.23}$$

The system (8.5.22), (8.5.23) will be the basis for our optimization problem. For the case $P = 0$, that implies $F_1 = F_2 = F_3 = 0$, from $(8.5.12)_{5,6,8,9}$, with

$$X_5 = c_3 s_2 M_2 + s_3 M_3, \quad X_6 = -s_3 M_2 + c_3 s_2 M_3, \tag{8.5.24}$$

we obtain

$$\begin{aligned}
(\cos\theta)' &= -\frac{X_5}{A_{22}}, \quad X_2' = 0, \\
X_5' &= \frac{X_2 \cos\theta}{A_{22}} - \frac{M_t}{A_{22}} X_6, \quad X_6' = \frac{M_t}{A_{22}} X_5.
\end{aligned} \tag{8.5.25}$$

The variable X_2 in (8.5.25) is given by $(8.5.14)_1$ and $\cos\theta = c_2 c_3$. By using the same procedure as in the case $P \neq 0$ we obtain

$$\sin^3\theta \left[\left(A_{22}\theta' \right)' + \frac{M_t^2}{A_{22}} \frac{\sin\theta}{(1+\cos\theta)^2} \right] = 0. \tag{8.5.26}$$

From (8.5.26) it follows that

$$\left(A_{22}\theta' \right)' + \frac{M_t^2}{A_{22}} \frac{\sin\theta}{(1+\cos\theta)^2} = 0. \tag{8.5.27}$$

Thus, equation (8.5.22) is valid *for all values of* P.

Equations (8.5.22), (8.5.23) are Euler–Lagrange equations of the following variational problem: determine the minimum of the functional I given as

$$I = \int_0^L \left(\frac{A_{22} \left(\Theta' \right)^2}{2} - \frac{M_t^2}{A_{22} \left(1 + \cos \Theta \right)} + P \cos \Theta \right) dS \qquad (8.5.28)$$

for $\Theta \in C^2 (0, L)$ and satisfying (8.5.23). The necessary condition for the minimum of I, that is,

$$\delta I \left(\theta, \delta \theta \right) = 0, \qquad (8.5.29)$$

is satisfied on the solution of (8.5.22).

Note that the integrand in functional (8.5.28) does not depend on t if $A_{22} = const.$ (i.e., the rod has constant cross section). Then, there is a Jacobi-type first integral of the Euler–Lagrange equations (8.5.22) that reads

$$\frac{A_{22} \left(\theta' \right)^2}{2} + \frac{M_t^2}{A_{22} \left(1 + \cos \theta \right)} - P \cos \theta = const. \qquad (8.5.30)$$

Also, the stability condition for the rod with constant cross section can be obtained from the second variation of (8.5.28). Calculating $\delta^2 I \left(\theta_0 = 0, \delta \theta \right)$, we obtain

$$\delta^2 I \left(\theta_0 = 0, \delta \theta \right) = \int_0^L \left[A_{22} \left(\delta \theta' \right)^2 - \left(\frac{M_t^2}{A_{22}} + P \right) \left(\delta \theta \right)^2 \right] dS. \qquad (8.5.31)$$

From (8.5.23) we conclude that $\delta \theta \left(0 \right) = \delta \theta \left(L \right) = 0$ so that (see [74])

$$\left(\frac{\pi}{L} \right)^2 \int_0^L \left(\delta \theta \right)^2 dS \le \int_0^L \left(\delta \theta' \right)^2 dS. \qquad (8.5.32)$$

By using (8.5.32), (8.5.31) we obtain that $\delta^2 I \left(\theta_0 = 0, \delta \theta \right) \ge 0$, if the condition $A_{22} \left(\frac{\pi}{L} \right)^2 \ge \left(\left(M_{cr}^2 / A_{22} \right) + P \right)$ is satisfied. Thus the critical load parameters M_{cr}, P_{cr} are determined from

$$A_{22} \left(\frac{\pi}{L} \right)^2 = \left(\frac{M_{cr}^2}{A_{22}} + P_{cr} \right). \qquad (8.5.33)$$

The stability boundary (8.5.33) agrees with the stability boundary obtained by other methods [23]. Next we use (8.5.28) with $A_{22} = const.$ to obtain an approximate solution to (8.5.22), (8.5.23). Let

$$t = \frac{S}{L}, \quad \lambda_1 = \frac{M_t L}{A_{22}}, \quad \lambda_2 = \frac{P L^2}{A_{22}}. \qquad (8.5.34)$$

Then, (8.5.22), (8.5.23) become

$$\ddot{\theta} + \lambda_1^2 \frac{\sin \theta}{\left(1 + \cos \theta \right)^2} + \lambda_2 \sin \theta = 0, \quad \theta \left(0 \right) = \theta \left(1 \right) = 0, \qquad (8.5.35)$$

where $(\cdot) = d(\cdot)/dt$. The functional (8.5.28) with $\bar{I} = IL^2/A_{22}$ reads

$$\bar{I} = \int_0^1 \left(\frac{\dot{\theta}^2}{2} - \lambda_1^2 \frac{1}{(1 + \cos\theta)} + \lambda_2 \cos\theta \right) dt. \qquad (8.5.36)$$

Let us assume an approximate solution (8.5.35) in the form

$$\Theta = C_1 t^4 + C_2 t^3 + C_3 t^2 + C_4 t + C_5. \qquad (8.5.37)$$

The boundary conditions imply that $C_5 = 0, C_4 = -(C_1 + C_2 + C_3)$. Thus, the function (8.5.36) becomes

$$\Theta = C_1 t^4 + C_2 t^3 + C_3 t^2 - (C_1 + C_2 + C_3)\, t. \qquad (8.5.38)$$

We determine C_1, C_2, and C_3 by the Ritz method. Let $\lambda_1 = 1, \lambda_2 = 10$. By substituting (8.5.38) into (8.5.36) and using the conditions $\partial \bar{I}/\partial C_1 = 0, \partial \bar{I}/\partial C_2 = 0$, we obtain

$$C_1 = 1.946, \quad C_2 = -3.892, \quad C_3 = 0.151. \qquad (8.5.39)$$

With these values the approximate solution to (8.5.35) reads

$$\Theta = 1.946 t^4 - 3.892 t^3 + 0.151 t^2 + 1.795 t. \qquad (8.5.40)$$

In Table 8.5.1 we compare the numerical solution θ to (8.5.35) with the approximate solution Θ given by (8.5.40).

Table 8.5.1

t	0	0.2	0.4	0.6	0.8	1.0
θ	0	0.337	0.544	0.544	0.337	0
Θ	0	0.337	0.543	0.543	0.337	0

8.6 Optimal Shape of a Simply Supported Rod (Lagrange's Problem)

The problem of determining the shape of the rod of greatest efficiency (a rod having minimum volume for given buckling load) was, for the case of a simply supported rod loaded by concentrated forces at its ends, formulated by Lagrange in 1773. The solution of the problem was obtained by many authors. Here we shall present a solution to Lagrange's problem obtained by use of the Pontryagin principle. Consider a rod shown in Figure 8.6.1.

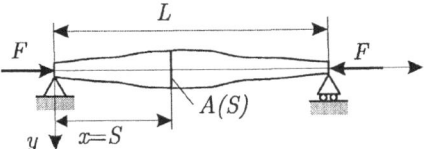

Figure 8.6.1

The differential equation determining the rod axis reads (see (8.2.7))

$$\frac{d}{dS}\left(EI\frac{d\theta}{dS}\right) + F\sin\theta = 0. \tag{8.6.1}$$

The boundary conditions corresponding to the rod shown in Figure 8.6.1 are

$$\frac{d\theta}{dS}(S=0), \quad \frac{d\theta}{dS}(S=L) = 0. \tag{8.6.2}$$

We assume that $I = \alpha A^2(S)$, where $\alpha = const.$, so that (8.6.2) becomes

$$\left(\alpha A^2(S)\,\theta'\right)' + F\sin\theta = 0. \tag{8.6.3}$$

The trivial solution to (8.6.3), (8.6.2) is

$$\theta_0 = 0. \tag{8.6.4}$$

Let $\theta = \theta_0 + \Delta\theta$. By substituting this into (8.6.3) we obtain (after linearization)

$$\left(\alpha A^2(S)\,\theta'\right)' + F\theta = 0, \tag{8.6.5}$$

where we omitted Δ in front of $\Delta\theta$. Introducing dimensionless parameters

$$t = \frac{S}{L}, \quad \lambda = \frac{F}{\alpha E L^2}, \quad a = \frac{A}{L^2} \tag{8.6.6}$$

in (8.6.5) we obtain

$$\left(a^2\dot{\theta}\right)^{\cdot} + \lambda\theta = 0, \tag{8.6.7}$$

subject to

$$\dot{\theta}(0) = 0, \quad \dot{\theta}(1) = 0, \tag{8.6.8}$$

where $\frac{d}{dt}(\cdot) = (\cdot)^{\cdot}$. The dimensionless volume of the rod is

$$w = \int_0^1 a(t)\,dt. \tag{8.6.9}$$

The Lagrange problem reads: determine the cross sectional area $a(t) \geq 0, t \in (0,1)$, such that for given w the smallest eigenvalue in (8.6.7), (8.6.8) is *maximal*. Alternatively, we may assume that λ is given and reformulate the Lagrange problem as: for given λ find $a(t) \geq 0, t \in (0,1)$, such that λ is the smallest eigenvalue of (8.6.7), (8.6.8) and w given by (8.6.9) is *minimal*.

We solve the second form of the Lagrange problem by using the Pontryagin maximum principle. We introduce new variables

$$x_1 = \theta, \quad x_2 = a^2\dot{\theta}, \tag{8.6.10}$$

so that (8.6.7) becomes

$$\dot{x}_1 = \frac{x_2}{a^2}, \quad \dot{x}_2 = -\lambda x_1, \tag{8.6.11}$$

subject to

$$x_2(0) = 0, \quad x_2(1) = 0. \tag{8.6.12}$$

Then, we formulate an optimization problem as: find $a(t) \geq 0$ such that

$$I = \int_0^1 a(t)\, dt \tag{8.6.13}$$

is minimal, subject to differential constraints (8.6.11). We form a Hamiltonian (see 7.6.6) as

$$H = a + p_1 \frac{x_2}{a^2} - p_2 \lambda x_1. \tag{8.6.14}$$

From (7.6.9) we obtain

$$\dot{p}_1 = -\frac{\partial H}{\partial x_1} = \lambda p_2, \quad \dot{p}_2 = -\frac{\partial H}{\partial x_2} = -\frac{p_1}{a^2}, \tag{8.6.15}$$

subject to (see (7.6.11))

$$p_1(0) = 0, \quad p_1(1) = 0. \tag{8.6.16}$$

The condition (7.6.10) reads

$$\frac{\partial H}{\partial u} = 1 - 2p_1 \frac{x_2}{a^3} = 0. \tag{8.6.17}$$

It can be seen that the solution of (8.6.15), (8.6.16) is given as

$$p_1 = -\frac{x_2}{2}, \quad p_2 = \frac{x_1}{2}. \tag{8.6.18}$$

There is a possibility to take $p_1 = x_2/2, p_2 = -x_1/2$, but this choice would not provide $a(t) \geq 0$. With (8.6.18) the condition (8.6.17) gives

$$a = \left(x_2^2\right)^{1/3}. \tag{8.6.19}$$

Returning to the system (8.6.11) we use (8.6.19) to obtain $(m = x_2)$

$$\ddot{m} + \frac{\lambda}{m^{1/3}} = 0, \tag{8.6.20}$$

subject to

$$m(0) = 0, \quad m(1) = 0. \tag{8.6.21}$$

Note that the optimality condition (8.6.19) reads

$$a^3 = m^2, \tag{8.6.22}$$

in agreement with [59]. By using $\ddot{m} = \frac{d\dot{m}}{dm}\dot{m}$ in (8.6.20) we get

$$\frac{1}{2}(\dot{m})^2 + \frac{3}{2}\lambda m^{2/3} = C = const. \tag{8.6.23}$$

We take the constant in (8.6.23) as $C = (3/2)\lambda C_0^{2/3}$, where C_0 is another constant. By using the symmetry of the problem, that is, $\dot{m}(t = 1/2) = 0$, we conclude that $C_0 = m(t = 1/2) = m_0$. Then, (8.6.23) leads to

$$\frac{(\dot{m})^2}{2} = \frac{3}{2}\lambda(m_0)^{2/3}\left[1 - \left(\frac{m}{m_0}\right)^{2/3}\right]. \tag{8.6.24}$$

Introducing a new variable

$$\nu = \left(\frac{m}{m_0}\right)^{1/3} \tag{8.6.25}$$

and separating variables in (8.6.24) we obtain

$$\sqrt{\frac{3}{\lambda}}(m_0)^{2/3}\frac{1}{2}\left[\arcsin\left(\frac{m}{m_0}\right)^{1/3} - \left(\frac{m}{m_0}\right)^{1/3}\sqrt{1 - \left(\frac{m}{m_0}\right)^{2/3}}\right] = t + C_1, \tag{8.6.26}$$

where C_1 is another constant. Applying the boundary conditions $\dot{m}(t = 1/2)$ $= 0$ and $m(1) = 0$ we obtain

$$C_1 = \sqrt{\frac{3}{\lambda}}(m_0)^{2/3}\frac{\pi}{4} - \frac{1}{2}, \quad (m_0)^{2/3} = \frac{2}{\pi}\sqrt{\frac{\lambda}{3}}. \tag{8.6.27}$$

Therefore (8.6.26) becomes

$$\frac{1}{\pi}\left[\arcsin\left(\frac{m}{m_0}\right)^{1/3} - \left(\frac{m}{m_0}\right)^{1/3}\sqrt{1 - \left(\frac{m}{m_0}\right)^{2/3}}\right] = t. \tag{8.6.28}$$

The constant m_0 is determined from the constraint. Thus by using (8.6.25) and (8.6.22) in (8.6.9) we get

$$w = 2\int_0^1 a_0\nu^2\frac{1}{\pi}\frac{2\nu^2}{\sqrt{1 - \nu^2}}d\nu = \frac{3}{4}a_0, \tag{8.6.29}$$

where $a_0 = a(t = 1/2)$. Also, from (8.6.29) we obtain

$$a_0 = \frac{4}{3}w, \tag{8.6.30}$$

so that $m_0 = (4w/3)^{3/2}$. Now by observing (8.6.22) we write (8.6.28) as

$$\frac{1}{\pi}\left[\arcsin\left(\frac{a}{a_0}\right)^{1/2} - \left(\frac{a}{a_0}\right)^{1/2}\sqrt{1-\left(\frac{a}{a_0}\right)}\right] = t = \frac{S}{L}. \tag{8.6.31}$$

Finally, (8.6.26), when substituted in (8.6.20), leads to

$$\lambda = \frac{3}{4}\pi^2\,(a_0)^2 = \frac{4}{3}\pi^2 w^2. \tag{8.6.32}$$

Thus, for given λ we determine a_0 from (8.6.32) and the optimal shape of the rod from (8.6.31). Often the solution is written in parametric form as follows. Take $m = m_0 \sin^3\theta$ and $w = 1$. Then (8.6.22) and (8.6.28) become [59]

$$\begin{aligned}
a &= \frac{4}{3}\sin^2\theta, \quad \theta - \frac{1}{2}\sin 2\theta = \pi t, \quad 0 \le t \le 1, \tag{8.6.33}\\
\lambda &= \frac{4}{3}\pi^2.
\end{aligned}$$

The method used here is strongly dependent on the fact that λ is a simple eigenvalue of the problem (8.6.7), (8.6.8) or (8.6.11), (8.6.12). For the case when λ is not a simple eigenvalue, the optimality condition (8.6.22) is different. We refer to [31], [99], and [101] for this case.

8.7 Optimal Shape of a Rod Loaded by Distributed Follower Force

In this section we study the problem of determining a Pflüger rod of greatest efficiency. A Pflüger rod is a simply supported rod loaded by a uniformly distributed follower–type load (see [86]). The uniformly distributed follower load is a nonconservative load. It is interesting, however, that the Pflüger rod loses stability by divergence, so that the stability analysis could be based on static (Euler) method. We shall formulate an optimization problem to determine the necessary conditions for the minimum of volume and determine the optimal distribution of the material along the rod axis. We follow the presentation of [16].

Consider a rod shown in Figure 8.7.1. The rod is simply supported at both ends with end C movable. The axis of the rod is initially straight and the rod is loaded by a uniformly distributed follower–type load of constant intensity q_0. We shall assume that the rod axis has length L and that it is inextensible.

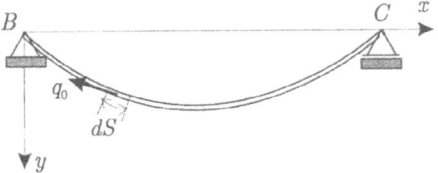

Figure 8.7.1

Let $x - B - y$ be a Cartesian coordinate system with the origin at the point B and with the x axis oriented along the rod axis in the undeformed state. The equilibrium equations are (see [14])

$$\frac{dH}{dS} = -q_x, \quad \frac{dV}{dS} = -q_y, \quad \frac{dM}{dS} = -V\cos\theta + H\sin\theta, \qquad (8.7.1)$$

where H and V are components of the contact force along the x and y axis, respectively, M is the bending moment, and θ is the angle between the tangent to the rod axis and the x axis. Also, in (8.7.1), q_x and q_y are components of the distributed forces along the x and y axis, respectively. Since the distributed force is tangent to the rod axis we have

$$q_x = -q_0\cos\theta, \quad q_y = -q_0\sin\theta. \qquad (8.7.2)$$

To the system (8.7.1) we adjoin the following geometrical relations:

$$\frac{dx}{dS} = \cos\theta, \quad \frac{dy}{dS} = \sin\theta, \qquad (8.7.3)$$

and constitutive relation

$$\frac{d\theta}{dS} = \frac{M}{EI}. \qquad (8.7.4)$$

In (8.7.3) and (8.7.4) we use x and y to denote coordinates of an arbitrary point of the rod axis and EI to denote the bending rigidity. The boundary conditions corresponding to the rod shown in Figure 8.7.1 are

$$\begin{aligned} x(0) &= 0, \quad y(0) = 0, \quad M(0) = 0, \\ y(L) &= 0, \quad M(L) = 0, \quad H(L) = 0. \end{aligned} \qquad (8.7.5)$$

The system (8.7.1)–(8.7.5) possesses a trivial solution:

$$\begin{aligned} H^0(S) &= -q_0(L - S), \quad V^0(S) = 0, \quad M^0(S) = 0, \\ x^0(S) &= S, \quad y^0(S) = 0, \quad \theta^0(S) = 0. \end{aligned} \qquad (8.7.6)$$

In order to formulate the minimum volume problem for the rod, we take the cross-sectional area $A(S)$ and the second moment of inertia $I(S)$ of the cross section in the form

$$A(S) = A_0 a(S), \quad I(S) = I_0 a^2(S), \qquad (8.7.7)$$

where A_0 and I_0 are constants (having dimensions of area and second moment of inertia, respectively). For the case of a rod with circular cross section we have the connection between A_0 and I_0 given by $I_0 = (1/4\pi)A_0^2$. Let $\Delta H, ..., \Delta\theta$ be the perturbations of $H, ..., \theta$ defined by

$$\begin{aligned} H &= H^0 + \Delta H, \quad V = V^0 + \Delta V, \quad M = M^0 + \Delta M, \\ x &= x^0 + \Delta x, \quad y = y^0 + \Delta y, \quad \theta = \theta^0 + \Delta\theta. \end{aligned} \qquad (8.7.8)$$

Then, by introducing the following dimensionless quantities

$$h = \frac{\Delta H L^2}{E I_0}, \quad v = \frac{\Delta V L^2}{E I_0}, \quad m = \frac{\Delta M L}{E I_0},$$

$$\xi = \frac{\Delta x}{L}, \quad \eta = \frac{\Delta y}{L}, \quad t = \frac{S}{L}, \quad \lambda = \frac{q_0 L^3}{E I_0}, \tag{8.7.9}$$

and by substituting (8.7.7) in (8.7.1)–(8.7.5) we arrive at the following nonlinear system of equations describing nontrivial configuration of the rod:

$$\dot{h} = -\lambda(1 - \cos\theta), \quad \dot{v} = \lambda\sin\theta,$$
$$\dot{m} = -v\cos\theta + [-\lambda(1 - t) + h]\sin\theta,$$
$$\dot{\xi} = 1 - \cos\theta, \quad \dot{\eta} = \sin\theta; \quad \dot{\theta} = \frac{m}{a^2}, \tag{8.7.10}$$

where $(\dot{\cdot}) = d(\cdot)/dt$. The boundary conditions corresponding to (8.7.10) are

$$\xi(0) = 0, \quad \eta(0) = 0, \quad m(0) = 0,$$
$$\eta(1) = 0, \quad m(1) = 0, \quad h(1) = 0. \tag{8.7.11}$$

Note that the system (8.7.10)–(8.7.11) has the solution $h(t) = 0, ..., \theta(t) = 0$ for all values of λ. Next we linearize (8.7.10) to obtain

$$\dot{h} = 0, \quad \dot{v} = \lambda\theta, \quad \dot{m} = -v - \lambda(1 - t)\theta,$$
$$\dot{\xi} = 0, \quad \dot{\eta} = \theta, \quad \dot{\theta} = \frac{m}{a^2}. \tag{8.7.12}$$

By using boundary conditions (8.7.11) in (8.7.12) we conclude that $h(t) = \xi(t) = 0$, and the rest of equations (8.7.12) could be reduced to

$$\ddot{m} + \frac{\lambda}{a^2}(1 - t)m = 0, \tag{8.7.13}$$

subject to

$$m(0) = m(1) = 0. \tag{8.7.14}$$

The system (8.7.13)–(8.7.14) constitutes a linear spectral problem. For the case when $0 < a(t) < \infty$, $t \in (0, 1)$ the eigenvalues $0 < \lambda_1 < \lambda_2 < \lambda_i < \infty$, are *simple* and the only accumulation point is at infinity. The theorem of Krasnoselskii,[21] when applied to our problem (see [89]), states that eigenvalues of the linear spectral problem (8.7.13)–(8.7.14) determine the bifurcation points of the non-linear system (8.7.10)–(8.7.11). Thus the bifurcation points of (8.7.13)–(8.7.14) are of the form $(0, \lambda_n)$, where λ_n are eigenvalues of the system (8.7.13)–(8.7.14).

---

[21]If the eigenvalues of the linearized problem are of odd algebraic multiplicity, then they determine the bifurcation points of the nonlinear problem. In our case, eigenvalues of the linearized problem are *simple* (algebraic and geometric multiplicities are equal to 1), and the conditions of Krasnoselskii theorem are satisfied.

Therefore, the rod of greatest efficiency is the one that has minimal volume for fixed lowest eigenvalue λ_1 of (8.7.13)–(8.7.14). Thus, for the rod of the greatest efficiency we have to minimize

$$V_0 = \int_0^1 a(t)dt, \tag{8.7.15}$$

subject to

$$\ddot{m} + \frac{\lambda_1}{a^2}(1-t)m = 0, \tag{8.7.16}$$

with boundary conditions (8.7.14). In (8.7.16) we consider λ_1 to be known. V_0 in (8.7.15) denotes the dimensionless volume of the rod.

To determine $a(t)$ such that V_0 is minimal, we shall again use the Pontryagin maximum principle. Let us rewrite optimization problem (8.7.15)–(8.7.16) as follows.

Find the control $u(t)$ such that $0 < u(t) < \infty$, $t \in (0,1)$, for which the optimality criterion

$$V_0 = \int_0^1 u(t)dt, \tag{8.7.17}$$

attains minimum value. The governing differential equations are

$$\begin{aligned} \dot{x}_1 &= x_2, \\ \dot{x}_2 &= -\frac{\lambda_1}{u^2}(1-t)x_1, \end{aligned} \tag{8.7.18}$$

subject to

$$x_1(0) = 0, \quad x_1(1) = 0. \tag{8.7.19}$$

For the system (8.7.17)–(8.7.18) the Hamiltonian (see (7.6.6)) reads

$$H = u + p_1 x_2 - p_2 \frac{\lambda_1}{u^2}(1-t)x_1, \tag{8.7.20}$$

where the variables p_1 and p_2 have to satisfy the following system of differential equations (see (7.6.9)):

$$\dot{p}_1 = -\frac{\partial H}{\partial x_1} = \frac{\lambda_1}{u^2}(1-t)p_2, \quad \dot{p}_2 = -\frac{\partial H}{\partial x_2} = -p_1, \tag{8.7.21}$$

subject to

$$p_2(0) = 0, \quad p_2(1) = 0. \tag{8.7.22}$$

The optimality condition (7.6.10) or (7.6.13) leads to

$$\frac{\partial H}{\partial u} = 1 + \frac{2\lambda_1}{u^3}(1-t)x_1 p_2 = 0. \tag{8.7.23}$$

By solving (8.7.23) for u we obtain

$$u = -\left[2\lambda_1(1-t)x_1p_2\right]^{1/3}. \tag{8.7.24}$$

Comparison of the boundary value problems (8.7.18)–(8.7.19) and (8.7.21)–(8.7.22) lead to the conclusion that

$$p_1(t) = x_2(t), \quad p_2(t) = -x_1(t). \tag{8.7.25}$$

Therefore, the control variable $u(t)$ given by (8.7.24) becomes

$$u = \left[2\lambda_1(1-t)x_1^2\right]^{1/3}. \tag{8.7.26}$$

Note that by using (8.7.25) in (8.7.20) we have $(\partial^2 H/\partial u^2) > 0$ so that the necessary condition for a minimum of H is satisfied. By substituting (8.7.26) into (8.7.18) we arrive at a single differential equation

$$\ddot{x} + \bar{\lambda}(1-t)^{1/3}x^{-1/3} = 0, \tag{8.7.27}$$

subject to

$$x(0) = 0, \quad x(1) = 0. \tag{8.7.28}$$

In (8.7.27) and (8.7.28) we used $x(t) = x_1(t)$, $\bar{\lambda} = (\lambda_1/4)^{1/3}$. Note that the end points $t = 0$ and $t = 1$ are singular points of the problem (8.7.27)–(8.7.28). To examine the local behavior of the solution near end points, we first transform the independent variable t to $\zeta = 1 - t$. Then (8.7.27) becomes

$$\ddot{x} + \bar{\lambda}\zeta^{1/3}x^{-1/3} = 0, \tag{8.7.29}$$

subject to

$$x(0) = 0, \quad x(1) = 0. \tag{8.7.30}$$

In (8.7.29) we again used $(\cdot) = d(\cdot)/d\zeta$. Suppose that the solution $x(\zeta)$ of (8.7.29)–(8.7.30) in the vicinity of $\zeta = 0$ behaves as

$$\begin{aligned}
x(\zeta) &\sim C\zeta^\alpha\left[1 + a_1\zeta + a_2\zeta^2 + a_3\zeta^3 + a_4\zeta^4 + \cdots\right], \\
\zeta &\to 0,
\end{aligned} \tag{8.7.31}$$

where C, α, and $a_i, i = 1, 2, ...,$ are constants. By substituting (8.7.31) in (8.7.29) we conclude that

$$\begin{aligned}
\alpha &= 1, \quad a_1 = -\frac{1}{2}\bar{\lambda}C^{-4/3}; \quad a_2 = -\frac{1}{36}\bar{\lambda}^2C^{-8/3}, \\
a_3 &= -\frac{7}{1296}\bar{\lambda}^3C^{-4}, \quad a_4 = -\frac{23}{15552}\bar{\lambda}^4C^{-16/3}.
\end{aligned} \tag{8.7.32}$$

In (8.7.32) the constant C remains undetermined. It is related to the first derivative of x at $\zeta = 0$, namely, $C = \dot{x}(0)$. Note that from (8.7.29) and (8.7.32) it follows that

$$\lim_{\zeta \to 0} \ddot{x}(\zeta) = -\frac{\bar{\lambda}}{C^{1/3}}. \tag{8.7.33}$$

To solve (8.7.29) numerically we shall write the first-order system of differential equations

$$\dot{x}_1 = x_2, \quad \dot{x}_2 = -\bar{\lambda}\zeta^{1/3}x_1^{-1/3}. \tag{8.7.34}$$

Then, we choose $x_2(0) = C$ and solve (8.7.34) as an initial value problem. The boundary condition $x_1(1) = 0$ will be satisfied by the shooting method. In the first integration step, the right hand-side of $(8.7.34)_2$ will be equated with (8.7.33).

We shall now formulate two variational principles for the boundary value problem (8.7.27)–(8.7.28). Let $\mathbf{X}$ be the following function space:

$$\mathbf{X} = \left\{ X(t) : X(t) \in C^2(0,1); X(0) = X(1) = 0 \right\}, \tag{8.7.35}$$

where $C^2(0,1)$ is the space of continuous functions mapping $(0,1)$ into real R, having continuous first and second derivatives. Next we consider the variational problem

$$\min_{X \in \mathbf{X}} I(X) = \int_0^1 \left[\frac{1}{2}\dot{X}^2 - \frac{3}{2}\bar{\lambda}(1-t)^{1/3}X^{2/3} \right] dt \tag{8.7.36}$$

for $X(t) \in \mathbf{X}$. It is easy to see that the minimizing element of (8.7.36), if it exists, satisfies (8.7.27)–(8.7.28). Therefore,

$$\delta I(x, \delta x) = 0. \tag{8.7.37}$$

Hamilton's variational principle (8.7.37) is also called the *primal* variational principle (see [8]). Calculating the second variation of (8.7.36) we obtain

$$\delta^2 I(x, \delta x) = \int_0^1 \left[(\delta \dot{x})^2 + \frac{1}{3}\bar{\lambda}(1-t)^{1/3}x^{-4/3}(\delta x)^2 \right] dt. \tag{8.7.38}$$

From (8.7.38) it follows that

$$\delta^2 I(x, \delta x) \geq \|\delta \dot{x}\|_{L_2}^2, \tag{8.7.39}$$

where $\|\delta \dot{x}\|_{L_2} = (\int_0^1 (\delta \dot{x})^2 \, dt)^{1/2}$. By combining (8.7.37) and (8.7.39) we conclude that on the solution $x(t)$ of the boundary value problem (8.7.27)–(8.7.28) the functional (8.7.36) attains a minimum (see discussion after (5.2.13)):

$$\min_{X \in \mathbf{X}} I(X) = I(x). \tag{8.7.40}$$

Now we proceed to construct the *dual variational principle* (see [8]). Let us denote by $F(t, X, \dot{X})$ the Lagrangian function of the variational principle (8.7.37), that is,

$$F(t, X, \dot{X}) = \frac{1}{2}\dot{X}^2 - \frac{3}{2}\bar{\lambda}(1 - t)^{1/3}X^{2/3}. \tag{8.7.41}$$

We get the canonical form of (8.7.27)–(8.7.28) by introducing the variable P as (see (1.8.1))

$$P = \frac{\partial F}{\partial \dot{X}} = \dot{X}, \tag{8.7.42}$$

and the function $H_0(t, X, P)$, also called the Hamiltonian, connected with F by the Legendre transformation (see (1.8.6))

$$H_0(t, X, P) = P\dot{X} - F = \frac{1}{2}P^2 + \frac{3}{2}\bar{\lambda}(1 - t)^{1/3}X^{2/3}. \tag{8.7.43}$$

The system (8.7.27)–(8.7.28) then becomes

$$\dot{X} = \frac{\partial K}{\partial P} = P, \quad \dot{P} = -\frac{\partial K}{\partial X} = -\bar{\lambda}(1 - t)^{1/3}X^{-1/3}. \tag{8.7.44}$$

In order to obtain the dual variational principle we first have to solve (8.7.43) for F. Then, in the expression so obtained, we substitute X in terms of $\dot{P}$ from (8.7.44)$_2$ and use the resulting expression in (8.7.36). After partial integration and application of boundary conditions, we obtain

$$G(P) = I(P, X(P)) = -\int_0^1 \left[\frac{1}{2}P^2 + \frac{1}{2}\bar{\lambda}^3\frac{1 - t}{\dot{P}^2}\right] dt. \tag{8.7.45}$$

Let $x(t)$ and $p(t)$ be the solution of canonical system (8.7.44). Then it is easy to show that the following dual variational principle holds:

$$\delta G(p, \delta p) = 0, \tag{8.7.46}$$

where the variation δp is $\delta p = P - p$ and P is an admissible function, that is, $P \in \mathbf{Y} = \{P : P \in C^1(0, 1); \int_0^1 [(1 - t)/\dot{P}^2] dt < \infty\}$. Note also that $I(x) = G(p)$. The elements of $\mathbf{Y}$ *need not satisfy any boundary condition*. However, in practical applications P is usually expressed as $P = \dot{X}$, where $X(t) \in \mathbf{X}$. Also, from (8.7.45) it follows that

$$\delta^2 G(p, \delta p) = -\int_0^1 \left[(\delta p)^2 + 3\bar{\lambda}^3\frac{1 - t}{\dot{P}^4}(\delta\dot{p})^2\right] dt \leq -\|\delta p\|_{L_2}^2. \tag{8.7.47}$$

By combining (8.7.46) and (8.7.47) we conclude that on $p(t)$ the functional (8.7.45) attains a maximum:

$$\max_{P \in \mathbf{Y}} G(P) = G(p). \tag{8.7.48}$$

From the results (8.7.37), (8.7.39), (8.7.46), (8.7.47) the following chain of inequalities could be derived:

$$G(P) \leq G(p) = I(x) \leq I(X). \tag{8.7.49}$$

We shall use (8.7.36) and (8.7.45) to obtain an error estimate of an approximate solution of (8.7.27), (8.7.28). This is achieved by using the inequality

$$\|\delta x\|_{L_\infty} = \sup_{t \in (0,1)} |\delta x(t)| \leq \frac{\|\delta \dot{x}\|_{L_2}}{\sqrt{2}}, \tag{8.7.50}$$

valid for any $\delta x(t)$ satisfying $\delta x(0) = \delta x(1) = 0$. From (8.7.49) and (8.7.50) we obtain

$$\|\delta x\|_{L_\infty} \leq [I(X) - G(P)]^{1/2}. \tag{8.7.51}$$

We present results of numerical integration of the system (8.7.18), (8.7.19). The value of λ_1 was taken as the lowest eigenvalue of the rod with constant cross section having unit volume. Precisely, we take $a(t) = 1$, $A_0 = \pi R_0^2$, $I_0 = \pi R_0^4/4$ in (8.7.7), where R_0 is the radius of the cross-section that we assume to be circular. Then, from (8.7.15) we conclude that $V_0 = 1$. With these values λ_1 becomes $\lambda_1 = 18.956266$ (see [14]).

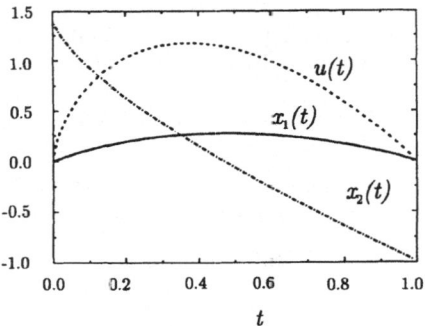

Figure 8.7.2

The numerical solution of (8.7.18), (8.7.19) is shown in Figure 8.7.2. Note that in accordance with (8.7.24) the cross section of the rod (in our notation u) is zero at both ends of the rod. By assuming that the optimal rod is also of circular cross section with the radius $r(t)$ we can take $u(t) = A(t)/A_0 = (r(t)/R_0)^2$. The radius $r(t)$ of the optimal rod as a function of dimensionless arc length is shown in Figure 8.7.3.

When the volume of the optimal rod is calculated from (8.7.17) we obtain

$$v_{opt} = \int_0^1 u(t)dt = 0.81051. \tag{8.7.52}$$

It is interesting to compare the critical force of the rod with constant cross section having the same volume as the optimal rod. The radius of the cross section of this rod is $r_0 = 0.90028R_0$. The corresponding dimensionless critical load is $\lambda_0 = 12.452807$.

We now use dual variational principles, described in the previous section, together with the Ritz method, to determine an analytical approximate solution of (8.7.27)–(8.7.28) for $\lambda_1 = 18.956266$, already used in numerical treatment. We assume the solution of the problem in the following form:

$$X(C_1, C_2, C_3, t) = m_3(t) = C_1 t(1 - t)(1 + C_2 t + C_3 t^2), \qquad (8.7.53)$$

where C_1, C_2, and C_3 are constants to be determined.

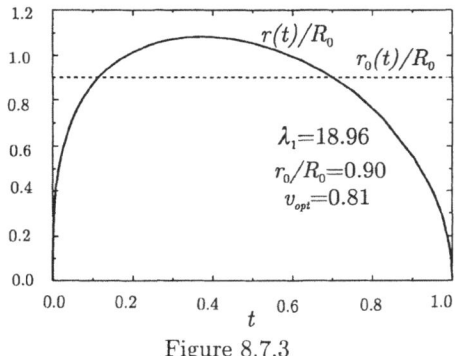

Figure 8.7.3

Since $x(t)$ in (8.7.27) represents the dimensionless moment (see (8.7.13)) we used this in writing (8.7.53). The index attached to m indicates the number of free constants that are to be determined. Note that $X(t)$ given by (8.7.53) is an admissible function since it satisfies the boundary conditions for all values of constants C_1, C_2, and C_3. By substituting (8.7.53) into (8.7.36) and minimizing with respect to C_1, C_2, and C_3 we obtain $C_1 = 1.3036, C_2 = -0.4563, C_3 = 0.2293$ so that (8.7.53) and the corresponding value of functional I become

$$m_3(t) = 1.3036t(1 - t)(1 - 0.4563t + 0.2293t^2), \quad I(m_3) = -0.4052. \quad (8.7.54)$$

For the functional G given by (8.7.45) we take the admissible function $P(t)$ in the form

$$P(D, D_1, D_2) = \dot{X}(D_1, D_2, D_3, t). \qquad (8.7.55)$$

Substituting (8.7.55) in (8.7.45) and maximizing with respect to D_1, D_2, and D_3, we obtain

$$
\begin{aligned}
P(D, D_1, D_2, t) &= D(1 + 2D_1 t + 3D_2 t^2) - Dt\left(2 + 3D_1 t + 4D_2 t^2\right), \\
D_1 &= 1.3128, \quad D_2 = -0.4646, \quad D_3 = 0.2353, \\
G(P) &= -0.4088.
\end{aligned}
\qquad (8.7.56)
$$

From (8.7.55), (8.7.56), and (8.7.51) we obtain the estimate of the error of the solution (8.7.54) as

$$\|\delta x\|_{L_\infty} \le [I(X) - G(P)]^{1/2} = 0.06. \qquad (8.7.57)$$

In Figure 8.7.4. we show the numerical $(m(t))$ and two approximate solutions $m_3(t)$ (given by (8.7.54)) and $m_2(t)$, obtained by minimization of the functional I for the trial function with two unknown constants, that is, $m_2(t) = C_1 t(1 - t)(1 + C_2 t)$. As could be seen from Figure 8.7.4 by increasing the number of constants in the trial function from two to three, we obtained convergence of approximate solutions to the exact one. Once we know the approximate solution $m_3(t)$ we can determine the approximate cross-sectional area of the rod from (8.7.26).

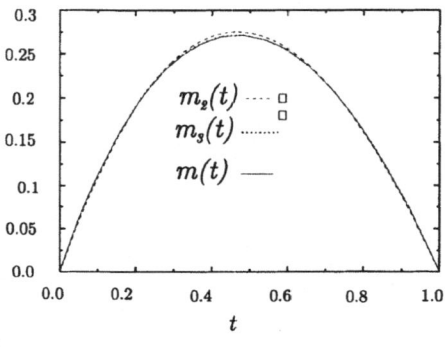

Figure 8.7.4

A generalization of the problem presented here was given in [20], where it was assumed that a compressive force F is acting at the moving end of the Pflüger rod. The resulting equation is (compare with (8.7.27))

$$\ddot{x} + \left[\bar{\lambda}_1(1 - t) + \bar{\lambda}_2\right]^{1/3} x^{-1/3} = 0, \qquad (8.7.58)$$

subject to

$$x(0) = x(1) = 0. \qquad (8.7.59)$$

In (8.7.58) and (8.7.59) we used $\bar{\lambda}_1 = (\frac{q_0}{\alpha EL}/4), \bar{\lambda}_2 = (\frac{FL^2}{EI_0}/4)$. The case treated by Keller [59] and in section 8.6 corresponds to $\bar{\lambda}_1 = 0, \bar{\lambda}_2 \ne 0$. If the solution of (8.7.58) and (8.7.59) is known, the cross section of the rod is determined form

$$a = \left\{2\left[4\bar{\lambda}_1(1 - t) + 4\bar{\lambda}_2\right] x^2\right\}^{1/3}. \qquad (8.7.60)$$

Equation (8.7.58) is the Euler–Lagrangian equation of the variational problem $\delta I_1 = 0$, where

$$I_1(X) = \int_0^1 \left\{\frac{1}{2}\dot{X}^2 - \frac{3}{2}\left[\bar{\lambda}_1(1 - t) + \bar{\lambda}_2\right]^{1/3} X^{2/3}\right\} dt, \qquad (8.7.61)$$

and X belongs to the space $\mathbf{X}$ defined by (8.7.35). The procedure of solving (8.7.58), (8.7.59) and after that determining the shape of the optimal rod (optimal cross section) according to (8.7.60) is similar to the previous case, and is omitted.

8.8 Optimal Shape of the Rotating Rod

Our intention in this section is to formulate an optimization problem for the rotating rod (see [17]). We start from the equations (8.4.1), (8.4.2), but we use different dimensionless quantities. Thus we introduce

$$t = \frac{S}{L}, \quad a = \frac{A}{L^2}, \quad y = \frac{\bar{y}}{L}, \quad w = \frac{W}{L^3},$$

$$m = \frac{M}{E\alpha L^3}, \quad v = \frac{V}{E\alpha L^2}, \quad \lambda^2 = \frac{\rho\omega^2 L^2}{E\alpha}, \tag{8.8.1}$$

and a new dependent variable

$$u = -\frac{v}{\lambda}. \tag{8.8.2}$$

From (8.4.1), (8.4.2) and (8.8.1), (8.8.2) we obtain

$$\dot{u} = \lambda a y, \quad \dot{m} = -\lambda u \cos\theta, \quad \dot{y} = \sin\theta, \quad \dot{\theta} = -\frac{m}{a^2}. \tag{8.8.3}$$

The boundary conditions corresponding to the rod shown in Figure 8.4.1 are

$$u(1) = 0, \quad m(1) = 0, \quad y(0) = 0, \quad \theta(0) = 0. \tag{8.8.4}$$

The system (8.8.3), (8.8.4) possesses a trivial solution, in which the axis of the rod remains straight for any value of the dimensionless rotation speed λ, in the form

$$u_0 = 0, \quad m_0 = 0, \quad y_0 = 0, \quad \theta_0 = 0. \tag{8.8.5}$$

To examine stability of the equilibrium configuration (8.8.5) we use the Euler method. Thus, we assume that

$$u = u_0 + \Delta u, \quad m = m_0 + \Delta m, ..., \quad \theta = \theta_0 + \Delta\theta, \tag{8.8.6}$$

where $\Delta u, ..., \Delta\theta$, denote perturbations of the corresponding variables. Then, by substituting (8.8.6) into (8.8.3) and linearizing the resulting expressions, we obtain (omitting Δ in front of perturbed variables)

$$\left(\frac{\dot{u}}{a}\right)^{\cdot} = \lambda\theta, \quad \left(\dot{\theta}a^2\right)^{\cdot} = \lambda u, \quad \dot{y} = \theta, \tag{8.8.7}$$

subject to

$$u(1) = 0, \quad \dot{u}(0) = 0, \quad \theta(0) = 0, \quad \dot{\theta}(1)a^2(1) = 0, \quad y(0) = 0. \tag{8.8.8}$$

For the bifurcation it is enough to consider

$$\left(\frac{\dot u}{a}\right)^{\cdot} = \lambda\theta, \quad \left(\dot\theta a^2\right)^{\cdot} = \lambda u, \tag{8.8.9}$$

subject to

$$u(1) = 0, \quad \dot u(0) = 0, \quad \theta(0) = 0, \quad \lim_{t\to1}\dot\theta(t)a^2(t) = 0. \tag{8.8.10}$$

By introducing new variables $u = w_1, \dot u = w_2, \theta = w_3, \dot\theta = w_4$ and the vector $\mathbf{w} = [w_1, w_2, w_3, w_4]$, the system (8.8.9), (8.8.10) can be written in compact form as[22]

$$\mathbf{F}(\lambda)\mathbf{w} = \frac{d}{dt}\left[\mathbf{A}\frac{d}{dt}\mathbf{w}\right] + \lambda\mathbf{B}\mathbf{w} = 0, \tag{8.8.11}$$

where $\mathbf{A}$ and $\mathbf{B}$ are given as

$$\mathbf{A} = \begin{bmatrix} 0 & 1/a & 0 & 0 \\ 0 & 0 & 0 & 0 \\ 0 & 0 & 0 & 0 \\ 0 & 0 & 0 & a^2 \end{bmatrix}, \quad \mathbf{B} = \begin{bmatrix} 0 & 0 & 1 & 0 \\ 0 & 0 & 0 & 0 \\ 0 & 0 & 0 & 0 \\ 1 & 0 & 0 & 0 \end{bmatrix}. \tag{8.8.12}$$

Note that the boundary condition (8.8.10)$_4$ that corresponds to $m(1) = 0$ is equivalent to $\dot\theta(1) = 0$ if $a(1) \neq 0$. However, if $a(1) = 0$ (as it will be in our analysis) the condition $m(1) = 0$ can be satisfied with $\dot\theta(1) \neq 0$. For fixed λ the existence of a nontrivial solution of (8.8.9), (8.8.10) is a necessary condition for the loss of stability. The dimensionless volume of the rod is given as

$$w = \int_0^1 a(t)dt. \tag{8.8.13}$$

We now state the following optimization problem.

For given λ determine $a(t) > 0$ for $t \in (0,1)$, so that λ is the smallest eigenvalue of (8.8.9), (8.8.10) and at the same time w given by (8.8.13) is minimal.

We call the rod with such $a(t)$ the *optimal rotating rod*. Thus, the optimal rotating rod is so shaped that any other rod with smaller volume w will buckle at a rotation speed that is smaller than λ. We note that λ must be an isolated eigenvalue in order for nonlinear equilibrium equations to have bifurcation points at λ, and we assume this to be true. Cox and McCarthy [32] pointed out that the assumption about isolated eigenvalues may be violated if the cross-sectional area $a(t)$ vanishes too severely when $t \to 1$. Thus, in principle, our assumption may be checked by the method similar to one presented in [32]. This rather delicate analysis is outside the scope of our presentation.

---

[22]The system (8.8.9) could also be writen as $\left(a^2\ddot y\right)^{\cdot\cdot} - \lambda^2 ay = 0$.

Let $x_1, ..., x_4$ be a set of dependent variables defined by

$$u = x_1, \quad \frac{\dot{u}}{a} = x_2, \quad \theta = x_3; \quad a^2 \dot{\theta} = x_4. \tag{8.8.14}$$

Then, the system (8.8.9), (8.8.10) becomes

$$\dot{x}_1 = ax_2, \quad \dot{x}_2 = \lambda x_3, \quad \dot{x}_3 = \frac{x_4}{a^2}, \quad \dot{x}_4 = \lambda x_1, \tag{8.8.15}$$

subject to

$$x_1(1) = 0, \quad x_2(0) = 0, \quad x_3(0) = 0, \quad x_4(1) = 0. \tag{8.8.16}$$

The problem of determining the shape $a(t)$ of the optimal rod may be stated as: determine the function $a(t) > 0$, $t \in (0,1)$, that minimizes the functional

$$I = \int_0^1 a(t)dt, \tag{8.8.17}$$

when the system is described by (8.8.15), (8.8.16). To solve the optimization problem we use the Pontryagin maximum principle. For system (8.8.15) the Hamiltonian function H is

$$H = a + p_1 a x_2 + p_2 \lambda x_3 + p_3 \frac{x_4}{a^2} + p_4 \lambda x_1, \tag{8.8.18}$$

where the variables $p_1, ..., p_4$ satisfy

$$\dot{p}_1 = -\frac{\partial H}{\partial x_1} = -\lambda p_4, \quad \dot{p}_2 = -\frac{\partial H}{\partial x_2} = -p_1 a,$$

$$\dot{p}_3 = -\frac{\partial H}{\partial x_3} = -\lambda p_2, \quad \dot{p}_4 = -\frac{\partial H}{\partial x_4} = -\frac{p_3}{a^2}, \tag{8.8.19}$$

subject to

$$p_1(0) = 0, \quad p_2(1) = 0, \quad p_3(1) = 0, \quad p_4(0) = 0. \tag{8.8.20}$$

The optimality condition (see (7.6.10) or (7.6.13)) leads to

$$\frac{\partial H}{\partial a} = 1 + p_1 x_2 - 2p_3 \frac{x_4}{a^3} = 0. \tag{8.8.21}$$

By solving (8.8.21) for a we obtain

$$a = \left\{ \frac{2p_3 x_4}{1 + p_1 x_2} \right\}^{1/3}. \tag{8.8.22}$$

Again we note that the solution of the boundary value problem (8.8.15), (8.8.16) leads to a solution of the boundary value problem (8.8.19), (8.8.20) if we make the following identification:

$$p_1(t) = x_2(t), \quad p_2(t) = -x_1(t), \quad p_3 = x_4, \quad p_4 = -x_3. \tag{8.8.23}$$

There is one more possibility to connect the solutions of (8.8.15), (8.8.16) and (8.8.19), (8.8.20). This is given by the following identification of dependent variables: $p_1 = -x_2$, $p_2 = x_1$, $p_3 = -x_4$, $p_4 = x_3$. However, this identification is not of interest since it does not provide $a(t) \geq 0$ in (8.8.22) and $(\partial^2 H/\partial a^2) > 0$ with H given by (8.8.18). Note that here we assumed that λ is a simple eigenvalue of (8.8.15), (8.8.16), so that there exist a single eigenvector with components $(x_1(t), ..., x_4(t))$. By using (8.8.23) the cross-sectional area $a(t)$ given by (8.8.22) becomes

$$a = \left\{ \frac{2(x_4)^2}{1 + (x_2)^2} \right\}^{1/3}. \tag{8.8.24}$$

Note also that with (8.8.23) substituted in (8.8.18) we have $(\partial^2 H/\partial a^2) > 0$ so that the necessary condition for the minimum of H is satisfied. From (8.8.24) and the boundary condition (8.8.16)$_4$, we conclude that

$$a(1) = 0. \tag{8.8.25}$$

Thus the optimal rod is tapered so that it has zero cross-sectional area and zero moment of inertia at its free end. Also, when the original variables (see (8.8.14)) are used in (8.8.24) we obtain

$$\frac{a^3 \dot{\theta}^2}{a^2 + \dot{u}^2} = \frac{1}{2}. \tag{8.8.26}$$

By using the boundary condition (8.8.10) in (8.8.26) we get

$$a(0)\dot{\theta}^2(0) = \frac{1}{2}. \tag{8.8.27}$$

We now transform (8.8.24). First we write it in the form

$$1 + (x_2)^2 = \frac{2(x_4)^2}{a^3} = \frac{2a^4\dot{\theta}^2}{a^3} = 2a\dot{\theta}^2. \tag{8.8.28}$$

From (8.8.28) we conclude that $a(t) \neq 0$ for $t \in (0, 1)$. Next, by differentiating (8.8.28) and by using (8.8.14), (8.8.15) we get

$$(a\dot{\theta}^2)^{\cdot} = \lambda\theta\frac{\dot{u}}{a} \tag{8.8.29}$$

as the optimality condition. Now we transform the system (8.8.7) as follows. Integrate (8.8.7)$_1$ to obtain

$$\left(\frac{\dot{u}}{a} \right) = \lambda \int_0^t \theta(\xi)d\xi, \tag{8.8.30}$$

where we used the fact that $\dot{u}(0) = 0$ and $a(0) \neq 0$ (see $(8.8.8)_2$ and $(8.8.26)$). Substituting $(8.8.30)$ into $(8.8.29)$, integrating, and using $(8.8.27)$, we obtain

$$a\dot{\theta}^2 = \lambda^2 \int_0^t \left[\theta(\xi) \int_0^\xi \theta(\zeta)d\zeta\right] d\xi + \frac{1}{2}. \qquad (8.8.31)$$

Since $\dot{y} = \theta$ (see $(8.8.7)_3$), the equation $(8.8.31)$ that represents the condition of optimality may be written as

$$a\ddot{y}^2 = \frac{1}{2}\left[1 + \lambda^2 y^2\right]. \qquad (8.8.32)$$

Finally, by differentiating $(8.8.7)_2$ and using $(8.8.30)$, we have

$$\left(\dot{\theta}a^2\right)^{\cdot\cdot} = \left(\ddot{y}a^2\right)^{\cdot\cdot} = \lambda\dot{u} = \lambda^2 a \int_0^t \theta(\xi)d\xi = \lambda^2 ay. \qquad (8.8.33)$$

Therefore, the optimal shape $a(t)$ of the rotating rod is determined from the solution of the system $(8.8.32)$, $(8.8.33)$:

$$\left(\ddot{y}a^2\right)^{\cdot\cdot} = \lambda^2 ay, \quad a\ddot{y}^2 = \frac{1}{2}\left[1 + \lambda^2 y^2\right], \qquad (8.8.34)$$

subject to

$$y(0) = 0, \quad \dot{y}(0) = 0, \quad \lim_{t\to 1}\ddot{y}(t)a^2(t) = 0, \quad \lim_{t\to 1}\left\{\left[\ddot{y}(t)a^2(t)\right]^{\cdot}\right\} = 0. \qquad (8.8.35)$$

We next analyze system $(8.8.34)$, $(8.8.35)$. First, we formulate three different variational principles corresponding to the system $(8.8.34)$, $(8.8.35)$. Also, we construct a conservation law corresponding to $(8.8.34)$, $(8.8.35)$.

(a) *The variational principle with two arguments.* Let $\mathbf{W}_1$ be the linear function space defined as

$$\mathbf{W}_1 = \left\{\mathbf{w} = (y, a) : y \in C^4(0, 1), \quad y(0) = \dot{y}(0) = 0, \right.$$
$$\left. a \in C^2(0, 1), \quad a \geq 0, \quad a(1) = 0\right\}. \qquad (8.8.36)$$

Consider the functional

$$I_1(y, a) = \int_0^1 F_1 dt, \qquad (8.8.37)$$

with the Lagrangian function

$$F_1 = a^2\ddot{y}^2 - \lambda^2 ay^2 - a. \qquad (8.8.38)$$

Suppose further that we want to determine the minimum of I_1 on $\mathbf{W}_1$. We claim that I_1 is stationary on the solution of $(8.8.34)$, $(8.8.35)$. To prove this, note that the condition of stationarity of I_1, that is, vanishing of the first variation δI_1, leads to the following Euler–Lagrangian equations:

$$\left(\ddot{y}a^2\right)^{..} = \lambda^2 ay, \qquad a\ddot{y}^2 - \frac{\lambda^2}{2}y^2 - \frac{1}{2} = 0, \tag{8.8.39}$$

and natural boundary conditions

$$\ddot{y}(1)a^2(1) = 0, \qquad \left\{\left[\ddot{y}(t)a^2(t)\right]^{.}\right\}_{t=1} = 0. \tag{8.8.40}$$

The system (8.8.39), (8.8.40) is equivalent to (8.8.34), (8.8.35).

(b) *The variational principle with one argument.* We can write system (8.8.34), (8.8.35) as a single differential equation of fourth order if we determine a from $(8.8.34)_2$ and then substitute the result in $(8.8.34)_1$. Thus we obtain

$$\left[\frac{1}{\ddot{y}^3}\left(1 + \lambda^2 y^2\right)^2\right]^{..} - 2\lambda^2 y \frac{\left(1 + \lambda^2 y^2\right)}{\ddot{y}^2} = 0, \tag{8.8.41}$$

subject to

$$y(0) = 0, \quad \dot{y}(0) = 0,$$

$$\frac{1}{4\ddot{y}^3(1)}\left[1 + \lambda^2 y^2(1)\right]^2 = 0, \quad \left(\left\{\frac{1}{4\ddot{y}^3(t)}\left[1 + \lambda^2 y^2(t)\right]^2\right\}^{.}\right)_{t=1} = 0. \tag{8.8.42}$$

Consider the space

$$\mathbf{W}_2 = \left\{y : y \in C^4(0,1); \quad y(0) = \dot{y}(0) = 0\right\} \tag{8.8.43}$$

and the functional

$$I_2 = \int_0^1 F_2 dt, \tag{8.8.44}$$

with Lagrangian function

$$F_2 = \frac{\left(1 + \lambda^2 y^2\right)^2}{\ddot{y}^2}. \tag{8.8.45}$$

Then the Euler–Lagrangian equation corresponding to $\delta I_2 = 0$ is equivalent to (8.8.41). Note that the natural boundary conditions for the minimization of I_2 on the set (8.8.43) are identical to (8.8.42).

(c) *The canonical formalism.* The variational principle $\delta I_2 = 0$ could be used to write (8.8.39), (8.8.40) in canonical form. We define a variable (a "momentum") p as

$$p = \frac{\partial F_2}{\partial \ddot{y}} = -2\frac{(1 + \lambda^2 y^2)^2}{\ddot{y}^3}. \tag{8.8.46}$$

Then, the Hamiltonian (see [107]) function is

$$H_2 = p\ddot{y} - F_2 = -\frac{3}{2^{2/3}} p^{2/3} \left(1 + \lambda^2 y^2\right)^{2/3}. \tag{8.8.47}$$

With (8.8.47) the canonical equations

$$\ddot{y} = \frac{\partial H_2}{\partial p}, \quad \ddot{p} = \frac{\partial H_2}{\partial y} \tag{8.8.48}$$

become

$$\ddot{y} = - \left[2\frac{\left(1 + \lambda^2 y^2\right)^2}{p}\right]^{1/3}, \quad \ddot{p} = -2^{4/3}\lambda^2 y \left[\frac{p^2}{\left(1 + \lambda^2 y^2\right)}\right]^{1/3}, \quad t \in (0, 1). \tag{8.8.49}$$

From (8.8.35) and (8.8.46) we obtain the boundary conditions corresponding to the system (8.8.49) as

$$y(0) = 0, \quad \dot{y}(0) = 0, \quad p(1) = 0, \quad \dot{p}(1) = 0. \tag{8.8.50}$$

Consider the space $\mathbf{W}_3$:

$$\mathbf{W}_3 = \left\{\mathbf{w} = (y, p) : y \in C^2(0, 1), \quad y(0) = 0, \quad \dot{y}(0) = 0, \right.$$

$$\left. p \in C^2(0, 1), \quad p(1) = 0, \quad \dot{p}(1) = 0\right\} \tag{8.8.51}$$

and the problem of determining the minimum on $\mathbf{W}_3$ of the functional

$$I_3 = \int_0^1 F_3 dt, \tag{8.8.52}$$

with

$$F_3 = \dot{y}\dot{p} - \frac{3}{2^{2/3}} p^{2/3} \left(1 + \lambda^2 y^2\right)^{2/3}. \tag{8.8.53}$$

It is easy to see that the condition $\delta I_3 = 0$ reproduces the system (8.8.49). Since F_3 does not depend explicitly on t we have a Jacobi-type first integral for (8.8.49) in the form

$$\dot{y}\dot{p} + \frac{3}{2^{2/3}} p^{2/3} \left(1 + \lambda^2 y^2\right)^{2/3} = const. \tag{8.8.54}$$

We now determine the constant in (8.8.54). By using the boundary conditions (8.8.50) it follows that

$$\dot{y}\dot{p} + \frac{3}{2^{2/3}} p^{2/3} \left(1 + \lambda^2 y^2\right)^{2/3} = C = \frac{3}{2^{2/3}} [p(0)]^{2/3}. \tag{8.8.55}$$

We shall use (8.8.55) to check numerical integration of the system (8.8.49). Namely, we shall, in each step of numerical integration, calculate the left-hand side of (8.8.55) and compare the value so obtained with the constant on the right-hand side.

Finally, we note that with (y, p) known, the cross-sectional area $a(t)$ is determined from the equations (8.8.32) and (8.8.46) so that

$$a = a(t) = \left[\frac{p^2}{2^5 (1 + \lambda^2 y^2)} \right]^{1/3}. \tag{8.8.56}$$

For the optimal rod we now derive another important relation. To this end we multiply $(8.8.49)_1$ by p and integrate to obtain

$$\int_0^1 \ddot{y} p \, dt = - \int_0^1 \left[2 \frac{\left(1 + \lambda^2 y^2\right)^2}{p} \right]^{1/3} dt \tag{8.8.57}$$

or

$$\int_0^1 \dot{y} \dot{p} \, dt = 2^{1/3} \int_0^1 p^{2/3} \left(1 + \lambda^2 y^2\right)^{2/3} dt. \tag{8.8.58}$$

Similarly, by multiplying $(8.8.49)_2$ by y, integrating, and using (8.8.58), we obtain

$$\int_0^1 \dot{y} \dot{p} \, dt = 2^{4/3} \int_0^1 \frac{\left(1 + \lambda^2 y^2\right)^{2/3}}{p^{1/3}} dt = 8w, \tag{8.8.59}$$

where we used (8.8.56) and (8.8.13). Now we use (8.8.58) and (8.8.59) in (8.8.55) to obtain

$$[p(0)]^{2/3} = \frac{20}{3} 2^{2/3} w. \tag{8.8.60}$$

By substituting (8.8.60) in (8.8.56) we finally get

$$w = \frac{3}{10} a(0). \tag{8.8.61}$$

The equation (8.8.60) is used to check the (overall) accuracy of numerical integration of the system (8.8.49), (8.8.50). It also shows that by choosing w we can determine $p(0)$ from (8.8.60), or if $a(0)$ is prescribed then the volume of the optimal rod can be determined from (8.8.61) as $a(0) = (10/3) w$.

The system (8.8.49), (8.8.50) is integrated using the Runge–Kutta double-precision procedure. Note that the point $t = 1$ is a singular point for the system (8.8.49), so that the equation $(8.8.49)_1$ cannot be satisfied at $t = 1$. Thus we proceed as follows: we constructed a sequence of numerical solutions (y_n, p_n), $n = 1, 2, ...$, with $y_n(0) = \dot{y}_n(0) = 0$ and $y_n(t) > 0, \dot{y}_n(t) > 0, p_n(t) < 0, \dot{p}_n(t) > 0$ for $t \in (0, 1)$ and $p_n(1) = -\varepsilon_n, \dot{p}(1) = \delta_n$ with the constants $\varepsilon_n > 0, \delta_n > 0$.

For each solution (y_n, p_n) the values of variables $p_n(0)$ and $\dot{p}_n(0)$ and the cross-sectional area $a_n = a_n(\varepsilon_n, \delta_n)$ are determined (a_n is determined according to (8.8.56)). Then, the optimal cross-sectional area $a(t)$ and corresponding initial values $p(0)$ and $\dot{p}(0)$ are obtained as $\lim a_n$, $\lim p_n(0)$, and $\lim \dot{p}_n(0)$, when $\varepsilon_n \to 0, \delta_n \to 0$, respectively.

We performed the calculations for $\lambda = \sqrt{10}$ and obtained the following values $p(0) = -0.195054200962$, $\dot{p}(0) = 0.37907$. With these values at $t = 1$ we obtained $p(1) = -3.744 \times 10^{-4}$, $\dot{p}(1) = 4.649 \times 10^{-4}$. The accuracy of integration was controlled by evaluation of the first integral (8.8.55) in each step of integration. The left-hand side in (8.8.55) was constant and equal to $\frac{3}{2^{2/3}} \left[0.195054^2 \right]^{1/3} = 0.6356$ up to 10^{-7}.

From (8.8.56) we obtain $a(0) = (p^2(0)/2^5)^{1/3} = 0.10595$. The dimensionless volume w of the optimal rod is determined using (8.8.13) and (8.8.56). For $\lambda = \sqrt{10}$ we have

$$w = \int_0^1 a(t)dt = 0.037111. \tag{8.8.62}$$

The condition (8.8.61) gives $w_{smooth} = \frac{3}{10}a(0) = 0.031781$. This value corresponds to the case when the conditions $p(1) = \dot{p}(1) = 0$ are satisfied exactly. It also shows that one should further increase the accuracy of numerical integration.

We note that in the neighborhood of the point $t = 1$ the functions $y(t)$ and $a(t)$ can be expanded in a series. Thus, suppose that

$$y(t) = b(1-t)^{-c}, \tag{8.8.63}$$

where b and c are constants. By substituting (8.8.63) into $(8.8.34)_2$ we obtain

$$a = \frac{\lambda^2}{2c^2(c+1)^2}(1-t)^4 + \frac{1}{2c^2b^2(c+1)}(1-t)^{4+2c} + \cdots. \tag{8.8.64}$$

For positive c the equation (8.8.64) leads to

$$a = \frac{\lambda^2}{2c^2(c+1)^2}(1-t)^4. \tag{8.8.65}$$

By using (8.8.65) in (8.8.40) we conclude that $0 < c < 5$. Numerical experiments show that $c = 2$, so that (8.8.65) leads to

$$a = \frac{\lambda^2}{72}(1-t)^4. \tag{8.8.66}$$

In numerical experiments we concluded that due to singularity at $t = 1$, the boundary conditions $p(1) = \dot{p}(1) = 0$ is difficult to satisfy with high accuracy (for arbitrary λ) with the smooth solutions of (8.8.49), (8.8.50). Thus, we examine the possibility of the existence of broken extremals with the discontinuity in $\dot{p}(t)$. To do this, we write the functional (8.8.52) as

$$I_4 = \int_0^1 F_4\,dt, \qquad (8.8.67)$$

where $F_4 = y\ddot{p} + \frac{3}{2^{2/3}}p^{2/3}\left(1 + \lambda^2 y^2\right)^{2/3}$. Broken extremals of (8.8.67) must satisfy the Weierstrass–Erdmann corner conditions at the points of discontinuity. For the functional (8.8.52) those conditions are given in section 6.6. For the functional (8.8.67) those conditions read (see [108, p. 231])

$$\left[\frac{\partial F_4}{\partial \ddot{p}}\right]_{t^*} = 0, \qquad \left[\frac{\partial F_4}{\partial \dot{p}} - \frac{d}{dt}\frac{\partial F_4}{\partial \ddot{p}}\right]_{t^*} = 0, \qquad (8.8.68)$$

where $[f]_{t^*} = (f)_{t^*+0} - (f)_{t^*-0}$ denotes the jump of f at $t = t^*$. In our case (8.8.68) becomes

$$[y(t)]_{t=t^*} = 0, \qquad [\dot{y}(t)]_{t=t^*} = 0. \qquad (8.8.69)$$

Thus, if $y(t)$ and $\dot{y}(y)$ are continuous, (8.8.69) is satisfied. The same conditions follow from (6.6.6), which is equal to (8.8.69)$_2$, while (8.8.69)$_1$ is assumed in deriving (6.6.6). Since $\dot{p}(t)$ represents transversal force, it can have value zero only at the end point $t = t^* = 1$, and the condition $[F_4]_{t^*=1} = 0$ is automatically satisfied since $t^* = 1, \delta t^* = 0$. Thus the broken extremal is characterized by the solution of

$$\ddot{y} = -\left[2\frac{\left(1 + \lambda^2 y^2\right)^2}{p}\right]^{1/3}, \qquad \ddot{p} = -2^{4/3}\lambda^2 y\left[\frac{p^2}{\left(1 + \lambda^2 y^2\right)}\right]^{1/3}, \qquad t \in (0,1),$$

$$(8.8.70)$$

subject to the boundary conditions

$$y(0) = 0, \quad \dot{y}(0) = 0, \quad \lim_{t \to 1-0} p(t) = 0, \quad \lim_{t \to 1-0} \dot{p}(t) \neq 0, \quad \dot{p}(1) = 0. \quad (8.8.71)$$

We performed numerical calculations with the same parameters as in the case of smooth solutions. Thus with $\lambda = \sqrt{10}$, we obtained the following values: $p(0) = -0.195054200962, \dot{p}(0) = 0.3791$ giving $p(1) = -6.56 \times 10^{-13}, \lim_{t \to 1-0}\dot{p}(t) = 5.69 \times 10^{-3}$. The volume in this case is $w_{brocken} = 0.037088$. This is the volume of the *optimal* rod for broken extremals since the possibility of the optimal solution with $p(t)$ discontinuous, that is, solution of (8.8.70) with $y(0) = 0, \dot{y}(0) = 0, \lim_{t \to 1-0}p(t) \neq 0, p(1) = 0, \lim_{t \to 1-0}\dot{p}(t) = 0$, leads to $w = 0.371133$, which is larger than the value obtained when $\dot{p}(1)$ is discontinuous.

Thus, the volume of the rod with $a(0) = 0.10595$ corresponding to a smooth extremal determined by the solution of (8.8.49), (8.8.50) has the value $w_{smooth} = \frac{3}{10}a(0) = 0.031781$, which is smaller than the volume corresponding to broken extremals and smooth extremal leads to the *optimal* rod. A cross section of the optimal rod $a(t)$, calculated according to (8.8.56), is shown in Figure 8.8.1.

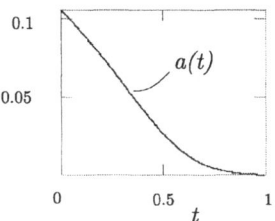

Figure 8.8.1

We show next that with the solution of (8.8.49), (8.8.50) for single λ we can determine the solution for any λ. Let (y, p) be the solution of the problem (8.8.49), (8.8.50) and let $a(t)$ be the corresponding cross-sectional area determined for the specified value of the dimensionless rotation λ. Let $(\hat{y}, \hat{p})$ and $\hat{a}(t)$ be the corresponding functions determined for the dimensionless rotation speed $\Lambda = \beta\lambda$, with the constant β given. By using (8.8.49), (8.8.50), and (8.8.56) it is easy to see that the following relations hold:

$$\hat{y} = \frac{1}{\beta}y, \quad \hat{p} = \beta^3 p, \quad \hat{a} = \beta^2 a. \tag{8.8.72}$$

Thus, with the solution for single λ, we have the solution for any Λ.

We now compare the volume of the optimal rod and the rod with constant circular cross section if both are stable up to the same angular velocity ω. Suppose that both rods are made of the same material, that is, that E, ρ are the same. Also, we assume that both rods lose stability at the same angular velocity ω. For the rod with constant circular cross section we have (see (8.4.5) and [14])

$$\hat{\lambda}^2 = \frac{\rho A \omega^2 L^4}{EI} = \frac{\rho \omega^2 L^4}{\alpha EA} = 12.362, \tag{8.8.73}$$

where $I = \alpha A^2$ with $A = const$. From (8.8.1) we have

$$\lambda^2 = \frac{\rho \omega^2 L^2}{E\alpha} = 10. \tag{8.8.74}$$

With (8.8.73), (8.8.74) we obtain

$$\frac{\lambda^2}{\hat{\lambda}^2} = \frac{A}{L^2} = a(0). \tag{8.8.75}$$

A rod with constant cross section A has the volume $W_{const.} = AL$, while the volume of the optimal rod is $W_{optimal} = wL^3$. By using (8.8.74) we obtain

$$W_{const.} = \frac{\lambda^2}{\hat{\lambda}^2}L^3, \tag{8.8.76}$$

so that

$$\frac{W_{optimal}}{W_{const.}} = \frac{\hat{\lambda}^2}{\lambda^2} w = 0.04589. \tag{8.8.77}$$

Therefore, the volume of the optimal rotating rod, in the example treated here, is just 4.59% of the volume of the rod that has same length and constant cross section! However, if we choose the same cross-sectional area at $t = 0$, then from (8.8.61) we conclude that the ratio $\frac{W_{optimal}}{W_{const.}} = \frac{3}{10}$. Note also that from (8.8.75) the dimensionless cross-sectional area of the rod with constant cross section becomes $a_{const.} = 0.80893L^2$. Therefore, for the circular cross section, the ratio of the radius of the optimal rod for $t = 0$, that is, $r(0)$, and the radius of the rod with constant cross section R is $(r(0)/R) = 0.362$. In Figure 8.8.2 we show the radius of the optimal rod and the rod of the constant cross section that lose stability at the same angular speed.

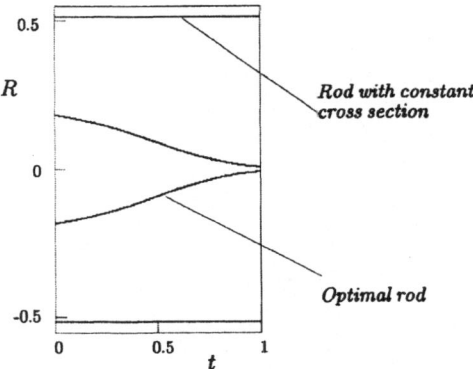

Figure 8.8.2

Finally, we analyze the functional (8.8.52) by the Ritz method. Suppose that the approximate solution is taken in the form

$$Y = C_1 t^3, \quad P = -C_2 (1 - t)^3. \tag{8.8.78}$$

The functional (8.8.52) then becomes

$$\begin{aligned}
I_3(C_1, C_2) &= \int_0^1 F_3 dt \\
&= \int_0^1 \left\{ 3C_1 t^2 3C_2 (1-t)^2 - \frac{3}{2^{2/3}} (C_2)^{2/3} \right. \\
&\qquad \left. \times (1-t)^2 \left(1 + \lambda^2 (C_1)^2 t^6\right)^{2/3} \right\} dt. \tag{8.8.79}
\end{aligned}$$

We plotted I_3 as a function of C_1 and C_2 in Figure 8.8.3. It shows saddle-type behavior (characteristic of a functional that is bilinear in generalized velocities). Minimization of (8.8.79) with respect to C_1 and C_2 leads to $C_1 = 3.6, C_2 = 0.21$.

To show that I_3 is a saddle-type functional we write it as (see (8.8.52), (8.8.53))

$$
\begin{aligned}
I_3 &= \int_0^1 \dot{y}\dot{p} - \frac{3}{2^{2/3}} p^{2/3} \left(1 + \lambda^2 y^2\right)^{2/3} dt \\
&= -\int_0^1 y\ddot{p} + \frac{3}{2^{2/3}} p^{2/3} \left(1 + \lambda^2 y^2\right)^{2/3} dt, \qquad (8.8.80)
\end{aligned}
$$

where we used (8.8.50).

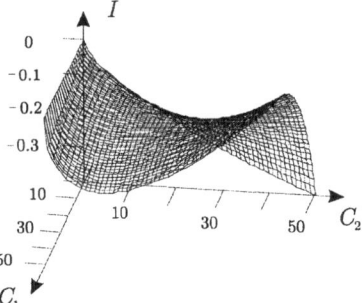

Figure 8.8.3

The functional (8.8.80) is of the type considered in [98, p. 219]. We write the function under the integral sign in (8.8.80) as $-y\ddot{p} - R$, where $R(y,p) = \frac{3}{2^{2/3}} p^{2/3} \left(1 + \lambda^2 y^2\right)^{2/3}$. Note that $-R$ is convex in p (i.e., $\partial^2 R / \partial p^2 > 0$) and concave in y (i.e., $\partial^2 R / \partial y^2 < 0$). Thus, we conclude that I_3 is a saddle functional. The Ritz-type of approximate solution shown in Figure 8.8.3, qualitatively, represents the functional correctly.

A possible generalization of the problem consists in introducing a constant axial force at the end of the rod. This load configuration is shown in the Figure 8.8.4. To the equilibrium and constitutive equations (8.4.1), (8.4.2) we adjoin the following boundary conditions

$$
\bar{x}(0) = 0, \quad \bar{y}(0) = 0, \quad V(L) = 0, \quad H(L) = -F_0, \quad M(L) = 0. \qquad (8.8.81)
$$

By using the dimensionless variables and parameters (8.8.1),

$$
\lambda_1 = \left(\frac{\rho \omega^2 L^2}{E\alpha}\right)^{1/2}, \quad \lambda_2 = \frac{F_0}{E\alpha L^2}, \qquad (8.8.82)
$$

and a new dependent variable

$$
u = -\frac{v}{\lambda_1}, \qquad (8.8.83)
$$

we obtain

$$\dot{u} = \lambda_1 a y, \quad \dot{m} = -\lambda_1 u \cos\theta + \lambda_2 \sin\theta, \quad \dot{y} = \sin\theta, \quad \dot{\theta} = -\frac{m}{a^2} \qquad (8.8.84)$$

subject to

$$u(1) = 0, \quad m(1) = 0, \quad y(0) = 0, \quad \theta(0) = 0. \qquad (8.8.85)$$

The system (8.8.84), (8.8.85) possesses a trivial solution, in which the axis of the rod remains straight for any value of the dimensionless rotation speed λ_1 and all values of the dimensionless compressive force λ_2 in the form

$$u_0 = 0, \quad m_0 = 0, \quad y_0 = 0, \quad \theta_0 = 0. \qquad (8.8.86)$$

The linearization of (8.8.84) about trivial solution (8.8.86) leads to

$$\left(\frac{u}{a}\right)^{\cdot} = \lambda_1\theta, \qquad \left(\dot{\theta}a^2\right)^{\cdot} = \lambda_1 u - \lambda_2\theta, \qquad \dot{y} = \theta, \qquad (8.8.87)$$

subject to (8.8.8). The dimensionless volume of the rod is, again, given by (8.8.13). If $a(t)$ in (8.8.87) is given, then the values of $(\lambda_1, \lambda_2) \in R^2$ for which (8.8.87), (8.8.8) have nontrivial solution define a set of curves C_n, $n = 1, 2, ..$ (see [5]), called the interaction curves. Let $(\lambda_1^1, \lambda_2^1)$ be a point on the lowest interaction curve C_1 (i.e., a curve corresponding to the first buckling mode). Then a straight line connecting the point $(0,0)$ with the point $(\lambda_1^1, \lambda_2^1)$ does not intersect any other interaction curve.

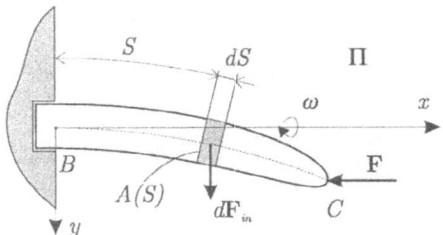

Figure 8.8.4

Suppose now that (λ_1, λ_2) are given. We define *the lightest compressed rotating rod* as the rod so shaped that any other rod of the same length and smaller volume will buckle under the given load characterized by (λ_1, λ_2). Thus, the problem of determining the shape of the lightest compressed rotating rod could be stated as follows.

Given (λ_1, λ_2), find $a(t) \in U$, where U is the space of continuous functions, having continuous first derivative on the interval $[0, L]$, such that, the boundary value problem, consisting of differential equations

$$\left(\frac{u}{a}\right)^{\cdot} = \lambda_1\theta,$$

$$\left(\dot{\theta}a^2\right)^{\cdot} = \lambda_1 u - \lambda_2\theta, \qquad (8.8.88)$$

subject to

$$u(1) = 0, \qquad \dot{u}(0) = 0,$$
$$\theta(0) = 0, \qquad \lim_{t \to 1} \dot{\theta}(t)a^2(t) = 0, \qquad (8.8.89)$$

has (λ_1, λ_2) as a point on the lowest interaction curve C_1 and that at the same time, the integral

$$V_0 = \int_0^1 a(t)dt \qquad (8.8.90)$$

is minimal.

Let $x_1, ..., x_4$ be a set of dependent variables defined by (8.8.14). Then, the system (8.8.88), (8.8.89) becomes

$$\dot{x}_1 = ax_2, \quad \dot{x}_2 = \lambda_1 x_3, \quad \dot{x}_3 = \frac{x_4}{a^2}, \quad \dot{x}_4 = \lambda_1 x_1 - \lambda_2 x_3, \qquad (8.8.91)$$

subject to

$$x_1(1) = 0, \quad x_2(0) = 0, \quad x_3(0) = 0, \quad x_4(1) = 0. \qquad (8.8.92)$$

Proceeding as in the previous case, we form the Hamiltonian function H as

$$H = a + p_1 ax_2 + p_2 \lambda_1 x_3 + p_3 \frac{x_4}{a^2} + p_4 [\lambda_1 x_1 - \lambda_2 x_3], \qquad (8.8.93)$$

where the variables $p_1, ..., p_4$ satisfy

$$\dot{p}_1 = -\frac{\partial H}{\partial x_1} = -\lambda_1 p_4, \quad \dot{p}_2 = -\frac{\partial H}{\partial x_2} = -p_1 a,$$
$$\dot{p}_3 = -\frac{\partial H}{\partial x_3} = -\lambda_1 p_2 + \lambda_2 p_4, \quad \dot{p}_4 = -\frac{\partial H}{\partial x_4} = -\frac{p_3}{a^2}, \qquad (8.8.94)$$

subject to

$$p_1(0) = 0, \quad p_2(1) = 0, \quad p_3(1) = 0, \quad p_4(0) = 0. \qquad (8.8.95)$$

The optimality condition $\min_a H(t, x_1, x_2, p_1, p_2, a)$ leads to

$$\frac{\partial H}{\partial a} = 1 + p_1 x_2 - 2p_3 \frac{x_4}{a^3} = 0. \qquad (8.8.96)$$

By solving (8.8.96) for a we obtain

$$a = \left\{ \frac{2p_3 x_4}{1 + p_1 x_2} \right\}^{1/3}. \qquad (8.8.97)$$

A further procedure follows the analysis presented before in this chapter, but with small differences. For example, instead of (8.8.34), (8.8.35) we obtain

$$\left(\ddot{y}a^2 \right)^{\cdot\cdot} = \lambda_1^2 ay - \lambda_2 \ddot{y}, \quad a\ddot{y}^2 = \frac{1}{2} \left[1 + \lambda_1^2 y^2 \right], \qquad (8.8.98)$$

subject to

$$y(0) = 0, \quad \dot{y}(0) = 0, \quad \lim_{t \to 1} \ddot{y}(t)a^2(t) = 0, \quad \lim_{t \to 1} \left\{ [\ddot{y}(t)a^2(t)]^{\cdot} \right\} = -\lambda_2 \dot{y}(1).$$

$$(8.8.99)$$

Note that we can eliminate a from (8.8.98) to obtain the single equation

$$\left[\frac{\left(1 + \lambda_1^2 y^2\right)^2}{\ddot{y}^3} \right]^{\cdot \cdot} = 2\lambda_1^2 y \frac{\left(1 + \lambda_1^2 y^2\right)}{\ddot{y}^2} - 4\lambda_2 \ddot{y}, \qquad (8.8.100)$$

with the boundary conditions

$$y(0) = 0, \quad \dot{y}(0) = 0,$$

$$\lim_{t \to 1} \frac{\left(1 + \lambda_1^2 y^2\right)^2}{\ddot{y}^3} = 0, \quad \lim_{t \to 1} \left\{ \left[\frac{\left(1 + \lambda_1^2 y^2\right)^2}{\ddot{y}^3} \right]^{\cdot} \right\} = -\lambda_2 \dot{y}(1).$$

$$(8.8.101)$$

We can construct a variational principle for (8.8.100), (8.8.101). Thus we consider the space

$$\mathbf{W}_5 = \left\{ y : y \in C^4(0,1), \quad y(0) = \dot{y}(0) = 0 \right\} \qquad (8.8.102)$$

and the functional

$$I_5 = \int_0^1 F_5 dt, \qquad (8.8.103)$$

with Lagrangian

$$F_5 = \frac{\left(1 + \lambda_1^2 y^2\right)^2}{\ddot{y}^2} + 4\lambda_2 \dot{y}^2. \qquad (8.8.104)$$

Then the Euler–Lagrangian equation corresponding to $\delta I_5 = 0$ is equivalent to (8.8.100). Note that the natural boundary conditions for the minimization of I_5 on the set (8.8.102) are identical to (8.8.101). The variational principle $\delta I_5 = 0$ could be used to write (8.8.100) in canonical form. We define a variable (a "hypermomentum"; see [107]) r as

$$r = \frac{\partial F_5}{\partial \ddot{y}} = -2 \frac{(1 + \lambda_1^2 y^2)^2}{\ddot{y}^3}. \qquad (8.8.105)$$

The "modified" momentum now becomes

$$p = \frac{\partial F_5}{\partial \dot{y}} - \frac{dr}{dt} = 8\lambda_2 \dot{y} + \left(2 \frac{(1 + \lambda_1^2 y^2)^2}{\ddot{y}^3} \right)^{\cdot}, \qquad (8.8.106)$$

so that the Hamiltonian function (see [107, p. 237]) is

$$H = p\ddot{y} + \dot{y}r - F_5 = \dot{y}\left(2\frac{(1+\lambda_1^2 y^2)^2}{\ddot{y}^3}\right)^{\cdot} - 3\frac{(1+\lambda_1^2 y^2)^2}{\ddot{y}^2} - 4\lambda_2 \dot{y}^2. \quad (8.8.107)$$

Since F_5 does not depend on t explicitly, we have

$$H = const. \quad (8.8.108)$$

With (8.8.105), equation (8.8.100) becomes

$$\ddot{y} = -2^{1/3}\frac{(1+\lambda_1^2 y^2)^{2/3}}{r^{1/3}},$$

$$\ddot{r} = -2^{4/3}\lambda_1^2 y\frac{r^{2/3}}{(1+\lambda_1^2 y^2)^{1/3}} + 8\lambda_2 2^{1/3}\frac{(1+\lambda_1^2 y^2)^{2/3}}{r^{1/3}},$$

$$t \in (0,1), \quad (8.8.109)$$

and

$$y(0) = 0; \quad \dot{y}(0) = 0; \quad r(1) = 0; \quad 8\lambda_2 \dot{y}(1) - \dot{r}(1) = 0. \quad (8.8.110)$$

Consider the space $\mathbf{W}_6$,

$$\mathbf{W}_6 = \{\mathbf{w} = (y,r) : y \in C^2(0,1), \quad y(0) = \dot{y}(0) = 0,$$
$$r \in C^2(0,1), \quad r(1) = 0, \quad 8\lambda_2 \dot{y}(1) - \dot{r}(1) = 0\}, (8.8.111)$$

and the problem of determining the minimum on $\mathbf{W}_6$ of the functional

$$I_6 = \int_0^1 F_6 dt, \quad (8.8.112)$$

with

$$F_6 = \dot{y}\dot{r} - \frac{3}{2^{2/3}}r^{2/3}\left(1+\lambda_1^2 y^2\right)^{2/3} - 4\lambda_2 \dot{y}^2. \quad (8.8.113)$$

It is easy to see that the condition $\delta I_6 = 0$ reproduces the system (8.8.109). Again, since F_6 does not depend explicitly on t, we have the Jacobi-type first integral for (8.8.109) in the form

$$\dot{y}\dot{r} + \frac{3}{2^{2/3}}r^{2/3}\left(1+\lambda_1^2 y^2\right)^{2/3} - 4\lambda_2 \dot{y}^2 = const., \quad (8.8.114)$$

which is equivalent to (8.8.108). We now determine the constant in (8.8.114). By using the boundary conditions it follows that

$$\dot{y}\dot{p} - 4\lambda_2 \dot{y}^2 + \frac{3}{2^{2/3}}r^{2/3}\left(1+\lambda_1^2 y^2\right)^{2/3} = C = \frac{3}{2^{2/3}}\left[r(0)\right]^{2/3}. \quad (8.8.115)$$

We shall use (8.8.115) to check numerical integration of the system (8.8.109).

Finally, we note that with (y, r) known, the cross-sectional area $a(t)$ is determined from the equations (8.8.98) and (8.8.105), so that

$$a = a(t) = \left[\frac{p^2}{2^5(1 + \lambda^2 y^2)} \right]^{1/3}. \qquad (8.8.116)$$

The optimal shape of the rod is shown in Figure 8.8.5 for a specific choice of parameters $\lambda_1 = \lambda_2 = \sqrt{10}$. The initial values of the variable r are $r(0) = -11.882130, \dot{r}(0) = 5.490713$. With these values the boundary conditions at $t = 1$ are satisfied as $r(1) = -6.339397 \times 10^{-14}, 8\lambda_2 \dot{y}(1) - \dot{r}(1) = 1.243450 \times 10^{-13}$. As is seen in Figure 8.8.5, the optimal rod is "cigar shaped" at the top. This shape is characteristic for the optimal rod loaded with the concentrated force only (see [59]). The optimal shape of the rotating rod without compressive force at the end has a thin end (see Figure 8.8.2). The dimensionless volume of the optimal rod is $w_{optimal} = 1.085$. We compare this value with the volume of the rod of constant circular cross section that is stable under the same compressive force and angular velocity. Thus in [14, p. 236] it is shown that the critical load parameters $\bar{\lambda}_1 = \frac{\rho A \omega^2 L^4}{EI}, \bar{\lambda}_2 = \frac{F_0 L^2}{EI}$ satisfy

$$2\bar{\lambda}_1 - \bar{\lambda}_2 \bar{\lambda}_1^{1/2} \sin \left(\sqrt{\bar{\lambda}_1 + \left(\frac{\bar{\lambda}_2}{2} \right)^2} + \frac{\bar{\lambda}_2}{2} \right)^{1/2}$$

$$\times \sinh \left(\sqrt{\bar{\lambda}_1 + \left(\frac{\bar{\lambda}_2}{2} \right)^2} - \frac{\bar{\lambda}_2}{2} \right)^{1/2}$$

$$+ \left(2\bar{\lambda}_1 + \bar{\lambda}_2^2 \right) \cos \left(\sqrt{\bar{\lambda}_1 + \left(\frac{\bar{\lambda}_2}{2} \right)^2} + \frac{\bar{\lambda}_2}{2} \right)^{1/2}$$

$$+ \left(2\bar{\lambda}_1 + \bar{\lambda}_2^2 \right) \cos \left(\sqrt{\bar{\lambda}_1 + \left(\frac{\bar{\lambda}_2}{2} \right)^2} + \frac{\bar{\lambda}_2}{2} \right)^{1/2}$$

$$\times \cosh \left(\sqrt{\bar{\lambda}_1 + \left(\frac{\bar{\lambda}_2}{2} \right)^2} - \frac{\bar{\lambda}_2}{2} \right)^{1/2} = 0. \qquad (8.8.117)$$

In the expressions for $\bar{\lambda}_1 = \frac{\rho A \omega^2 L^4}{EI}$ and $\bar{\lambda}_2 = \frac{F_0 L^2}{EI}$ the moment of inertia must be expressed as $I = \alpha A^2$. Using this and (8.8.82) we obtain

$$\bar{\lambda}_1 = \lambda_1^2 \frac{L^2}{A} = \lambda_1^2 \frac{1}{w_{const}}, \quad \bar{\lambda}_2 = \lambda_2 \left(\frac{L^2}{A} \right)^2 = \lambda_2 \frac{1}{(w_{const})^2}, \qquad (8.8.118)$$

where w_{const} is the dimensionless volume of the rod with constant cross section. Using (8.8.118) with $\lambda_1 = \lambda_2 = \sqrt{10}$ in (8.8.117), we obtain $w_{const} = 1.585$.

Thus $w_{optimal}/w_{const} = 1.085/1.585 = 0.684$. We note that for the classical case (see [91], [59]), that is, $\lambda_1 = 0, \lambda_2 = 10$, we obtain $w_{optimal} = 1.743455$, while the solution of (8.8.117) yields $w_{const} = 2.012$. Thus, $w_{optimal}/w_{const} = 1.743/2.012 = 0.866 = \sqrt{3}/2$, as is known (see [91, p. 106]). Thus, in the special case of a rod loaded by compressive force only ($\lambda_1 = 0, \lambda_2 \neq 0$), our results reduce to those obtained before analytically. Numerical solution for the case $\lambda_1 = \sqrt{10}, \lambda_2 = 0$ confirms our results obtained in the first part of this section with one important difference. Namely, the case $\lambda_1 = \sqrt{10}, \lambda_2 = 0$ has $\lim_{t \to 1} \dot{a} = 0$ while for the case $\lambda_1 = \sqrt{10}, \lambda_2 \to 0$ we have $\lim_{t \to 1} \dot{a} = -\infty$, that is, the rod is "cigar shaped" as in the case presented in section 8.6.

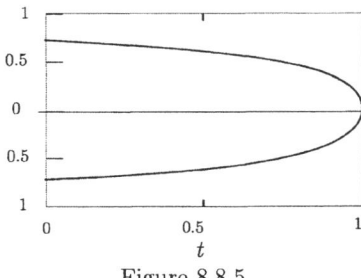

Figure 8.8.5

8.9 Optimal Shape of a Rod Loaded by a Force and a Torque

Next we determine the optimal shape of the compressed and twisted rod analyzed in section 8.5. We shall follow the presentation of [50].

Let $\theta = \theta_0 + \Delta\theta$. By substituting this into (8.5.22) and neglecting the higher order terms, we obtain

$$\left(A_{22}\Delta\theta'\right)' + \left[\frac{M_t^2}{A_{22}} + P\right]\Delta\theta = 0, \tag{8.9.1}$$

subject to

$$\Delta\theta(0) = 0, \quad \Delta\theta(L) = 0. \tag{8.9.2}$$

The existence of a nontrivial solution to (8.9.1), (8.9.2) is a necessary condition for the existence of a nontrivial solution to (8.5.22), (8.5.23). If the eigenvalues of the boundary value problem (8.9.1), (8.9.2) are simple, then this condition is also sufficient. The eigenvalues of (8.9.1), (8.9.2) are simple if $A_{22}(S) \neq 0$ and $\frac{M_t^2}{A_{22}(S)} + P \neq 0$ for $S \in (0, L)$. Thus $A_{22}(S) > 0$ guarantees that the eigenvalues of (8.9.1), (8.9.2) are simple.

Let us assume that $A_{22}(S)$ (the bending stiffness) is given as

$$A_{22}(S) = \alpha E A^2(S), \tag{8.9.3}$$

where $\alpha = const.$, E is the modulus of elasticity, and $A(S)$ is the cross-sectional area. With (8.9.3), equation (8.9.1) becomes

$$\left(A^2 \Delta\theta'\right)' + \frac{1}{\alpha E}\left[\frac{M_t^2}{\alpha E A^2} + P\right]\Delta\theta = 0. \tag{8.9.4}$$

The volume of the rod is given as $V = \int_0^L A(S)\, dS$. If $A(S)$ in (8.9.4) is given, then the values of $(M_t^2, P) \in R^2$ for which (8.9.4),(8.9.2) has nontrivial solution define a set of interaction curves. Let $((M_t^2)_1, (P)_1)$ be a point on the lowest interaction curve so that a straight line connecting the point $(0,0)$ with the point $((M)_1, (P)_1)$ does not intersect any other interaction curve.

Suppose now that (M_t^2, P) is given. As before, we define *the lightest rod* as the rod having the shape that any other rod of the same length (in our case, equal to L) and smaller volume will buckle under a given load characterized by (M_t^2, P). Thus, the problem of determining the shape of the lightest rod could be stated as follows.

Given (M_t^2, P), find $A(S) \in U$ such that

$$\left(A^2 \Delta\theta'\right)' + \frac{1}{\alpha E}\left[\frac{M_t^2}{\alpha E A^2} + P\right]\Delta\theta = 0 \tag{8.9.5}$$

subject to

$$\Delta\theta(0) = 0, \quad \Delta\theta(L) = 0 \tag{8.9.6}$$

has (M_t^2, P) as a point on the lowest interaction curve C_1 and such that at the same time, the integral

$$V = \int_0^L A(S)dS \tag{8.9.7}$$

is minimal.

The set U consists of admissible cross-sectional area functions. We assume that it is the space of continuous functions, having continuous first derivative on the interval $[0, L]$, that is, $U = C^1[0, L]$. Thus, we do not impose any restriction on the minimal admissible cross section.

Introducing the dimensionless quantities

$$a = \frac{A}{L^2}, \quad v = \frac{V}{L^3}, \quad \lambda_1 = \frac{M_t}{\alpha E L^3}, \quad \lambda_2 = \frac{P}{\alpha E L^2}, \quad t = \frac{S}{L}, \tag{8.9.8}$$

in (8.9.5), (8.9.6) we obtain

$$\left(a^2 \dot{u}\right)' + \left[\frac{\lambda_1^2}{a^2} + \lambda_2\right]u = 0 \tag{8.9.9}$$

subject to

$$u(0) = 0, \quad u(1) = 0, \tag{8.9.10}$$

where $\Delta\theta = u, (\cdot)^{\cdot} = \frac{d}{dt}(\cdot)$. The dimensionless volume now reads

$$v = \int_0^1 a(t)\,dt. \tag{8.9.11}$$

Now we state the optimization problem as follows. Find the cross-sectional area $a(t)$ so that $0 < a(t) < \infty$, $t \in (0,1)$, for which the optimality criterion

$$v = \int_0^1 a(t)\,dt \tag{8.9.12}$$

attains minimum value and the constraints are given in the form (8.9.9), (8.9.10).
    With new variables

$$x_1 = u, \quad x_2 = a^2\dot{u}, \tag{8.9.13}$$

the governing differential equations become

$$\dot{x}_1 = \frac{x_2}{a^2}, \quad \dot{x}_2 = -\left[\frac{\lambda_1^2}{a^2} + \lambda_2\right]x_1, \tag{8.9.14}$$

subject to

$$x_1(0) = x_1(1) = 0. \tag{8.9.15}$$

For the system (8.9.12)–(8.9.14) the Hamiltonian function H could be easily constructed as

$$H = a + p_1\frac{x_2}{a^2} - p_2 x_1\left[\frac{\lambda_1^2}{a^2} + \lambda_2\right], \tag{8.9.16}$$

where the generalized momenta p_1 and p_2 have to satisfy

$$\dot{p}_1 = -\frac{\partial H}{\partial x_1} = p_2\left[\frac{\lambda_1^2}{a^2} + \lambda_2\right], \quad \dot{p}_2 = -\frac{\partial H}{\partial x_2} = -\frac{p_1}{a^2}, \tag{8.9.17}$$

subject to

$$p_2(0) = p_2(1) = 0. \tag{8.9.18}$$

The optimality condition leads to

$$\frac{\partial H}{\partial a} = 1 - 2p_1\frac{x_2}{a^3} + 2\lambda_1^2 p_2\frac{x_1}{a^3} = 0. \tag{8.9.19}$$

Solving (8.9.19) for a^3, we obtain

$$a^3 = 2\left[p_1 x_2 - \lambda_1^2 p_2 x_1\right]. \tag{8.9.20}$$

The solution p_1, p_2 of the system (8.9.17), (8.9.18) could be obtained from the solution x_1, x_2 of the system (8.9.14), (8.9.15) if we set

$$p_1 = x_2, \quad p_2 = -x_1. \tag{8.9.21}$$

With (8.9.21), equation (8.9.20) becomes

$$a^3 = 2\left[x_2^2 + \lambda_1^2 x_1^2\right]. \tag{8.9.22}$$

By using (8.9.21), the second derivative of H with respect to a becomes

$$\frac{\partial^2 H}{\partial a^2} = 6\frac{x_2^2}{a^4} + 6\lambda_1^2 \frac{x_1^2}{a^4} > 0. \tag{8.9.23}$$

Thus, H attains a minimum. Differentiating (8.9.22) using (8.9.14) and integrating, we obtain

$$a = -\frac{2}{3}\lambda_2 x_1^2 + C, \tag{8.9.24}$$

where C is an arbitrary constant. By observing (8.9.15)$_1$ we obtain $C = a(0)$. With this value substituted in (8.9.24), it follows that

$$\lambda_2 x_1^2 = \frac{3}{2}\left[a(0) - a\right]. \tag{8.9.25}$$

Substituting (8.9.25) into (8.9.22) and solving the result for $\lambda_2 x_2^2$, we get

$$\lambda_2 x_2^2 = \lambda_2 \frac{a^3}{2} - \frac{3\lambda_1^2}{2}\left[a(0) - a\right]. \tag{8.9.26}$$

Note that (8.9.25) implies $a(0) \geq a(t)$. We shall analyze the symmetric deformation mode, so that $\dot{u}(1/2) = 0$. This condition implies

$$x_2(1/2) = 0. \tag{8.9.27}$$

By substituting (8.9.27) into (8.9.26) calculated at $t = 1/2$, we obtain

$$\frac{3\lambda_1^2}{2}a(0) = \frac{3\lambda_1^2}{2}a(1/2) + \frac{\lambda_2}{2}a^3(1/2). \tag{8.9.28}$$

Finally, by using (8.9.28) in (8.9.26), we have

$$\lambda_2 x_2^2 = \frac{3\lambda_1^2}{2}\left[a - a(1/2)\right] + \frac{\lambda_2}{2}\left[a^3 - a^3(1/2)\right]. \tag{8.9.29}$$

Now, from (8.9.29), we conclude that $a \geq a(1/2)$. Therefore,

$$a(1/2) \leq a(t) \leq a(0). \tag{8.9.30}$$

Now we make an important assumption. Namely, in the analysis that follows, we assume that the couple $M_t \neq 0$, so that

$$\lambda_1 > 0. \tag{8.9.31}$$

Next we show that $a(t) > 0, t \in [0, 1/2]$. Suppose $a(t_1) = 0$ for $t_1 \in [0, 1/2]$. From $(8.9.13)_2$ we conclude that $x_2(t_1) = 0$. By using this in (8.9.26) we have that $a(0) = 0$. This, together with (8.9.30) implies $a(t) = 0, t \in [0, 1/2]$, which contradicts (8.9.22). Combining this with the symmetry assumption, we conclude that $a(t) > 0, t \in [0, 1]$. Therefore, the boundary value problem (8.9.9), (8.9.10) has simple eigenvalues. We determine now the cross-sectional area of the optimal rod.

(a) Suppose that $\lambda_2 = 0$. From (8.9.25) we obtain

$$a(t) = a(0) = const. \tag{8.9.32}$$

Using this in (8.9.9), (8.9.10) we have

$$\ddot{u} + \frac{\lambda_1^2}{a^4(0)} u = 0, \tag{8.9.33}$$

subject to

$$u(0) = 0, \quad u(1) = 0. \tag{8.9.34}$$

Thus

$$a(0) = \left(\frac{\lambda_1}{\pi}\right)^{1/2}. \tag{8.9.35}$$

(b) Suppose that $\lambda_2 > 0$. By differentiating (8.9.25), we obtain

$$2\lambda_2 x_1 \dot{x}_1 = -\frac{3}{2}\dot{a}. \tag{8.9.36}$$

In (8.9.36) we substitute $\dot{x}_1$ from $(8.9.14)_1$, and in the resulting expression we substitute x_1, x_2 from (8.9.25), (8.9.26), respectively. The result is

$$\dot{a} = -\frac{2}{a^2\sqrt{3}} \sqrt{[a(0) - a]\{\lambda_2 a^3 - 3\lambda_1^2[a(0) - a]\}}. \tag{8.9.37}$$

In writing (8.9.37) we choose the minus sign for $\dot{a}$ in accordance with (8.9.30). The connection between $a(0)$ and $a(1/2)$ follows from (8.9.28) and reads

$$a(0) = a(1/2)\left[1 + \frac{\lambda_2}{3\lambda_1^2}a^2(1/2)\right]. \tag{8.9.38}$$

By separating variables in (8.9.37) we obtain, after integration,

$$\frac{1}{2} = \frac{\sqrt{3}}{2} \int_{a(1/2)}^{a(0)} \frac{a^2 da}{\sqrt{[a(0) - a]\{\lambda_2 a^3 - 3\lambda_1^2[a(0) - a]\}}}, \tag{8.9.39}$$

where $a\,(0)$ is given by (8.9.38). For given λ_1, λ_2 we determine $a\,(1/2)$ and $a\,(0)$ from (8.9.38), (8.9.39) and then with $a\,(0)$ known, we separate variables and integrate (8.9.37) to obtain

$$t = -\frac{\sqrt{3}}{2} \int_{a(0)}^{a} \frac{a^2\,da}{\sqrt{[a\,(0) - a]\,\{\lambda_2 a^3 - 3\lambda_1^2\,[a\,(0) - a]\}}}. \tag{8.9.40}$$

Finally, from (8.9.40) we determine the shape of the optimal rod.

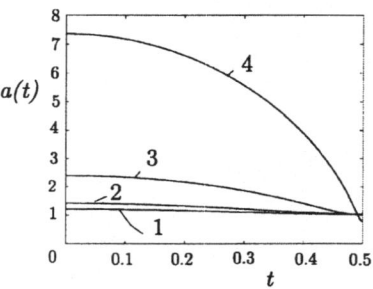

Figure 8.9.1

Now we present the results obtained by numerical integration of corresponding equations. First, we choose $\lambda_1 = \pi$ and then determine $a\,(0)$ and $a\,(1/2)$ for a few values of λ_2 by solving the system (8.9.38), (8.9.39). The results are shown in Table 8.9.1. It is evident from Figure 8.9.1 and Table 8.9.1 that the rod has the smallest cross section in the middle.

Table 8.9.1

(λ_1, λ_2)	$(\pi, 1)$	$(\pi, 5)$	$(\pi, 10)$	$(\pi, 40)$	$(\pi, 400)$
$a\,(0)$	1.042	1.213	1.419	2.400	7.357
$a\,(1/2)$	1.008	1.029	1.040	1.010	0.786

This cross section is different from zero, and that guarantees that the eigenvalues of (8.9.9), (8.9.10) are simple. The shape of the optimal rod, that is, $a\,(t)$ for $t \in (0, 1)$, follows from (8.9.40). The cross section of the optimal rod, shown in Figure 8.9.1, is given for $\lambda_1 = \pi$ and for four values of λ_2. Namely, λ_2 takes values $\lambda_2 = 5, \lambda_2 = 10, \lambda_2 = 40, \lambda_2 = 400$, and the corresponding curves in Figure 8.9.1, are denoted as 1, 2, 3, 4, respectively.

The behavior of the optimal rod near the point $t = 1/2$ is shown in Figure 8.9.2. Note that in accordance with (8.9.37), (8.9.38) the cross section of the optimal rod satisfies $\dot{a}(0) = 0$ and $\dot{a}(1/2) = 0$. From Figure 8.9.2 and Table 8.9.1 an interesting property of the optimal rod follows. Namely, if we fix λ_1 and increase λ_2, the cross section of the optimal rod at the middle point of the rod $(t = 1/2)$ first increases and then decreases.

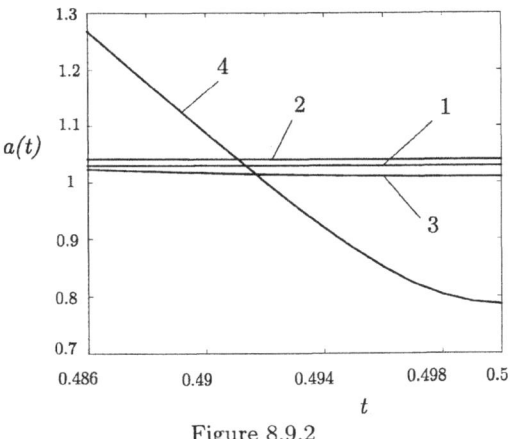

Figure 8.9.2

8.10 Variational Principle for Small Deformation Imposed on Large Deformation of a Rod

In this section we shall formulate a variational principle for small deformations imposed on large ones for elastic rods. In the context of three-dimensional elasticity for the theory of small deformations imposed on large ones, see [70]. The motivation for the principle that we will formulate is given in [110], where mechanical systems with a finite number of degrees of freedom were treated (see also section 1.4.3).

Consider the elastic rod shown in Figure 8.2.1. Suppose that the rod has constant cross section so that the differential equation (8.2.7) becomes

$$\frac{d^2\theta}{dt^2} + \lambda \sin\theta = 0, \qquad (8.10.1)$$

where $\lambda = FL^2/EI$ and $t = S/L$, subject to

$$\frac{d\theta}{dt}(0) = 0, \quad \theta(1) = 0. \qquad (8.10.2)$$

Equation (8.10.1) is of the type (1.4.78) with $a_{ij} = 1, i, j = 1, q = \theta, [jm, i] = 0, Q_i = -\lambda \sin\theta = -\partial\Pi/\partial\theta$ with $\Pi = 1 - \lambda \cos\theta$. Suppose that for given $\lambda = \lambda_0$ the solution to (8.10.1), (8.10.2) is denoted by θ_0, so that

$$\frac{d^2\theta_0}{dt^2} + \lambda_0 \sin\theta_0 = 0. \qquad (8.10.3)$$

We assume that θ_0 is small (in bifurcation, theory from trivial configuration $\theta_0 = 0$; see (8.2.9) and (8.3.8)). Suppose further that λ is perturbed, so that

$\lambda = \lambda_0 + \Delta\lambda$, and $\Delta\lambda$ is small. The solution of (8.10.1), (8.10.2) for this new λ we denote by $\theta = \theta_0 + \xi$. The function $\xi(t)$ represents a perturbation that measures the additional deformation to which the rod is subjected. We call it an *imposed* deformation, and we assume that it is small. By substituting $\lambda = \lambda_0 + \Delta\lambda$ and $\theta = \theta_0 + \xi$ in (8.10.1), (8.10.2) and neglecting the terms of the order higher than linear in $\Delta\lambda$, θ_0, and ξ, we obtain[23]

$$\frac{d^2\xi}{dS^2} + \lambda_0 \left(\cos\theta_0\right)\xi = 0, \tag{8.10.4}$$

subject to

$$\frac{d\xi}{dt}(0) = 0, \quad \xi(1) = 0. \tag{8.10.5}$$

The system (8.10.4), (8.10.5) determines the perturbation ξ. In [110] and in section 1.4.3 a variational principle was formulated that reproduces (in the context of analytical mechanics of a system with finite number of degrees of freedom) *both* equations of basic motion and equations for perturbations. Following this idea in [18], a variational principle reproducing (8.10.1), (8.10.2) *and* (8.10.4), (8.10.5) is formulated. It is easy to see that the Euler–Lagrangian equations for the functional (compare with (1.4.92))

$$I_1(\theta, \xi, \lambda) = \int\limits_0^1 \left[\frac{d\theta}{dt}\frac{d\xi}{dt} - \lambda\xi\sin\theta\right]dt, \tag{8.10.6}$$

where $\theta \in \mathbf{X}$ and $\xi \in \mathbf{X}$, with

$$\mathbf{X} = \left\{\theta: \quad \frac{d\theta(0)}{dt} = 0, \quad \theta = 0\right\}, \tag{8.10.7}$$

are

$$\frac{d^2\theta}{dt^2} + \lambda\sin\theta = 0, \qquad \frac{d^2\xi}{dt^2} + \lambda(\cos\theta)\xi = 0. \tag{8.10.8}$$

Thus, on the solution of (8.10.1), (8.10.2), (8.10.4), (8.10.5), the functional (8.9.6) is stationary.

The functional (8.6.10) has a value equal to zero on the solution (8.10.1), (8.10.2). To see this, integrate the first term under the integral sign partially and use boundary conditions to obtain

$$I_1(\theta, \xi, \lambda) = \int\limits_0^1 \left[-\frac{d^2\theta}{dt^2} - \lambda\sin\theta\right]\xi\,dt. \tag{8.10.9}$$

---

[23] In the case when $\lambda_0 > \lambda_{bifur}$, where λ_{bifur} is a value of λ at the bifurcation point $(\lambda_{bifur}, \theta_0)$ equation (8.10.4) holds for arbitrary θ_0, since in that case $\Delta\lambda = 0$, that is, we ask for different solution for the same value of λ_0.

Thus, $I_1(\theta_0, \xi, \lambda) = 0$. Further, we can interpret I_1 in the context of a weak solution of the problem (8.10.1), (8.10.2). Namely, a weak solution of (8.10.1) is defined as a function $\theta \in \mathbf{X}$ for which (see [81], [33])

$$\int\limits_0^1 \left[\frac{d\theta}{dt} \frac{d\phi}{dt} + \lambda \phi \sin \theta \right] dt = 0 \qquad (8.10.10)$$

for all $\phi \in \mathbf{Y}$. Here $\mathbf{X}$ are $\mathbf{Y}$ linear spaces of functions to which the solution belongs and linear spaces of test functions (if $\mathbf{X}$ is the space of continuous functions satisfying (8.10.2) having the first and second derivative continuous in the interval $(0,1)$, then (8.10.10) defines a classical solution). Therefore, the functional I_1 is the functional defining the weak solution of (8.10.1). In it *both* functions, θ from the solution set and ϕ from the set of test functions, are subject to variation.

Note also that the Lagrangian in (8.10.6) does not depend explicitly in time, so that the following Jacobi-type first integral exists (see (1.4.45)):

$$\frac{d\theta}{dt} \frac{d\xi}{dt} + \lambda \xi \sin \theta = const. \qquad (8.10.11)$$

The variational principle (8.10.6) could be used to study the bifurcation of the trivial solution of (8.10.1), (8.10.2). In that case $\theta_0 = 0, \Delta\lambda \neq 0$, and the procedure is presented in [18].

For the problem (8.3.4), (8.3.5) we have

$$\frac{d^2\phi}{dt^2} + \lambda \left[\sin \phi - \frac{\lambda}{2}(\alpha - \beta) \sin 2\phi \right] = 0, \qquad (8.10.12)$$

subject to

$$\frac{d\phi(0)}{dt} = 0, \quad \frac{d\phi(1)}{dt} = 0. \qquad (8.10.13)$$

We consider the functional

$$I_2 = \int\limits_0^1 \left\{ \frac{d\phi}{dt} \frac{d\xi}{dt} - \lambda \xi \left[\sin \phi - \frac{\lambda}{2}(\alpha - \beta) \sin 2\phi \right] \right\} dt. \qquad (8.10.14)$$

It is easy to see that the Euler–Lagrangian equations for (8.10.14) are

$$\frac{d^2\phi}{dt^2} + \lambda \left[\sin \phi - \frac{\lambda}{2}(\alpha - \beta) \sin 2\phi \right] = 0,$$

$$\frac{d^2\xi}{dt^2} + \lambda[\cos \phi - \lambda(\alpha - \beta) \cos 2\phi]\xi = 0, \qquad (8.10.15)$$

with boundary conditions for ξ:

$$\frac{d\xi(0)}{dt} = 0, \quad \frac{d\xi(1)}{dt} = 0. \qquad (8.10.16)$$

The first integral for (8.10.15) is

$$\frac{d\phi}{dt}\frac{d\xi}{dt} + \lambda\xi\left[\sin\phi - \frac{\lambda}{2}(\alpha - \beta)\sin 2\phi\right] = const., \qquad (8.10.17)$$

and we have

$$I_2 = \int_0^1 \left\{\frac{d\phi}{dt}\frac{d\xi}{dt} - \lambda\xi\left[\sin\phi - \frac{\lambda}{2}(\alpha - \beta)\sin 2\phi\right]\right\} dt = 0 \qquad (8.10.18)$$

on the solution of (8.10.12).

Finally, we consider the rotating rod problem described by (8.4.10), (8.4.11):

$$\frac{d^2u}{dt^2} - \lambda\sin\theta = 0, \qquad \frac{d^2\theta}{dt^2} - \lambda u\cos\theta = 0, \qquad (8.10.19)$$

subject to

$$\frac{du(0)}{dt} = 0, \quad u(1) = 0, \quad \theta(0) = 0, \quad \frac{d\theta(1)}{dt} = 0. \qquad (8.10.20)$$

Let η and ξ be the perturbations of u and θ, respectively. Then the equations for (ξ, η) read

$$\frac{d^2\eta}{dt^2} - \lambda\xi\cos\vartheta = 0, \quad \frac{d^2\xi}{dt^2} - \lambda[\eta\cos\vartheta - u\xi\sin\vartheta] = 0 \qquad (8.10.21)$$

and satisfy

$$\frac{d\eta(0)}{dt} = 0, \quad \eta(1) = 0, \quad \xi(0) = 0; \quad \frac{d\xi(1)}{dt} = 0. \qquad (8.10.22)$$

Consider the functional

$$I_2 = \int_0^1 \left[\frac{d\vartheta}{dt}\frac{d\xi}{dt} + \frac{du}{dt}\frac{d\eta}{dt} + \lambda\eta\sin\vartheta + \lambda u\xi\cos\vartheta\right] dt. \qquad (8.10.23)$$

The Euler–Lagrangian equations for the functional (8.10.23) read

$$\frac{d^2u}{dt^2} - \lambda\sin\theta = 0, \qquad \frac{d\theta}{dt^2} - \lambda u\cos\theta = 0,$$

$$\frac{d^2\xi}{dt^2} - \lambda[\eta\cos\theta - u\xi\sin\theta] = 0, \qquad \frac{d^2\eta}{dt^2} - \lambda\xi\cos\theta = 0. \qquad (8.10.24)$$

Thus, we reproduce both (8.10.19) and (8.10.21). Finally, the system (8.10.24) possesses a first integral of the type

$$\frac{d\theta}{dt}\frac{d\xi}{dt} + \frac{du}{dt}\frac{d\eta}{dt} - \lambda\eta\sin\theta - \lambda u\xi\cos\theta = const. \qquad (8.10.25)$$

Also, the relation

$$I_2 = \int_0^1 \left[\frac{d\theta}{dt}\frac{d\xi}{dt} + \frac{du}{dt}\frac{d\eta}{dt} + \lambda\eta\sin\theta + \lambda u\xi\cos\theta \right] dt = 0 \qquad (8.10.26)$$

holds on the solution of the boundary value problems defined by (8.10.19), (8.10.20) and (8.10.21), (8.10.22).

Bibliography

[1] B. Abraham-Shrauner, Lie group symmetries and invariants of the Henon-Heils equations. *J. Math. Phys.*, **31**, 1627–1631 (1990).

[2] H. Airault, Existence and construction of quadratic invariants for the equation $(\frac{d^2}{d\xi^2})v = \lambda(\xi)(\frac{d}{d\xi})v + F(\xi, v)$. *Int. J. Non-Linear Mechanics*, **21**, 197–203 (1986).

[3] H. Airault, Polynomial invariants for the equation $\frac{d^2}{d\xi^2}v = \lambda(\xi)\frac{d}{d\xi}v + F(\xi, v)$. *Int. J. Non-Linear Mechanics*, **21**, 331–339 (1986).

[4] K.-H. Anthony, A new approach to thermodynamics of irreversible processes by means of Lagrange-formalism. In *Disequilibrium and Self-Organisation* (C.W. Kilmister editor) pp. 75–92. D. Reidel, Dordrecht, the Netherlands, 1986.

[5] S. S. Antman, *Nonlinear Problems of Elasticity.* Springer-Verlag, New York, 1995.

[6] P. Appell, *Traité de mécanique rationnelle, Tome deuxiéme.* Gauthier-Villars, Paris, 1953.

[7] V. I. Arnold, *Mathematical Methods of Classical Mechanics.* Springer-Verlag, New York, 1980.

[8] A. M. Arthurs, *Complementary Variational Principles.* Clarendon Press, Oxford, UK, 1980.

[9] T. M. Atanackovic, The brachistrochrone for a material point with arbitrary initial velocity. *American J. Physics*, **46**, 1274–1275 (1978).

[10] T. M. Atanackovic and B. S. Baclic, A new time invariant for heat conduction with finite wave speed. *Zeitschrift für Ang. Mathematik und Mechanik (ZAMM)*, **60**, 168–169 (1980).

[11] T. M. Atanackovic, On conservation laws for continuous bodies. *Acta Mechanica*, **38**, 157–167 (1981).

[12] T. M. Atanackovic, Variational principles for column buckling. *IMA Journal of Appl. Math.*, **27**, 221–228 (1981).

[13] T. M. Atanackovic and M. Achenbach, Stability of an extensible rotating rod. *Continuum Mech. and Thermodynamics*, **1**, 81–95 (1989).

[14] T. M. Atanackovic, *Stability Theory of Elastic Rods*. World Scientific, River Edge, NJ, 1997.

[15] T. M. Atanackovic, On the rotating rod with shear and extensibility. *Continuum Mech. and Thermodynamics*, **9**, 143–153 (1997).

[16] T. M. Atanackovic and S. S. Simic, On the optimal shape of Pflüger column. *Eur. J. Mech., A/Solids*, **18**, 903–913 (1999).

[17] T. M. Atanackovic, On the optimal shape of the rotating rod. *Juornal of Applied Mechanics (Transactions of ASME)* **68**, 860–864 (2001).

[18] T. M. Atanackovic, On a Vujanovic-type variational principle with application to rod theory. *Q. Jl. Mech. Appl. Math.*, **54**, 1–11 (2001).

[19] T. M. Atanackovic and V. B. Glavardanov, Buckling of a twisted and compressed rod. *International Journal of Solids and Structures*, **39**, 2987–2999 (2002).

[20] T. M. Atanackovic, Stability bounds and optimal shape of elastic rods. In *Modern Problems of Structural Stability* (A. Seyranian and I. Elishakoff, editors), pp.1–56. Springer-Verlag, Wien, 2002.

[21] B. S. Baclic and B. D. Vujanovic, The Hamilton-Jacobi method for arbitrary rheo-linear dynamical systems. *Acta Mechanica*, **163**, 51–79 (2003).

[22] D. C. Barnes, Some remarks on a result by Atanackovic, *IMA Journal of Appl. Math.*, **29**, 271–273 (1982).

[23] P. B. Beda, A. Steindl, and H. Troger, Postbuckling of a twisted prismatic rod under terminal thurst. *Dynamics and Stability of Systems*, **7**, 219–232 (1992).

[24] M. A. Biot, *Variational Principles in Het Transfer*. Clarendon Press, Oxford, UK, 1970.

[25] F. Brezzi, J. Descloux, J. Rappaz, and B. Zwahlen, On the rotating beam: Some theoretical and numerical results. *Calcolo*, **21**, 345–367 (1984).

[26] A. E. Bryson and Y. C. Ho, *Applied Optimal Control*, Hemisphere, New York, 1975.

[27] J. Cisło, J. T. Lopuszański, and P. C. Stichel, On the inverse variational problem in classical mechanics. In *Particles, Fields and Gravitation* (J. Rembieliński, editor), pp. 219–225. American Institute of Physics, Woodbury, New York, 1998.

[28] Ph. Clément and J. Descloux, A variational approach to a problem of rotating rods. *Arch. Rational Mech. Anal.*, **114**, 1–13 (1991).

[29] L. Collatz, *Eigenwertaufgaben mit Technischen Anwendungen*, Geest and Portig, Leipzig, 1963 (Russian translation, Nauka, Moscow, 1968).

[30] R. Courant, *Methods of Mathematical Physics, Vol. II, Partial Differential Equations.* Interscience Publishing, New York, 1962.

[31] S. J. Cox and M. L. Overton, On the optimal design of columns against buckling. *SIAM J. Mathematical Analysis*, **23**, 287–325 (1992).

[32] S. J. Cox, and C. M. McCarthy, The shape of the tallest column. *SIAM J. Mathematical Analysis*, **29**, 547–554 (1998).

[33] R. F. Curtain and A. J. Pritchard, *Functional Analysis in Modern Applied Mathematic*, Academic Press, New York, 1977.

[34] Dj. S. Djukic, On a generalized Lagrange equations of the second kind. *PMM,* **37**, 156–159 (1973).

[35] Dj. S. Djukic, A procedure for finding first integrals of mechanical systems with gauge-variant Lagrangians. *Int. J. Non-Linear Mechanics*, **8**, 479–488 (1973).

[36] Dj. S. Djukic, Conservation laws in clasical mchanics for quasi-coordinates. *Arch. Rational Mech. Analysis*, **56**, 79–98 (1974).

[37] Dj. Djukic and B. D. Vujanovic, Noether's theory in classical nonconservative mechanics. *Acta Mechanica*, **23**, 17–27 (1975).

[38] Dj. S. Djukic and T. Sutela, Integrating factors and conservation laws on non-conservative dynamical systems. *Int. J. Non-Linear Mechanics*, **19**, 331–339 (1984).

[39] V. V. Dobronravov, *Foundations of Mechanics of nonholonomic Systems (Osnovi Mehaniki Negolonomnih Sistem).* Vishaja Skhola, Moscow, 1970.

[40] V. V. Eliseyev, The nonlinear dynamics of elastic rods. *Prikl. Matem. Mekhan.*, **52**, 635–641, 1988.

[41] N. W. Evans, On Hamiltonian systems of two degrees of freedom with invariants in the momenta of the form $p_1^2 p_2^2$. *J. Math. Phys.*, **31**, 600–604 (1990).

[42] Mei Fengxiang, A Field method for solving the equations of motion of nonholonomic systems. *Acta Mechanica Sinica*, **5**, 260–268 (1989).

[43] Mei Fengxiang, Parametric equations of nonholonomic nonconservative systems in the event space and the method of their integration. *Acta Mechanica Sinica*, **6**, 160–168 (1990).

[44] Mei Fengxiang, Generalized Whittaker equations and field method in generalized classical mechanics. *Applied Mathematics and Mechanics (Engish Edition)*, **11**, 569–576 (1990).

[45] Mei Fengxiang, On one method of integration of the equations of motion of nonholonomic systems with the higher order constraints. *PMM*, **55**, 691–695 (1991).

[46] Mei Fengxiang, A field method for integrating the equations of motion of nonholonomic controllable systems. *Applied Mathematics and Mechanics (Engish Edition)*, **13**, 181–187 (1992).

[47] Mei Fengxiang, On the integration methods of non-holonomic dynamics. *Int. J. Non-Linear Mech.*, **35**, 229–238 (2000).

[48] Mei Fengxiang, Nonholonomic mechanics, *Appl. Mech. Rev.*, **53**, 283–305 (2000).

[49] I. M. Gelfand and S. V. Fomin, *Variational Calculus*. FM, Moscow, 1961 (in Russian).

[50] V. B. Glavardanov and T. M. Atanackovic, Optimal shape of a twisted and compressed rod. *Eur. J. Mech. A/Solids*, **20**, 795–809 (2001).

[51] N. J. Günter and P.G.Leach, Generalized invariants for the time-dependent harmonic oscillator. *J. Math. Phys.*, **18**, 572–576 (1977).

[52] G. Hamel, *Theoretische Mechanik*, Springer-Verlag, Berlin, 1949.

[53] W. Heisenberg, *Philosophic Problems of Nuclear Sciences*. Fawcet Publications, Inc., Greenwich, CT, 1966.

[54] M. R. Hestens, *Elements of the Calculus of Variations*. McGraw-Hill, New York, 1956.

[55] H. Josephs and R. L. Huston, *Dynamics of Mechanical Systems*, CRC Press, Boca Raton, FL, 2002.

[56] E. Kamke, *Differentialgleichungen, Gevönliche Differentialgleichungen*. 6 Auflage, Nauka, Moscow, 1971 (in Russian).

[57] T. R. Kane, *Dynamics*, Holt, Rinehart & Winston, New York, 1968.

[58] R. S. Kaushal, S. C. Mishra, and K. C. Tripathy, Construction of the second constant of motion for two-dimensiona classical problems. *J. Math. Phys.*, **26**, 420–427 (1985).

[59] J. Keller, The shape of the strongest column, *Arch. Rational Mech. Anal.*, **5**, 275–285 (1960).

[60] J. Kevorkian and J. D. Kole, *Perturbation Methods in Applied Mathematics*. Springer-Verlag, New York, 1980.

[61] D. E. Kirk, *Optimal Control Theory,* Prentice-Hall, Englewood Cliffs, NJ, 1970.

[62] V. V. Kuznetsov, and S. V. Levyakov, Complete solution of the stability problem for elastica of Euler's column. *Int. J. Non-Linear Mechanics,* **37**, 1003–1009 (2002).

[63] P. G. Leach, Invariants and wave functions for some time-dependent harmonic oscilltor-type Hamiltonians. *J. Math. Phys.,* **18**, 1902–1907 (1977).

[64] P. G. L. Leach, R. Martens, and S. D. Maharaj, Self-similar solutions of the generalized Emden-Fowler equation. *Int. J. Non-Linear Mechanics,* **27**, 575–582 (1992).

[65] H. Leipholz, *Stability of elastic systems,* Sijthoff & Noordhoff, Alphen aan den Rijn, 1980.

[66] H. R. Lewis Jr., Class of exact invariants for classical quantum time dependent harmonic oscillator. *J. Math Phys.,* **9**, 1976–1986 (1968).

[67] A. Libai, Equations for the nonlinear planar deformation of beams. *Journal of Appl. Mech. (transactions of ASME),* **59**, 1028–1030 (1992).

[68] A. I. Lurie, *Analytical Mechanics.* FM, Moscow, 1961 (in Russian).

[69] D. Mangeron and S. Deleanu, Sur une classe d'équations de la méchanique analytique au sens de I. Tzénoff. *Dokl. Bolgar. Akad. Nauk,* **15**, 9–12 (1962).

[70] J. E. Marsden and T. J. R. Hughes, *Mathematical Foundation of Elasticity,* Prentice-Hall, Englewood Cliffs, NJ, 1983.

[71] A. J. McConnell, *Applications of Tensor Analysis.* Dover, New York, 1957.

[72] I. V. Meshcerskij, *Collection of Problems in Theoretical Mechanics.* Nauka, Moscow, 1975 (pp. 378–379) (in Russian).

[73] A. Miele, *Theory of Optimum Aerodynamic Shapes,* Academic Press, New York, 1965.

[74] D. S. Mitrinovic, *Analytic Inequalities.* Springer-Verlag, Berlin, 1970.

[75] I. Ju. Neimark and N. A. Fufaev, *Dynamics of Nonholonomic Systems.* Nauka, Moscow, 1967 (in Russian).

[76] A. H. Neyfeh, *Perturbation Methods.* Wiley, New York, 1973.

[77] J. Nielsen, *Vorlesungen über elementare Mechanik.* Springer-Verlag, Berlin, 1935.

[78] E. Noether, Inariante Variationsprobleme. *Nach. Ges. Wiss. Göttingen,* Heft **2**, 235–257 (1918).

[79] H. N. Núñez-Yépez, J. Delgado, and A. L. Salas-Brito, Variational equations of Lagrangian systems and Hamilton's principle. In *Contemporary Trends in Nonlinear Geometric Control Theory and Its Applications* (A. Anzaldo-Meneses et al., editors), pp. 405–422. Proceedings of the International Conference, México City, México, 2000. World Scientific, Singapore, 2002.

[80] F. Odeh and I. A. Tadjbakhsh, A nonlinear eigenvalue problem for rotating rods. *Arch. Rational Mech. Anal.*, **20**, 81–94 (1965).

[81] J. T. Oden and J. N. Reddy, *Variational Methods in Theoretical Mechanics*. Springer-Verlag, Berlin, 1983.

[82] J. G. Papastavridis, *Analytical Mechanics, A Comprehensive Treatise on the Dynamics of Constrained Systems*. Oxford University Press, Oxford, UK, 2002.

[83] L. A. Pars, *An Introduction to the Calculus of Variations*. Heinemann, London, 1962.

[84] L. A. Pars, *A Treatise on Analytical Dynamics.* Heinemann, London, 1968.

[85] I. A. Pedrosa, Canonical transformations and exact invariants for dissipative systems. *J. Math Phys.*, **28**, 2662–2664 (1987).

[86] A. Pflüger, *Stabilitätsprobleme der Elastostatik.* Springer-Verlag, Berlin, 1975.

[87] L. S. Pontryagin, V. G. Boltjanskij, R. V. Gamerklidze, and E. F. Mishchenko, *The Mathematical Theory of Optimal Processes.* Wiley, New York, 1962.

[88] W. H. Press, B. P. Flannery, S. A. Teukolsky, and W. T. Vetterling, *Numerical Recipes, The Art of Scientific Computing.* Cambridge University Press, Cambridge, UK, 1986.

[89] P. Rabier, *Topics in One-Parameter Bifurcation Problems.* Springer-Verlag, Berlin, 1985.

[90] P. V. Ranganathan, Invariants of a certain non-linear N dimensional dynamical system. *Int. J. Non-Linear Mech.*, **27**, 43–50 (1992).

[91] J. Ratzersdorfer, *Die Knickfestigkeit von Stäben und Stabwerken.* Springer-Verlag, Wien, 1936.

[92] G. D. Riccia, On the Lagrange representation of a system of Newton equations. In *Dynamical Systems and Microphysics,* (A. Avez, A. Blaquière, and A. Marzollo, editors), pp. 281–292. Academic Press, New York, 1982.

[93] R. M. Rosenberg and C. S. Hsu, On the geometrization of normal vibration of nonlinear systems having many degrees of freedom. In *Analytical Methods in the Theory of Nonlinear Vibrations*, Vol. 1. ISNV, Kiev, 1961.

[94] A. P. Sage, and C. C. White, III, *Optimum Systems Control* (second ed.), Prentice-Hall, Englewood Cliffs, NJ, 1977.

[95] R. M. Santilli, *Foundations of Theoretical Mechanics I, The Inverse Problem in Newtonian Mechanics*. Springer-Verlag, New York, 1978.

[96] W. Sarlet, Symmetries, first integrals and inverse problem of a Lagrangian mechanics, *J. Phys. Appl. Math. Gen.*, **14**, 2227–2238 (1981).

[97] W. Sarlet and L. Y. Bahar, Quadratic integrals for linear nonconservative systems and their connection with the inverse problem of Lagrangian dynamics. *Int. J. Non-Linear Mech.*, **16**, 271–281 (1981).

[98] M. J. Sewell, *Maximum and Minimum Principles*. Cambridge University Press, Cambridge, UK, 1987.

[99] A. P. Seyranian, A solution of the problem of Lagrange. *Sov. Physics Doklady*, **28**, 550–551 (1983).

[100] A. P. Seyranian, New solutions of Lagrange's problem. *Physics Doklady*, **342**, 182–184 (1995).

[101] A. P. Seyranian and O. Privalova, The Lagrange problem on optimal column: Old and new results. Moscow State Lomonosov University. Preprint 60 (2000). Also in *Structural Optimization* (in press).

[102] S. C. Sinha and C. C. Chou, Approximate eigenvalues for system with variable parameters. *Journal of Appl. Mechanics (Transaction of ASME)*, **46**, 203–205 (1979).

[103] V. V. Stepanov, *Course in Differential Equations*. Moscow University Edition, Moscow, 1953 (in Russian).

[104] T. Sutela and B. D. Vujanovic, Motion of nonconservative dynamical system via a complete integral of a partial differential equation. *Tensor* (NS), **38**, 303–310 (1982).

[105] K. R. Symon, The adiabatic Invariant of the linear and nonlinear oscillator. *J. Math Phys.*, **11**, 1320–1330 (1970).

[106] J. L. Synge, On the geometry of dynamics, *Phil. Transactions of the Royal Soc. of London, Ser. A*, **CCXXVI**, 31–106 (1926).

[107] B. Tabarrok and W. L. Cleghorn, Application of principle of least action to beam problems. *Acta Mechanica*, **142**, 235–243 (2000).

[108] V. A. Troickii, and L. V. Petruhov, *Shape Optimization of Elastic Bodies.* Nauka, Moscow, 1982 (in Russian).

[109] I. Tzénoff, On a new form of the equations of analytical dynamics. *Dokl. Akad. Nauk SSSR,* **89**, 21–24 (1953).

[110] B. Vujanovic, Synge's disturbed equations as a variational problem and their first integrals. *Bulletin de l'Académie Royale de Belgique,* **51**, 692–698 (1965).

[111] B. Vujanovic, A group-variational procedure for finding first integrals of dynamical systems. *Int. J. Non-Linear Mech.,* **23**, 269–278 (1970).

[112] B. D. Vujanovic, Conservation laws of dynamical systems via D'Alembert's variational principle. *Int. J. Non-Linear Mechanics,* **13**, 185–197 (1978).

[113] B. D. Vujanovic, On the field momena method in mechanics. *Tensor* (NS), **33**, 117–122 (1979).

[114] B. D. Vujanovic, On a gradinet method in nonconservative mechanics. *Acta Mechanica,* **34**, 167-179 (1979).

[115] B. D. Vujanovic, On the integration of the nonconservative Hamilton's dynamical equations. *Int. J. Engng. Sci.,* **19**, 1739–1747 (1981).

[116] B. D. Vujanovic and A. M. Strauss, Linear and quadraic first integrals of a forced linearly damped oscillator with a single degree of freedom. *Journal of the Acoustic Society of America,* **69**, 1213–1214 (1981).

[117] B. D.Vujanovic, Conservation laws and Hamilton-Jacobi-like method in nonconservative mechanics. In *Dynamical Systems and Microphysics Geometry and Mechanics* (A. Avez, B. Blaquiere, and A. Morazollo, editors), pp. 293–301. Academic Press, New York, 1982.

[118] B. D. Vujanovic, A field method and its applications to the theory of vibrations. *Int. J. Non-Linear Mech.,* **21**, 381–396 (1984).

[119] B. D. Vujanovic and A. M. Strauss, Study of motion and conservation laws of nonconservative dynamical systems. *Hadronic Journal,* **7**, 163–185 (1984).

[120] B. D. Vujanovic and A. M. Strauss, Application of a field method to the theory of vibration. *Journal of Sound and Vibration,* **114**, 375–378 (1987).

[121] B. D. Vujanovic and A. M. Strauss, Applications of Hamilton-Jacobi method to linear and nonlinear vibration theory. *J. Math. Phys.,* **29**, 1604–1609 (1988).

[122] B. D. Vujanovic and S. E. Jones, *Variational Methods in Nonconservative Phenomena.* Academic Press, New York, 1989.

[123] B. D. Vujanovic, Conservation laws of rheolinear dynamical systems with one and two degrees of freedom. *Int. J. Non-Linear Mech.*, **27**, 309–322 (1992).

[124] B. D. Vujanovic, Application of Hamilton–Jacobi method to the study of rheolinear oscillators. *Acta Mechanica*, **93**, 179-190 (1992).

[125] B. D. Vujanovic, Application of the field-momentum method to rheonomic dynamics. *Int. J. Non-Linear Mech.*, **29**, 515–528 (1994).

[126] B. D. Vujanovic, Conservation laws and reduction to quadratures of the generalized time-dependent Duffing equation. *Int. J. Non-Linear Mech.*, **30**, 783–792 (1995).

[127] B. D. Vujanovic, T. Kawaguchi, and S. Simic, A class of conservation laws of linear time-dependent dynamical systems. *Tensor* (NS), **58**, 243–252 (1997).

[128] G. B. Whitham, *Linear and Non-Linear Waves*. Wiley, New York, 1974.

[129] B. D. Vujanovic, A. M. Strauss, S. E. Jones, and P. P. Gillis, Polynomial conservation laws of the generalized Emden–Fowler equation. *Int. J. Non-Linear Mech.*, **33**, 377–384 (1998).

[130] E. T. Whittaker, *A Treatise on the Analytical Dynamics of Particles and Rigid Bodies* (fourth ed.), Cambridge University Press, Cambridge, UK, 1965.

[131] L. C. Young, *Calculus of Variations and Optimal Control Theory*. Saunders, Philadelphia, 1969.

Index